Explorations in Classical Sociological Theory

Seeing the Social World

KENNETH ALLAN
University of North Carolina

PINE FORGE PRESS
An Imprint of Sage Publications, Inc.
Thousand Oaks • London • New Delhi

For information:

Pine Forge Press
2455 Teller Road
Thousand Oaks, California 91320
E-mail: order@sagepub.com

Sage Publications Ltd.
1 Oliver's Yard
55 City Road
London EC1Y 1SP
United Kingdom

Sage Publications India Pvt. Ltd.
B-42, Panchsheel Enclave
Post Box 4109
New Delhi 110 017 India

Printed in the United States of America

Library of Congress Cataloging-in-Publication Data

Allan, Kenneth, 1951-
Explorations in classical sociological theory : seeing the social world / Kenneth Allan.
p. cm.
Includes bibliographical references (p.) and index.
ISBN 1-4129-0572-9 (pbk.: acid-free paper)
1. Sociology—Philosophy—History—19th century—Textbooks.
2. Sociology—Philosophy—History—20th century—Textbooks. I. Title: Seeing the social world. II. Title.
HM445.A45 2005
301′.01—dc22

2004019257

This book is printed on acid-free paper.

05 06 07 08 10 9 8 7 6 5 4 3 2 1

Acquisitions Editor:	Jerry Westby
Production Editor:	Denise Santoyo
Copy Editor:	Teresa Herlinger
Typesetter:	C&M Digitals (P) Ltd.
Indexer:	Kathy Paparchontis
Cover Designer:	Michelle Lee Kenny

Contents

Acknowledgments

My thanks go out to all my students who have helped me focus in on important issues in teaching theory. I especially thank several of my graduate students who aided in various stages of the manuscript: Liz Shimkus, Damion Raba, Aly Esler, and Josh Kelley. My editors at Pine Forge, Jerry Westby and Ben Penner, have given wonderfully creative assistance, and my copyeditor, Teresa Herlinger, is a true wordsmith whose eyes never failed to see that to which I was blind; many, many thanks. It's been a joy working with you. I'm also thankful for the constructive comments I received from the book reviewers: Bruce Day, Margaret Hunter, Amy Chasteen-Miller, Jieming Chen, Ralph Pyle, and especially Richard Garnett; the book is substantially better because of their work. My thanks especially to my undergraduate professors: to David McKell for showing me the heart of sociology; to Douglas Deagher who gave me a B- on my first theory exam and later gave me the opportunity to study the classics firsthand; and to Richard E. Skeen, from whom I learned that teaching is a subversive activity. And to Steve Kroll-Smith, who gave me the freedom and support necessary to find my voice, my continuing gratitude. I am particularly grateful to my wife and best friend, Jen, without whose patience and support this project would never have been completed.

Photo Credits

Special thanks to all the rights holders of the following theorists' photographs:

Chapter 2, Herbert Spencer: © Bettman/Corbis

Chapter 3, Karl Marx: © Bettman/Corbis

Chapter 4, Emile Durkheim: © Bettman/Corbis

Chapter 5, Max Weber: © Granger Collection

Chapter 6, Georg Simmel: © Granger Collection

Chapter 7, George Herbert Mead: © Granger Collection

Chapter 8, Harriet Martineau, Charlotte Perkins Gilman, and W.E.B. Du Bois, © Bettman/Corbis

Chapter 10, Talcott Parsons: © Granger Collection

Prelude

Two personal experiences formed the motivation for this book. The first occurred when I was a graduate student. In my last year of graduate school, I attended a workshop on teaching theory to undergraduates. Several of the most prominent theorists in the United States were on the panel, two of them world class. I was very excited about the presentations, not only because of the stature of the people presiding, but also because I had known for years that I wanted to teach theory. I felt that this workshop was going to be one of the most important of my career. Yes, I know that research is the guiding force in our discipline, and I have always looked forward to paper presentations as well. But teaching is special. In teaching, you can change the way someone sees the world. And teaching theory can impact students' perspectives most of all, because it is all about how people think.

I got to the workshop early and eagerly secured a front-row seat. By the time the presentations began, the place was packed. Apparently, there were a lot of people interested in teaching theory to undergraduates. I was excited. I was about to hear two of the most accomplished theorists in the world tell me how to convey theoretical ideas to the minds of students. My notepad was open and my pen was on the paper when the first speaker began.

The first words out of his mouth were, "You can't teach theory to undergraduates." Say again?!? I really hope my face didn't convey the shock I felt. I was stunned. It was like the wisdom of the ages boiled down to "you can't do it." I sat there with pen on paper and the pen never moved for the entire hour. I made it my goal to prove that professor wrong.

The next experience occurred just a few years ago. I had been teaching both undergraduate and graduate theory for quite some time, and I was having a ball. To me, the ability to form and discuss ideas is something that is fundamentally human. Some people can spend hours watching football or talking about politics; for me it's theory, the pinnacle of our ideas and thought. If I'm involved in talking about ideas and theory, the world could probably pass away and I'd barely notice. I know, it's pretty sad, and it actually leaves me few people to talk to. But for the moment, I have you, so let's get back to my story.

As I said, I had been teaching theory for some years and having a great time. After class one day, as I was erasing the board, a student came up to ask some questions. Just before she left she said, "You know, I got into sociology because I love it. And every class I've taken has been really exciting. But this class makes me want to quit." Ouch! There's nothing like finding out you're a failure at one of the things you love best. Don't get me wrong, I had (and have!) plenty of success stories. But this woman's comment went straight to my heart.

However, I knew what motivated her comment; and after teaching theory myself I finally knew what had prompted the remarks of the professor at the workshop. Theory is hard work. Well, actually, *thinking* is hard work and theory is one of the hardest forms of thinking. It's generally pretty abstract, filled with specialized terms, and it requires us to think differently than we normally do. Human beings are meant to think concretely, and it's socially necessary for us to think in patterned ways. Theoretical thinking disturbs all that. It's actually painful to make your brain think differently, which is what we try to do in a theory class.

Not only is it painful to think theoretically, theory books are difficult. Part of that is unavoidable. As I said, thinking theoretically is difficult, and it just stands to reason that theory books would be tough to read. Also, part of what we are doing in theory textbooks is condensing a library's worth of material into one relatively short chapter. Think about that when you're reading the chapter on Marx in this book: it took Marx four or five long books to say what I'm going to cover in one chapter. I also believe that there is another, perhaps more basic reason why most theory books are difficult: they are themselves works of theory. They aren't necessarily designed as teaching aids.

Ever since my student's comment, I have been working at making theory more accessible to students. This textbook is the product of those efforts, and as a result, it's a little different than most—it is a teaching book first and a theory book second. One thing I try to do as much as possible is to write in a conversational tone. Too often our texts are dry and don't convey a sense of excitement about the material. Because I'm interested in inviting you to think about your social world, I want the book to be inviting. I include a lot of personal examples and insights. In fact, I actually had a professor who had read one of the earlier versions of a chapter say that she had learned as much about me as she had about Simmel! I'm not sure that's really true, but I tell quite a few stories in these pages to convey my meaning.

Another difference is that I focus on the perspectives of the individual theorists. Most theory books categorize the theorists into one of three perspectives—functionalism, conflict theory, or interactionism—or they force the theory into a paradigm, most notably science. In either case, what the theorist is trying to say gets slighted. More importantly, the reason to have multiple voices is that the social world is much more complex than functionalism, interactionism, conflict theory, or science. So, I spend time helping you understand the perspective of each theorist. We can categorize Herbert Spencer, Émile Durkheim, and Charlotte Gilman as evolutionary functionalists, but it is a very different thing to see the world through Spencer's versus Gilman's eyes, or Durkheim's versus Spencer's, and so on.

One other difference is that I emphasize the historical eras in which the theorists lived and the one in which we live. What we consider classical social theory was produced out of the fervor of modernity. The great minds of the time were captivated by the massive social changes they observed. They considered the possibilities of their era in both positive and negative terms. You and I are also living in a time that is filled with substantial social changes. Many contemporary theorists think this time period is as distinct from modernity as modernity was from traditional society. So part of what I'm going to do is examine what each theorist says about "modern" societies—the societies of their day—and then I'm going to explore the notion that maybe we are living in just such a transition period. Are we now post- or late-modern rather than modern? I really don't know, but it's exciting to think about. We could be living at the threshold of a whole new world.

And, finally, I am infatuated with the poetry of the theorists. The theorists in this book, when pushed to their limits by a significant problem, were able to capture multiple layers of meaning in a single phrase. That's part of what makes them classic: we can read them again and again and come away with new insights each time. There is, then, both power and poetry in classical theory. And these phrases, whether we agree with them or not, have the power to excite our own thinking. As a result, this book is filled with selected quotations. I really encourage you to take time to read and think about them; don't just skip over them. They might just inspire you as they do me.

Here's the bottom line: I wrote this book in the hopes of exciting and disturbing your thinking. I sincerely hope to invite you into a space where we can think new thoughts and see new things. In the end, I hope you leave our discussion with more questions than answers. They say you can lead a horse to water but you can't make him drink. Like many old sayings, this one isn't true. All you have to do is get the horse thirsty. I hope, after all is said and done, that you are thirsty for more after reading this book.

A Note to Students: Throughout the text, important terms are in **bold** typeface. You will find these terms defined in the glossary section of this book. Most of the bolded words are technical terms that have very specific meanings and are particularly relevant for understanding a particular theorist's theory. Some of these concepts actually carry across all of the theorists in this book. Religion is a good example. In these cases, I include each theorist's definition, so that you can begin to compare and contrast what the classic theorists are saying about these important concepts.

CHAPTER 1

Imagining Society

Imagination is more important than knowledge.

Albert Einstein

I came to sociology by accident. After two careers, I decided to go back to school. It was important for me to get into a profession in which I could help people and make a difference. So I picked psychology as my major. But while taking a Sociology of Education class for liberal arts requirements, I heard a sociologist speak about his involvement in Nicaragua. Nicaragua had been undergoing economic, political, and social reforms since 1979. Previously the country had known 40 years of dictatorship under the Somoza family. But during the 1980s, the United States government supported a rebel group (the Contras) that plunged Nicaragua into a civil war. During this politically and militarily risky time, this professor adopted a village. Two to three times a year he took a group of students down to build houses, dig ditches, bring medical supplies, and care for the community. Here was an individual that was making a difference. His academic knowledge wasn't sterile; it affected his life and those around him. And that was exciting to me.

Two things struck me about the presentation. First, here was a man who walked the talk. He didn't hold knowledge in an unfruitful way; what he knew made him responsible. He cared about the human condition, and that wasn't something that I had come across at the university before. The second thing that struck me about the presentation was his view of Karl Marx. I have always hated oppression. But as a good American I had been told to hate Marx, too; he was the father of communism and communism was bad (I grew up during the Cold War). But this man whom I respected because of his fight against human suffering told us that he was motivated by the theories of Marx. That day in class I found out that Marx told a fascinating story about economic oppression. This story takes elements from the past, along with social factors in modernity, and predicts what will happen in society. It tells how (not just why) people are economically oppressed.

I was hooked. This teacher was a man who made a difference and knew *why* he *had* to make a difference. I wondered if it had something to do with sociology. The professor certainly claimed that it did. For two weeks, I met with my professors in psychology and went door-to-door in the sociology department talking to everybody I could. At the end of those two weeks, I changed my major to sociology. Yep, I was hooked on sociology. But what I didn't realize at the time was that I was hooked on theory as well.

I had an important "theory moment" on my first day of graduate school. I had just moved back to Southern California, having spent a good deal of my adult life away. I was on my way to class, driving on the freeway, and noticed a car ahead of me. The car was chopped, or lowered so that the body was only a few inches off the ground. It had wide tires and rims. The side windows were tinted and there was a conspicuous symbol painted on the back window. The car was painted a mesmerizing purplish color that changed as the sun hit it from different angles. As I pulled closer to the car I could see "dingle balls" hanging from the back window. I thought, "Things haven't changed a bit." I smiled to myself with that comfort that comes from nostalgia, until I pulled directly next to the car and looked at the driver. He was a white kid! In fact, there were two white kids in the car, which left out the possibility that he was driving his friend's car (at least in my mind it did). It was a white kid driving what I had thought was a Chicano's car.

There may be a number of lessons that can be drawn from this experience—like don't stereotype—but for me the experience left me with questions and problems. I grew up in Southern California, and when I was in high school and early college, a car like the one I saw would *only* have been driven by a Chicano. The cultural boundaries surrounding identity were clear. White kids fixed their cars in certain ways and wore certain clothes, same with Chicanos and blacks. The kids I hung out with weren't really prejudiced. We would interact with one another, party with one another, and have friends in the different groups. But we all clearly claimed our identities through culture, and the cultures were all different, with distinct boundaries. This car blew all that cultural understanding out the window. Apparently, things had changed.

But why and how did they change? These questions are why I called that experience on the freeway a theory moment. Questions, problems, and oddities are the cauldron out of which theory is born. It wasn't until a couple of years later that I actually figured out the answers to those questions. It took awhile in school before I came across theories that spoke of cultural fragmentation and the effects that increases in markets, advertising, and communication and transportation technologies can have on culture.

But I was thrilled when I found my answer. Using theory, I was able to see what was hidden. Theory is that which lifts the veil and connects the dots. It lifts the veil because it can show us what is going on beneath the surface. Like the way Marx's theory exposes the workings of a system (capitalism) that we grew up taking for granted, theory connects the dots because it can help us make sense of the social world around us. Seeing through the prisms of theory can be a wonderful experience. Even the mundane can become fascinating—like standing in a line.

I used to hate standing in line. Actually, I probably still do, but I have also been enthralled by it ever since I discovered a theory called *ethnomethodology* (a perspective that looks for how social order is produced in face-to-face interactions). Ethnomethodology let me see that lines, or queues, are ongoing social accomplishments. We take the ability to line up for granted, but how do we accomplish it? The other day I stood in line for an hour for show tickets. When I got through the line and got my tickets, I noticed that the line was just as long as it was when I got in it, and it was in almost the exact same form. The line wound around this large room like a snake and then went out a door. Now here's the interesting thing: there was no one there telling us how to line up, and there were no ropes to guide us. The new queue contained all different people, yet the line looked exactly the same as when I had been in it. How did we do it, and then how did *they* do it again? Theory gives us the eyes to be able to connect those dots.

Foundations of Theory

Perspectives: Have you ever thought that you and your parents live in separate worlds? I know I did when I was younger. And you know what? You're right. Your parents may listen to your favorite band and say, "Why are they screaming? That's not even music." They might then point to the "beautiful voice" of their favorite singer and tell you, "Now that's the way singing is supposed to sound; that's music!" It's kind of amazing, but people can look at the same thing and see, or hear, very different worlds.

Think about this story: There once was a group of blind men who examined an elephant, and each, by his individual experience of the animal, drew very different conclusions about the nature of an elephant. One determined that an elephant is like a very smelly rope (based on the tail); another said that an elephant is like a tree trunk (based on a leg); yet another concluded that an elephant is like a brick wall (based on the side); and the last one said that an elephant is like a hose (based on the trunk). Forrest Gump may be right, that life is like a box of chocolates, but the world is like an elephant—what you *see* is what you get, and *what you see depends on where you stand.*

Because human reality is a cultural reality, *perspectives* are an essential and unavoidable part of our existence. Joel Charon (2001) explains it this way:

> Perspectives sensitize the individual to see parts of reality, they desensitize the individual to other parts, and they guide the individual to make sense of the reality to which he or she is sensitized. Seen in this light, a perspective is an absolutely basic part of everyone's existence, and it acts as a filter through which everything around us is perceived and interpreted. There is no possible way that the individual can encounter reality "in the raw," directly, as it really is, for whatever is seen can be only part of the real situation. (p. 3)

In other words, we never directly experience the world; we encounter it through our perspectives. For a trained sociologist, theory is our perspective—it is a way of seeing and not seeing the world.

Theories are useful precisely because they do call our attention to certain elements in the social environment. I never realized what an amazing social achievement lining up outside a theater is until I understood ethnomethodology—and I never understood how people in different religions could all be convinced they are right until I read Durkheim. In sociology, we have a large number of theoretical perspectives that tell us what to see so that we gain insight into the social world; among them are functionalism, conflict theory, interactionist theory, critical theory (including feminism, race, and queer theory, postmodernism, and so on), exchange theory, rational choice theory, dramaturgy, ethnomethodology, structuration, network theory, ecological theory, social phenomenology, and on and on.

All of these perspectives are based on assumptions and values, and they contain ideas, concepts, and language. They tell us how to see the social world and how to talk about it. Each of these becomes particularly important to the theorist. Sets of words help order the human world in general. Think about when and what you eat, for example. Biologically we are all driven to eat. But what and when we eat is socially and linguistically ordered. Thus, restaurants like Denny's serve "breakfast all day." What is meant by that phrase is that certain foods that are normally associated with eating in the morning are available for "lunch" or "dinner." The sociological perspective has specific sets of words associated with it as well, and these words order and make the world comprehensible in a specific way. Thus, part of what you will be learning in this book are sets of words—languages that help to organize the world in a specific manner.

Perspectives also contain **assumptions**: things that we suppose to be true without testing them. All theories and theorists make certain assumptions. They form the bedrock upon which theory can be built. Two things are extremely important when thinking about assumptions: assumptions are never proven or disproved and without them it is impossible to theorize (or even live). They thus inform everything that a theorist says. Let's take human nature, for example. Almost every theorist makes certain assumptions about what constitutes human nature. If a theorist assumes that human nature is individualistic and self-seeking, what then is she going to think about social structures? Chances are she will think that social structures are good and necessary to create social order. On the other hand, if a theorist thinks that human nature is intrinsically good and altruistic, then social structures can be a bad thing that oppresses people. The idea of human nature cannot be tested (can you think why?), yet it is impossible to theorize without it, and once we assume something about human nature, it influences every aspect of our thoughts. (By the way, functionalists generally assume that human nature is egocentric and in need of social control, and Marxists generally assume that human nature is social and thus structures can be oppressive.) *Remember: every theory is a perspective and is founded upon and contains assumptions, values, concepts, and languages.*

The Sociological Perspective

Context: One of the things that sociologists assume is that humans are social creatures. In other words, in order to understand human behavior, it must be seen as socially embedded. Ever since your Introduction to Sociology class, you have been told that to think sociologically is to see the social factors that influence human behavior. Let's use an illustration to understand sociological thinking. Let's pretend you're flying over the United States in an airplane at 25,000 feet. What do you see? Chances are you see some of the land divided up into large squares or rectangles of green, gray, and brown. You see small and large clusters of human-made buildings. You see long lines that seem to connect the clusters and some of the colored patches. Notice that some of the clusters have more of the lines going into them than others. What can you surmise from these observations? What are you seeing? You are seeing population centers that are connected by roadway infrastructures. You can begin to deduce which of the clusters (cities) are most important because they are the ones that have the most and largest connections. We can also see that most of the population lives in cities, not rural settings. People, then, tend to be densely packed together. We can also see that the United States has tremendous amounts of natural resources from which to draw, because there is an abundance of rural land, and it is bordered on two sides by ocean.

Now let's go down a few thousand feet. We can see other kinds of connections now, like telephone and digital lines. We see how within the cities there are other more dense clusters of larger buildings. The larger buildings would be the most important buildings in the society. What kind of buildings are they? And where do the lines of communication and transportation go? Which buildings and what cities receive the most information and materials? These configurations would be different in different kinds of societies.

If we go even further down, into one of the large buildings, we find that people are organized spatially. There are some people crowded together in relatively small rooms and there are others by themselves in spacious rooms. We can again look at how the communication is structured in this building and see that most of the information is sent to the rooms with the fewest number of people. Moreover, we can see that people are spatially organized by certain physical characteristics—certain kinds of people tend to be in the spacious or crowded rooms. We also notice that the people in the roomy areas tend to have framed pieces of paper from other institutions up on their walls, and the people in the cramped rooms don't. We might also see charts that denote the organization of the company, with lines connecting some people and not others.

Let's drop down even further and we can hear people speak. We notice that how people speak is different in the spacious offices than in the congested rooms. We can also see how people interact. We can see them in exchanges, flirtations, conflicts, dyads, triads, and so on. We can see them present selves and manage identities. And we can see that the people in the spacious offices get more deference from others.

Then we can look at one person, John Smith in cubby B. How are we to understand him? How does he feel? What kinds of things are important to him? What are his dreams and disappointments? How are we to think sociologically about John? It is important to keep in mind that the sociological perspective is always a contextual point of view.

Levels of analysis: Obviously, I've set us up for the answer, but it is an important point—a position that all our news and talk programs, movies, and social values move us away from seeing. Sociologists generally don't see the individual as an individual. We understand individuals in terms of their connections. Part of how we have to think about John is through the various levels that we just talked about: the **macro, meso,** and **micro** levels. The theories that we are going to be looking at will challenge us to think on all three of those levels. The macro level is like the plane at 25,000 feet. It is concerned with those factors that influence human behavior that exist completely beyond the influence of the individual. The facts that the United States is bordered by two oceans and has tremendous natural resources are things that are outside our control, but have important consequences for our nation and for us as individuals (Weber's geo-political theory helps us understand these effects). The same is true about most of the population living in cities. Large numbers of people living in urban settings happens mostly because of industrialization—something with which neither you nor I had anything to do, but it has had amazing consequences for us (Simmel's theory about the metropolis serves us here).

We began to look at meso theorizing when we moved to the level of the city and building, though both those are embedded in macro-level connections (for example, the largest buildings would be the ones that express the most important societal institution). At this level, we see bureaucratic elements such as the organizational chart, status by credential, rational communication channels, and so forth. We also can observe some macro-level elements, such as race and gender, playing themselves out in the organization. And at the **interaction** (micro) level, we also see the macro-level structure of language being used and modified and created. As we move down these levels, we can see increasing control of the individual (**agency**), but the sociological point of view understands agency and the individual as embedded—circumscribed and influenced—within larger contexts.

Each of our theorists is going to provide opportunities for us to think at these levels. Some are quite exclusive to one or the other, such as Mead at the micro and Marx at the macro. But most present us with a mix. Part of what we as theorists need to be able to do is to think flexibly—to be able to move from one level to another and to see connections among the levels. We need to think sociologically. We need to understand whatever it is we are observing in terms of its embeddedness.

Assumptions Concerning Society

Once we've assumed the sociological perspective, we have a very important question to answer. Ironically, it is a question that most of us, even sociologists, haven't taken the time to think about, let alone answer. That's actually pretty common when it comes to assumptions—assumptions work best when they are taken for granted. However, good theory is based on clear thinking. We can't get away from assumptions, but we can make them knowingly.

Probably the most fundamental issue in sociology is deciding how society exits. This is a question of ontology. The word **ontology** literally means "the study of being," and in philosophy ontology refers to the study of how things exist. The main ontological question for philosophers is, how do abstract entities exist? We don't want to get caught up in philosophical issues, but I do want to bring this issue of ontology home in a way that we can understand. Let me ask you a question: what's the difference between a rock and a brick? Well, there are a lot of differences, but can we think of them being ontologically distinct? In other words, is their mode of existence different? Yes, the ontological source for rocks and bricks is not the same. If humans never walked the face of the earth, the rock would still exist, but the brick would not. The brick exists only because of the tool context that humans create—it needs humans to exist, whereas the rock does not.

Society as an object: Obviously, society needs humans to exist: if you took away all the humans, society would cease to exist. So, that's not exactly the ontological issue with which we are concerned. We're going to take for granted the existence of humans, but even so we are left with the question of social ontology. The question is best put like this: after human beings create society, how does it continue to exist? I know the question still might sound a bit odd, but bear with me. There are basically two kinds of responses to this question. The first, and actually the most common, says that society exists as an object, something that continues outside of the individual or group and has independent effects. In the words of Émile Durkheim, society exists as a reality *sui generis,* in and of itself. Society, then, is a thing in the environment that can be studied and about which objective claims can be made. Making the assumption of objectivity means that we can *discover* things about society because it exists apart from us, *there* in the environment.

This idea of society as an object generally came out of the Enlightenment. In fact, the roots of sociology are grounded in this period—without the Enlightenment there would be no such thing as sociology. The **Enlightenment** is a European intellectual movement that began around the time Isaac Newton published *Principia Mathematica* in 1686, though the beginnings go back to Bacon, Hobbes, and Descartes. Among the most important Enlightenment thinkers are Jean-Jacques Rousseau, Voltaire, David Hume, and Adam Smith. The Enlightenment is so called because the people creating this intellectual revolution felt that the use of reason and logic would enlighten the world in ways that fate and faith could not. The principle targets of this movement were the church and the monarchy, and the ideas central to the Enlightenment were progress, empiricism, and tolerance.

The ideas of progress and empiricism are especially important. Prior to the Enlightenment, the idea of progress wasn't particularly important, especially technical progress. The reason for this is that the dominant worldview was religious. In general, religion sees the world in terms of fate, faith, and revelation. Any change under a religious system comes not because of human effort and progress, but, rather, because God has revealed some new truth. Purely religious systems thus tend toward status quo and honoring tradition. On the other hand, the belief in *progress* puts human beings at the center (humanism) and asserts that people can take control of their lives and the environment in order to make things better. This humanistic system privileges change, and change is defined as progress.

The idea of progress puts human beings in the driver's seat, and it implies that the universe and the social worlds are empirical. That is, **empiricism** is the assumption that there are no spiritual or unseen forces in back of the physical or social worlds, but, rather, real things exist as facts that are discernable through at least one of the five senses. Together, the ideas of progress and empiricism helped to form a uniquely modern philosophical orientation: positivism.

One of the easiest ways to understand positivism is to compare it to fatalism. Fatalism, or having a fatalistic attitude, has a negative connotation today, but it wasn't always so. Fatalism refers to a belief in the Fates. In antiquity, the Fates were believed to be goddesses who oversaw destiny and determined the course of human existence. This kind of idea is found in Christianity as well in the doctrine of predestination. A less specific form of fatalism is practiced today by anyone that believes there is some spiritual force in back of the universe. You can see that fatalism isn't necessarily negative. Another way to put this is that positivism is positive in a very specific way.

Positivism is the opposite of fatalism: Positivism assumes that the universe is empirical (without spiritual force), operates according to law-like principles, and that humans can discover those laws and use them to understand, control, and predict the forces that influence their lives. Fatalism puts a spiritual force at the center of existence; positivism puts humanity at the center of existence. Thus, it is positive in a humanistic sense. **Positivism,** then, is a philosophy that confines itself to sense data, denies any spiritual forces or metaphysical considerations, and emphasizes the ability of human beings to affect their own fate, generally through **science.**

The philosophy of positivism began with Auguste Comte. Comte is generally referred to as the "father of sociology" because he is credited with coining the term. However, Comte wasn't alone in his use of the term *sociology.* In fact, Harriet Martineau, whom we will be studying later in this book, actually used the term in some personal letters before Comte's text. If the dates of these writing matter, we should then say that sociology began with a woman and that Martineau is the mother of sociology. Nevertheless, it is probably the case that the term was in general use at the time and that both Comte and Martineau were simply using words with which a certain group of intellectuals was familiar (Lengermann & Niebrugge-Brantley, 2000, p. 292).

Comte (1896) argued that each branch of knowledge must pass through three different and mutually exclusive stages. The theological or fictitious stage is the natural beginning of knowledge. In this stage, humans seek the essential nature of things; the first and final causes of all effects. This is the search for absolute knowledge. The next stage is the metaphysical or abstract stage. It is really only a modification of the first: rather than seeing supernatural beings in back of all things, the mind hypothesizes abstract forces that produce all phenomena, such as the four basic elements (earth, fire, water, and air). The final stage is the scientific or positive stage. In this stage, reason and observation lead humanity to discover the natural laws of the universe. Comte felt that this was the final and fixed stage for human knowledge.

Comtean philosophy regards all phenomena as subject to invariant natural laws. Thus, the business of positivistic science is not the search for first or final causes, such as the search for the ultimate cause of the universe or the beginning cause of human society. This kind of search is considered a defining characteristic of the first two stages, and the pursuit of this kind of causality is considered fruitless and occupied with unsolvable questions. Science's job, and thus the job of sociology according to this viewpoint, is to discover these invariant laws and reduce them to the smallest possible number. "Our real business is to analyze accurately the circumstances of phenomena, and to connect them by the natural relationship of succession and resemblance" (Comte 1896, p. 6). As you probably have already guessed, the beliefs of progress, empiricism, and positivism are the cornerstones of science.

Scientific theory: Science is a knowledge system that is particularly oriented toward controlling the universe through technology. Like all perspectives, science is founded on a set of assumptions. It assumes that the universe is empirical, that it operates according to law-like principles, and that humans can discover those laws though rigorous investigation. Science also has very specific goals, as do most knowledge systems. Through discovery, scientists want to explain phenomena, predict and control phenomena, and accumulate knowledge. Scientists have done a pretty good job of fulfilling these goals. We can flip a switch and turn on the lights, and we can take an analgesic to get rid of a headache. Of course, what lies in back of these discoveries and applications is theory.

In one sense, we all have theories. We have common or folk theories about the way things work, especially about the social world. In fact, we need to have these theories in order to function in the human world. So, we may have a theory of gender that explains to us why Tom behaves the way he does ("he's a guy") and allows us to predict and control his behaviors to some degree (some people really don't like it when Tom doesn't act like a guy). But *scientific* theory is a particular type. For one thing, scientific theories are intended to be testable, and when scientists do test their theories, they try to prove them wrong. In contrast, we generally don't put our folk understandings to the test, and we certainly don't like it when we are proven wrong. Scientific theories also have other qualities, as described below.

Scientific theory is a formal and logically sound argument explaining some empirical phenomenon in general or abstract terms. Such theories are formal in the sense that they are written down and all the "hidden baggage" (like assumptions) explored and made explicit. Logic, in the sense that we are using it here, pertains to making an argument. An argument is the presentation of a particular course of reasoning designed to give others a basis for thinking in a certain manner. Thus, arguments are always aimed at presenting a conclusion, and the points offered in support of the conclusion are called premises. There is an important relationship between a conclusion and its premises: an argument is considered valid in the case that it is impossible for its premises to be logically true while its conclusion is false. The point here is that, in scientific theory, the associations among the premises or concepts must make sense and must be related to each other by the rules of argumentation.

Scientific theory is made up of at least three elements: concepts, definitions, and relationships. **Concepts** are the ideas we use to understand what we are looking at, and in science, concepts must be abstract. Think again about the goals of science. Prediction and control always require generalized knowledge, because it must be usable in many settings. In order for knowledge to be generalized out of a single empirical setting, it must be abstract. For example, here is a quote from Marx that is tied too strongly to an empirical state: "Modern industry has established the world market, for which the discovery of America paved the way." As is, the statement can't be used to predict anything because America can only be discovered once. One way of making this statement more abstract is to change "discovery of America" to "geographic expansion." We then can make a usable, testable statement (this is called a proposition): The greater is the level of geographic expansion, the greater will be the level of market development. This is a very important point: *in order to fulfill the goals of science, theory must be abstract and cannot be tied to the context.* So, the goals and assumptions of science necessitate a certain kind of theory.

Further, because a concept is an idea and not an object, we must be very careful to stake out the parameters of the concept through explicit and uniform definitions. A rock, for example, is an object and we can usually tell where it begins and ends. But the case isn't as clear when we are talking about gender. Where does masculinity begin and end? What will count as masculine and what will not? Also, science strives to construct its knowledge in such a way that it is a public activity; that is, the very methods employed to construct the knowledge are explicit and known. Accordingly, scientific theory should contain explicit definitions of all the concepts used so that the knowledge constructed from the theory can be tested and replicated by others.

One further element that we must consider is that scientific theory contains **relational statements.** These are statements that explain the relationships among and between the concepts. The relationship statements contain the variability of a theory. In other words, scientists know how a phenomenon will change because they understand the relationships among their concepts. Think again about the goals of scientific theory and you'll see that these relational statements are necessary. Explanation, prediction, and control imply something dynamic, not static (otherwise there wouldn't be anything to predict). For example, let's take two concepts, education and income. By themselves they simply allow us to identify qualities that appear to be associated with two different entities. However, if we could put them together or relate them to each other in some way, then we might be able to make some predictions. Something like, the greater is the level of education, the greater will be the level of income. (That is certainly what has prompted many of us to attend school.) The phrases "the greater is" and "the greater will be" are relationship statements. In this case, the relationship is positive, that is, they both vary in the same direction in relation to each other (up or down).

There are a couple of basic ways in which these kinds of theoretical statements can be made. These relationships may be modeled—we can draw a diagram that shows the relationships spatially—or they may be stated in propositional form. In a **dynamic model,** the relationships are denoted symbolically: + (positive) – (negative). A **proposition** is a concise statement that proposes relationships among two or more concepts. The "greater is"–"greater will be" statements in the previous paragraph are propositional statements. For those of our theorists who use a scientific approach, we will be using both models and propositions to express theory.

Getting back to the goals of science, I want to note that the issue of social control, in terms of using theory, is a contested idea. It sometimes strikes people as morally wrong to "control" society or individuals. However, most of the early sociological thinkers, like Durkheim, were convinced that a scientific approach could be used to prevent or mitigate the ill effects of social unrest, revolution, and early capitalism. Clearly, if society is objective and behaves according to universal laws, then figuring out how it works and controlling it to eradicate such things as racism and inequality is desirable. It can be compared to medicine controlling germs and human bodies to eliminate polio. The hope of this kind of progress is a defining feature of the Enlightenment and sociology as a science. This perspective is summed up by Jonathan H. Turner (1993), a strong advocate for scientific sociology:

> Sociology can be a natural science; it can develop concepts that denote the basic properties of the social universe; it can develop abstract laws that enable us to understand the dynamics of the social universe. . . . Without the conviction that laws of human organization can be discovered, sociological theory will be arrested. (pp. 5, 13)

Society as a web of signification: Almost since the beginning, the notion of the human sciences has had its detractors. A good example, and one that influenced Max Weber, is William Dilthey. Dilthey argued that the social sciences are intrinsically different from the natural sciences. The biggest difference of course is the subject matter. On the one hand, the natural sciences study physical objects; but on the other hand, the human sciences study something that is defined by autonomy or free will. In other words, one of the things that makes people human is the ability to knowingly choose. While atoms may respond to different external pressures, they can't individually choose their own paths, but people can. This means that the kind of knowledge that can be produced is different. To explain a social event in the scientific sense is to assume an external world (society) that works in a mechanical fashion, according to its own laws, that in turn determines human action. Do you see the problem? If humans are defined by the presence of free will and choice, then you can't assume that there are these external forces (like society) that take away that choice.

There is yet another problem in addition to the dilemma of agency: using scientific theory in sociology assumes that society is an empirical object. Sociologists who make this assumption believe that society is a thing that can be studied like any other thing in the universe. On the other hand, some sociologists assume that society is made up of "webs of signification" or meaning. *And meaning, by definition, is not empirical.* Let's take paper money, for example. There are some empirical things about the bill, like ink and paper. But the empirical elements aren't important to us at all. The meaning, or that which the paper and ink signify, is what's important. And the meaning of money, and every other social object, changes over time and across situations (think about Roman currency or inflation). The important thing is to see that it is the meaning that is primary for humans and that becomes our reality, not the actual thing itself.

Thus, the interpretist perspective makes a problem out of what most objective sociologists take for granted. For example, most sociologists take for granted that gender exists, and then they try to describe how it works to produce inequality. Those who do not assume that society is an object think that gender is an achievement, that it is created. It is itself the problem to be explained: How do we produce gender in face-to-face encounters? How is it brought up? Who brings it up? What does it mean when it is brought up? How does the meaning change? From this perspective, human behavior isn't determined or released by institutional structures; it emerges from social interaction. And because society is a web of signification (meaning), then society emerges as well.

So, interpretists generally focus on meaning and argue that symbolic and subjective meanings are the most important features of social life. But—and here is the important part—meaning does not exist as an intrinsic feature of any sign, symbol, or object. It has to be attributed or added. This feature of meaning, the fact that it has to be attributed, is why I said that meaning is by definition non-objective. Take for example the swastika. For us living in the wake of Nazism, the swastika means bigotry, racism, and the belief in white supremacy. But it hasn't always meant that. The symbol has been found in ancient Chinese culture, Native American culture, and early Christian culture. It has meant goodwill and peace and has symbolized the Christ figure. The meaning does not live in the symbol; it has to be attributed to it. Meaning does not exist in social categories either. We have to create them, and we can create them any way we want. For example, we usually think of gender as having two categories (male and female), but some societies have a third category (the berdache) that is neither male nor female.

Even the meanings of the categories that we do use have changed through time and are different across cultures. African Americans, for example, were at one time in the United States considered less than fully human (more precisely, "African Americans" didn't exist at all). Today that characterization is ridiculous. But what defines a black person today? Racially, an individual may be considered to be black here in the United States but be looked down on and rejected as "half-caste" in Africa. Langston Hughes considers this question of meaning: "Why is Negro blood so much more powerful than any other kind of blood in the world? If a man has Irish blood in him, people will say, 'He's *part* Irish.' . . . But if he has just a small bit of colored blood in him, BAM—*He's a Negro!*" In addition, remembering what we said about gender, race doesn't even have to be a meaning that is produced as a result of interaction at all. What race means at any given time is dependent upon the way people in face-to-face encounters interact around the idea. It can be something that has positive connotations, like appreciating differences, or it can be an attitude that people adapt to denigrate others.

Let's consider this issue of meaning in a different way. Sociologists talk a lot about institutions like education and the economy. But are they empirical (remember that empirical means that you can touch, taste, hear, see, or smell it)? Not really. Think about it this way—if I were to ask you to point to a tree, you would be able to do it without much trouble. But if I were to ask you to point to the institution of education or institutional racism, you would have difficulty doing it. You could point to behaviors, but to actually point to the institution is impossible. Your difficulty indicates that these things don't really exist as separate, objective entities; they only exist in and as the result of interactions. We have to imagine that there is such a thing as the economic institution, and we have to interact around that imagination. As we interact and behave as if a structure is real, it becomes real. It emerges out of our interaction. And, from this perspective, the same is true with everything in our lives. It isn't the empirical, objective world that constitutes human reality; our reality is a meaningful one and thus by definition it is not empirical.

Interpretive theory: Using science as a touchstone, we can see that this conceptualization of the social world demands a different kind of theory. If we assume that society is an object, a thing that exists outside of individuals and interactions, then we can use and anticipate the benefits of scientific theory. If the social world is empirical and operates according to law-like principles, then humans can discover those laws and use them to predict and control society. It follows that theory is a logical argument using abstract concepts and relationships to explain, predict, and control social behaviors. If, on the other hand, society is a web of meanings that are locally produced and understood, then theory cannot be very abstract (lifted out of the local context), and it cannot contain relationships of the kind noted above because the human world doesn't operate according to universal and unchanging laws—people are free-willed, creative, and spontaneous.

When society is seen as symbolic, the kind of theory developed to make sense of it is interpretive theory. *Interpretive theory* is itself an interpretation, a grounded way of understanding the social world of others. Here theory's principal role is to simply provide a "lexicon of valuable concepts, terms, and ways of talking about human behavior" (Wuthnow, 1995, p. 2). This kind of theory is never far from the local context; the concepts that are used are more tied to the setting than they are to scientific theory. And the likelihood of knowledge accumulation is doubtful. As Clifford Geertz (1973) says, "progress [in knowledge] is marked less by a perfection of consensus than by a refinement of debate. What gets better is the precision with which we vex each other" (p. 29). The major goal of such theory is to understand the social worlds of others, to gain "access to the conceptual world in which our subjects live so that we can, in some extended sense of the term, converse with them" (p. 24).

Theories at this end of the spectrum often use analogies, types, analytical frameworks, and/or thick descriptions. To use an **analogy** is to explain through resemblance. Analogies may be borrowed from theater, art, literature, law, or almost any human enterprise. So to illuminate a social world to an outsider, we can use something that the outsider already understands. For example, we may describe the kind of dance that occurs in a mosh pit as "ritualized violence." Here we are using an analogy from religion. Calling the violence ritual means that it conveys symbolic meaning and is not carried out physically, and it implies that the dance is a way through which high levels of emotional energy are created.

Types are usually thought of as ideal categories. As Weber (1949) argues, ideal types have "the significance of a purely ideal *limiting* concept with which the real situation or action is *compared* and surveyed for the explication of certain of its significant components" (p. 93). In other words, these types or concepts work as heuristic devices, frameworks from which to gain some understanding of the situation. We can understand types through the analogy of a yardstick: types are abstract schemes against which we can compare and thus measure a variety of social phenomena, just like we use a yardstick to measure distance in inches and feet. An ideal type is the product of pure logic and not value, as "it has nothing to do with any type of perfection other than a purely *logical* one" (Weber, pp. 98–99). So in constructing an ideal type, the researcher isn't trying to say what would be perfect for the situation, but rather that this measure exists only as a logical idea and not in the social world at all.

Analytical frameworks are sets of concepts that are used to understand an empirical phenomenon. Generally, the concepts used do not have a high level of abstraction but correspond rather intuitively to the situation at hand. Because the purpose of the framework is to analyze or understand and not to predict or control, there are never any relationships explicated within it. The concepts are used as sensitizing devices. They tell the researcher what to look for in the setting. For example, Talcott Parsons (1949) created an analytical framework for "voluntaristic action." Parsons argued that individuals make free choices concerning goals and means in environments that restrict choices. In order to understand why people behave the way they do, Parsons suggested that we look at the actor as goal oriented, the normative constraints (norms, values, and beliefs), the situational constraints (such as the individual's personality system), and the availability of means. The normative and situational constraints influence the choice of goals and means, but not in any predictive way. But the framework does tell the researcher what to pay attention to—goals, means, and constraints. And the framework tells the researcher what kinds of questions to ask—such as, how do the available norms influence the individual's choice of goals?

Thick descriptions are descriptions of social phenomena that are microscopic in detail. Clifford Geertz (1973), borrowing from Gilbert Ryle, argues for thick descriptions through the problem of the eye twitch versus the wink. The difference between the two cannot be photographed, but the differences in meaning are vast. But there's more: the twitch or the wink may be parodied, in which the physical action is neither a wink nor a twitch. Further still, the parody may be practiced—it is then not a wink or twitch or parody. In order to understand the differences between the wink and the twitch and all of the ways that they may be framed, thick descriptions must be used to get at the meaning of the behavior to the actors in the situation. Geertz refers to the thick descriptions and their layers of meaning as "piled-up structures of inference and implication" (p. 7).

Table 1.1 Two Theoretical Paradigms—Science and Interpretation

	Scientific Sociology	*Hermeneutic Sociology*
Assumptions	• The social world is empirical • The social world operates according to law-like principles • Those principles can be discovered and used by humans	• The social world is symbolic • The social world is produced through on-going interpretive moves
Goals	• Explain phenomena • Predict phenomena • Control phenomena • Cumulate knowledge	• Understand diverse social worlds
Theoretical Elements	• Abstract concepts • Explicit definitions • Logical relationships	• Analogies, types, analytical frames, thick descriptions

Bent Flyvbjerg (2001) gives a nice summation of the benefits of this method of theorizing:

> This approach leaves ample scope for readers to make different interpretations and to draw diverse connections. . . . For readers who stick from beginning to end with the minutiae of a case narrative told in this manner the payback is likely to be an awareness of the issues under study that cannot be obtained from "maps," that is, summaries, concepts, or theoretical formulas. (p. 86)

Keep these two extremes (objective, interpretist) in mind when reading our theorists. Remember that the differences between these two perspectives really hinge on assumptions about society. Is society an object? Or is society a web of signification? *What we assume about society determines the way we theorize about society.* I've outlined the differences between these two approaches in Table 1.1

Assumptions Concerning Values

Remember that we mentioned that perspectives contain values? Because theory is a perspective, it implies that all theories contain values. Scientific theory is characterized as being **non-evaluative**. That is, science is not supposed to tell people how to live; it is simply supposed to discover what exists and to develop technology to control the empirical world. However, most philosophers of science today recognize that taking the objectified stance is itself a value orientation. For example, is the earth something that should be respected and should we attempt to live in harmony with it? Or, is the earth simply an object that needs to be controlled for human benefit? Despite what you may feel about ecology, both points of view place a value or meaning on the earth; neither is less of a value orientation than the other.

This intrusion of values on knowledge is even more an issue for the social sciences. Why do we study gender or race relations? Why does someone want to study the problem of date rape or spousal abuse? Neither date rape nor spousal abuse existed 100 years ago. The behaviors existed but they didn't exist as date rape or spousal abuse. They existed as part of the accepted culture of gender—"boys will be boys," men had the right of access, and men had the right and obligation to discipline their wives. What changed the existence of these behaviors and made them a social problem worth investigating? Many factors went into defining these issues, but one of the chief factors is that we began to value women more in our society. The same is true about race and most of the other things with which sociologists are concerned. Thus, values lie at the very heart of the questions we ask of society.

Marx's theory is obviously value based. He doesn't like capitalism and wants to bring about a revolution. But Durkheim's theories are equally value based, though we usually don't recognize it. Durkheim wants to use a form of social engineering to create a morally integrated society. Both of these positions are based on value assumptions. One assumes that society has some deep-seated problems and values social change (Marx); the other assumes that society works like an organism and values integration (Durkheim). Now at first blush, these distinctions sound like the differences between conflict and functional theory, and to a certain extent that is true. However, not all conflict theorists are critical in the way that I am talking about. For example, among contemporary conflict theorists, neither Randall Collins nor Ralf Dahrendorf would be considered a critical theorist. Collins and Dahrendorf simply want to describe and understand how conflict works.

In general, what is at issue here is the way the theorist feels about society. A theorist may think about the way conflict works in a society, but see it as simply part-and-parcel of the way humans do business, as does Simmel. This theorist isn't critical of the conflict; she merely wants to understand it as an element of human behavior. On the other hand, someone else may think that a particular system is oppressive and that conflict is the effect of one **class** trying to rule another. This thinker is critical of the conflict and wants to rid society of it.

The critical perspective: **Critical theory** intentionally takes a condemnatory view of society. *Condemnatory* is a harsh word, and I hesitate to use it. However, it conveys both a value orientation and judgment, and in that sense it is appropriate. According to the *Merriam-Webster Unabridged Dictionary,* to *condemn* is to pronounce as ill-advised, reprehensible, wrong, or evil, typically after definitive judgment; it is to declare guilt and to make manifest faults. Critical theory is explicitly value based and is censorious of current social relations. In general, the critical perspective has four interrelated features: value-rationality, deconstructionism, praxis, and utopian vision. The term *value-rationality* comes directly from Weber. One way to understand what it means is to compare it to another of Weber's terms, *instrumental rationality.* Instrumental rationality is concerned with the most efficient relationship between means and ends. The driving concern in instrumental rationality is utilitarian calculation. Science uses instrumental rationality; this is especially evident in its claim to be value-free. Science isn't concerned with what ought to be; science is concerned with what is and how to best predict or control it. Value-rationality, on the other hand, is vitally and explicitly concerned with finding out and arguing for what is good for people.

Jürgen Habermas, a contemporary social theorist, says that this kind of theory is focused on emancipation—the work of bringing about freedom, growth, and improvement. As Habermas would point out, this kind of theory is engaged in deconstruction. What I mean by the term **deconstruction** is the process through which the taken-for-granted assumptions and ideologies that undergird most of society are dismantled. This deconstruction removes the gloss that hides conditions of domination and constraint. Deconstruction in critical theory began with Karl Marx. Marx exposed capitalism as a system of class oppression, wherein the bourgeoisie dominate and constrain the proletariat. Marx also showed how systems of ideology, like religion, work to hide the power of the elite. This work of deconstruction continues today. For example, for years traditional family relations in this country were taken for granted and assumed to be natural. Part of the work of feminist theory has been to uncover or deconstruct the system of patriarchy that is the source for the traditional roles of husband (head of the household) and wife (homemaker).

Another feature of this kind of theory is **praxis.** The idea of praxis contains a couple of issues, but the most important for our purpose is the notion of transformative action. To be concerned with human emancipation and what is right for humanity, and to dismantle the myths, discourses, and ideologies that hide and construct oppressive systems, isn't enough. Critical theory is overtly concerned with bringing about change—change in and through the theorists' own behaviors, as well as change in the behaviors of the oppressed and the elite. These changes in behavior are meant to bring about social change, and most critical theorists have an idea of what society should change into. This sense of utopian vision is, of course, linked to value-rationality and deconstruction. The critical theorist is one who deconstructs with a purpose—a purpose informed by a value system that privileges one element of society over another.

The descriptive perspective: A theorist may be critical of society or may simply want to describe society as it works. The descriptive point of view is of course closely linked to the goals of science, but the continuum I am developing here isn't a replication of the object–meaning continuum. Science embodies both the objective and descriptive points of view; but it is not the case that theorists who adopt the interpretive point of view are necessarily critical. In fact, as we will see, the critical or interpretive perspective isn't generally found in classical theory, but it is an earmark of contemporary theory.

We can understand our theorists on this critical–descriptive continuum as well (remember that the poles are considered ideal types). Marx, Martineau, Gilman, and Du Bois, for example, are clearly critical of society. They feel that there are certain structures in society that oppress human beings and rob them of their full potential (for Marx it's class, for Martineau and Gilman it's gender, and for Du Bois it's race). Weber and Simmel aren't explicitly critical, but neither do they simply describe society. Weber argues that modernity creates a kind of iron cage out of which individuals and groups will have a hard time escaping. And Simmel is particularly interested in the outsider's experience and the alienating effects of modernity and objective culture (as with gender, for example).

Durkheim and Spencer are both descriptive, though there are deeply held ideas of what constitutes the best society. Both Schutz and Mead are merely concerned with how things work and are more descriptive. It is interesting to note that the poles of our continuum contain theorists who are concerned with consciousness. Marx and Du Bois are very concerned with how people think, especially how they think about themselves. But they see the culture and structure of domination as negatively impacting consciousness (for Marx it is false consciousness and alienation, and for Du Bois it is the issue of black double-consciousness). Schutz and Mead, on the other hand, are interested in how language and interactions affect how people think and feel, apart from any criticism of society.

So, what is society? Is it an object or a web of meanings? Is it filled with awesome structures that overshadow individual choice, or is it only seen in local contexts and mundane conversations? And, what shall we *do* about society? Do ideologies and power relationships permeate society? Shall we strike out quixotically and seek to destroy the devils? Or, shall we approach society as an automobile engine, delving into its hidden mechanisms and figuring out how it works? I don't know the answers, but these are exciting questions. With each question we turn society and catch a different facet of the human rainbow. So, what do you think? What is society and what do you want to do about it?

The Beginnings of Theory and Learning to Theorize

Thinking: Understanding theory is more than simply knowing what Durkheim or Martineau said. It is more than content; theory is something that you do. A theory course shouldn't simply give you new terms or more information about famous people; it should give you eyes to see and ears to hear. When you get done with this book, I don't want you to simply be able to tell me what Schutz's concept of "reciprocity of perspectives" means; I want you to be able to see the world through Schutz's eyes. Of course, a person could spend a good number of years studying Schutz or Marx or any of these people, so we're just going to be able to enter into a conversation with them in this book. But I hope that conversation whets your appetite. I hope you end up wanting to theorize about your world—I hope in the end you are hungrier to lift the veil and connect the dots.

> I have seldom seen one of these young men, once he is well caught up, in a condition of genuine intellectual puzzlement. And I have never seen any passionate curiosity about a great problem, the sort of curiosity that compels the mind to travel anywhere and by any means, to re-make itself if necessary, in order *to find out.* (C. W. Mills, 1959, p. 103)

Theory, or more properly, theorizing, is an active disposition of the mind. Theorizing is a dynamic relationship that exists between the individual and his or her environment. In other words, theory is about thinking, and thinking is purposive action. Thinking doesn't just happen—we have to intentionally think. It's a behavior; it's action. In order to think, we have to pay attention to what is in our brain and the world around us. Thinking is what theory is all about. Theory is the content, action, and product of our thought.

Theory is the content of our thought in that it influences what we see and don't see. That's why there is a difference between sociological/social theory and psychological/individual theory—the content is different. So, what you think is extremely important. I want to help you think sociologically and socially. That really is a difficult task given the cultural preference for individualistic thinking in our society.

Passionate curiosity: Theory is a product of thought because we generally form theory in response to questions or problems. As Mead tells us, we think the most when we are faced with a problem to solve, like where to put the TV and the speakers and the sofa so we can experience our Home Theater to its fullest potential. Generally, adults spend much of their time walking around on automatic pilot—we've solved most of the mundane problems we are faced with and function on routine. But when something is different or problematic in our environment, we think about it. For example, the other day I was driving to work, a route I've taken more times than I can count. My route was routine. However, on this day there was an accident. When I saw the backup of cars, I began to think. I took into account the possible delay, when I needed to be at work (parking is terrible at our university), and potential alternative roads. What I came up with was a simplistic theory about the way traffic works.

The centrality of questions and problems to theory is at the heart of the quote by C. W. Mills (if you skipped it, go back and read it—it's important). Mills is arguing against a certain kind of sociology, one that is driven by data and methodology. His "young men" (and most were men) were graduate students and new sociologists who were enamored with statistical techniques for understanding society. He refers to them as research technicians rather than social researchers. While his argument is salient, I want us to focus on how he characterized theory: a condition of genuine intellectual puzzlement and passionate curiosity—a condition so intense that a woman or man is willing to *remake the mind in order to find out.* Theorists are driven by such curiosity and puzzlement—they are driven by questions.

Social theory as we know it was born out of the ferment of revolution, social change, and science. People began to think that social things mattered and that society was a thing in itself (apart from individuals and apart from God) because of some very serious social changes. These changes took place over an almost 200-year period, beginning with the Enlightenment, but became acute during the nineteenth century. The beginning of the nineteenth century saw the end of the French Revolution, a revolution that had lasted close to 20 years. The effects of the revolution were felt throughout Europe and the United States. The French Revolution had followed on the heels of the American Revolution and people became very concerned over the issues of social integration and disintegration: what holds a society together; what makes it fall apart? The concerns of these revolutions—inequality and social order—also became part of the **collective consciousness**.

Another social change that happened at about the same time was the Industrial Revolution. The Industrial Revolution had a number of profound effects. It moved masses of people from the countryside into the cities (urbanization). It also changed the face of work forever—there was now a machine between a person and the work of her or his hands. It produced new social classes (the bourgeoisie and proletariat) and created a new gulf between classes, one based on the ownership of the means of production. Industrialization also facilitated two new emotional orientations and ideologies. Some people felt that industrialization would bring hope and equality (through products for the masses and the trickledown effect); others felt that it would bring despair and pronounced inequality (factory work and elite capitalism). Industrialized capitalism helped change the configuration of society forever. Because of capitalism and nationalism (the idea of the **nation-state** was another new notion at this time), religion was pushed to the side; it no longer was the heart of social life. Politics and the economy came to dominate.

I'm not interested at the moment in detailing the history of these changes; my point isn't to have you learn the historical milieu out of which sociology and theory arose (though from a sociology of knowledge standpoint it is of interest). The thing I'm most concerned with is that you see that massive social changes occurred and these created *genuine intellectual puzzlement and passionate curiosity* in the minds of people who could see past their own private concerns. These people re-constructed their minds, and the way we think about society, in order to find out what these changes meant.

What I want you to see is that we theorize in response to a puzzle, a mental itch. Concerns, problems, issues, and tensions are what drive theorizing. Why does crime exist? Why does the crime rate change from one society (or time) to another? Why are certain types of people more likely to commit crimes than others? Why do women make on average 70% of what a man makes? How does or can society change? How do movies and television influence how we think about ourselves? What is the link between media and violence? In order to answer any of these questions, and the millions of others we can list, we must use theory.

Let me ask you, what are your questions? Is there something that burns in your mind such that you are willing to "to travel anywhere and by any means, to re-make [yourself] if necessary, in order *to find out*"? I hope so. To think is the most human of all vocations. Theory is thinking—it's actually *hard* thinking, and it can only begin with a question. Charles Lemert (2000) points out something interesting about theory: According to the *Oxford English Dictionary,* the word *theory* comes from the Greek word *theoros,* which means "one who travels in order to see things" (p. 289). Questions are important because over the next several hundred pages of this book, I'm going to be giving you answers, answers that have come to us from some of the most brilliant minds that have ever lived. But in order to appreciate the answers, we have to be captured by the questions. Part of what I hope to do in this book is to excite your mind to *think*—to think sociologically, critically, reflexively, and theoretically about your world. It's an amazing world, and theory is an amazing way to get into it. In some ways, I hope you end up with more questions than answers after reading this book.

In order to ask questions, you have to pay attention. As social theorists, we must immerse ourselves in social life. We must feel it and taste it; it must become a sensual experience for us. As we immerse ourselves, we must stand reflexively inside our self, and we must attempt to stand outside our self as well. To stand reflexively within our self means to experience our point of view in a critical manner. What

I mean by critical is that we have to question things that we take for granted. For example, most white people are unaware of privilege. They are unaware that they aren't routinely pulled over by police because of their race. Most teenagers and young adults aren't aware of the lack of institutional responsibility that befalls their status position, not yet having to take on the responsibilities of family and career. And most of us aren't aware of how we form queues or lines. But more than that, most of us are unaware of the myths, ideologies, discourses, and social structures in back of things that we take for granted every day—things like a driver's license or city planning or newscasts.

So, pay attention as you watch TV, or stand in a line, or go on a date, or shop at the mall, or drive or walk down the street, or when you see a homeless person. Ask questions, and you've begun the theorizing process. I would love to have everyone stop right here and dialog about questions (What questions about society drive you?). The reason that I would like to stop is that it generally isn't good to answer questions that haven't yet been asked. That's what theory is—answers to questions. But truth be told, there is benefit in learning about the theorists even before you have your own questions. One of the reasons for this is that the work of many theorists focuses on a few themes: knowing their concerns can sensitize us to asking questions about the most important social factors.

Themes in Classical Theory

In this book, we are going to place emphasis on the individual perspectives of the theorists; but in doing so, I'm not implying that there aren't commonalities. Many of our thinkers were concerned with a handful of issues, most of them being defined by the culture that grew out of the Enlightenment. Thinkers in the Enlightenment were very concerned with such things as religion, human rights, economic relations, government, and so on. Taken together, these issues constituted what came to be known as **modernity**. We can think of modernity, in the social sense, as the time period brought on by politics surrounding the nation-state, industrialized capitalism, urbanization, massive changes in communication and transportation technologies, the dominance of science and technology, the spread of rational organizational technologies (bureaucracy), rapid and increasing commodification brought about through markets, and so on. The issue of modernity and its social changes was uppermost in the minds of most of our early theorists. Many came up with typologies to understand the differences, such as Durkheim's organic and mechanical solidarity or Simmel's organic and rational group membership.

Part of what I'm going to do in this book is trace what our theorists have to say about modernity. By doing so, what we'll see is that certain themes develop and get played out differently depending on the theorist. The reason I want to pick out certain issues revolving around modernity is that there are many contemporary theorists who argue that we are no longer in the modern era. The time period that we are living in now has been variously termed **postmodern**, late-modern, postindustrial, and so forth. I want us to seriously consider the possibility that we are living in an age of change as profound as the Enlightenment and the shift from traditional to modern society. Substantial transformations occurred because of this historical shift, not only in terms of social structures and arrangements, but also in terms of our individual experiences. The implication, of course, is that if we are living in a postmodern society, then not only is the society that we live in different than that of your parents and grandparents, but your experience of yourself as an individual is different as well. Thus, in each of the chapters, I will conclude our discussion by pulling out a few features that the theorist seems to think are defining characteristics of modernity and comparing them to what post- or late-modern thinkers are saying. As we go along, then, I will be expanding our understanding of what the ideas of modernity and postmodernity mean. I want us to be captivated and provoked by the possibilities of our era.

In addition to seeing that historical periods are important, I also want to call your attention to other themes that the theorists visit over and over. These themes won't necessarily get pulled into our discussion of modernity and postmodernity, but they are extremely important to note and trace. Among the more significant of these themes are religion, equality/oppression, empiricism, culture, and economic relations. The concern with religion is a direct result of the Enlightenment. One of the targets of Enlightenment philosophers was religion. They were particularly critical of it because it is the opposite of logic, rationality, and progress—in other words, everything that would make an individual or society enlightened. But they were also curious about the purpose that religion served and how it related to other social structures; for example, the idea of the separation of church and state came out of the Enlightenment.

A kindred concern is culture. This modern interest in culture is of course linked with the Enlightenment's concern with what counts as knowledge, truth, and reality, as well as the defining characteristics of human nature. I'm stretching the meaning of culture here a bit, because what we mean by culture today is somewhat different than the way it was used in the past. Nevertheless, in looking back at our theorists, we can see a strong and abiding interest in those things that we would today gather under the category of culture—things like Marx's species being and false consciousness, Weber's notion of rationality and disenchantment, and Durkheim's idea of the collective consciousness. Things shift a bit as we move through our theorists and thus through time. Weber began to be concerned with meaning, but Schutz centered his entire theory on the question of meaning. Meaning, and thus culture, is at the heart of Mead's theory, and Simmel was clearly concerned with the preponderance of what he calls "objective culture." Du Bois talks about double consciousness, and for Parsons, culture is the joystick of the cybernetic hierarchy of control.

Another theme that clearly stands out is equality and oppression. Of course, most of us know that Marx was concerned with class oppression, but he and Engels were also concerned about gender inequality. Simmel as well wrote extensively about gender issues, as did of course Martineau and Gilman (recent additions to our pantheon of classical theorists). In addition, Weber was extremely concerned about things like race and gender but in more abstract terms: he referred to them as status groups. And Du Bois was explicitly concerned with race. What I want us to see here is that the Enlightenment's concern for equality and tolerance threads its way in many different forms throughout our theorists' work. And, interestingly enough, the ideas of equality and tolerance get tangled in fascinating ways according to postmodern thinking.

Three additional themes worth mentioning are the economy, empiricism, and social cohesion and change. We'll see that many of our theorists talk about capitalism, markets, money, and the division of labor. We will also see that there is an argument for, or at least acceptance of the idea of, empiricism with all of our earlier theorists. But beginning with Weber, we see a move away from empiricism and toward interpretist schemes. Mead even offers us a different approach to truth and reality called pragmatism. When we consider postmodern issues, we will see that there are strong arguments against empiricism and even the certainty of meaning that Schutz and Mead seem to assume.

I mentioned earlier that most of our thinkers were concerned with the enormous social changes that were occurring at the time; in their theories, this concern became the recurring theme of social cohesion and change. The idea of social cohesion refers to those factors that hold society together. With the loss of traditional authority and the de-centering of religion, the question of social cohesion became paramount. Within the question of what holds society together is the question of change: How do societies change? Can societies change without falling apart, or is revolution and upheaval necessary? We will revisit these themes as we move through the theorists and reflect on a few of them in the concluding chapter.

This book is an invitation to you to see your world theoretically—it is an amazing experience to have eyes to see and ears to hear. Theory is what supplies those eyes and ears for sociologists. In the remainder of this book, we will be looking at individual theorists, people who have changed the way we look at our world. There are three things that I want you to learn from the theorists that we will be considering. The *content* of their theory is one of them, of course; and the content includes concepts and relationships (potentially). But I also want us to experience their *perspectives.* The way these people saw the world is just as important as what generally passes as their theory. I want you to be able to see the world as Mead or Du Bois or Marx. The world changes as you step into each of their shoes. More importantly, in back of both the theory content and perspectives are the *questions,* and exposing ourselves to their questions will help us see and question the social world for ourselves.

> Theory is our collective memory, the brain center in which we store the basic elements of what we have learned and the strategies we have available to carry us into the future. (Collins, 1988, p. 8)

Summary

- All knowledge systems, including sociology and theory, are based on perspectives. Perspectives are founded upon assumptions and values, and they contain ideas, concepts, and language. Perspectives help us to see certain things, but they also blind us to others.

• Sociology as a whole is built upon a contextualizing perspective. Everything that sociologists study, whether it's an institution such as the state (macro level), an organization like your university (meso level), or an individual (micro level), is assumed to be social. That is, everything human is in the main created, influenced, and maintained by—and has its purpose in—social relations.

• The most basic assumption that sociologists make concerns the way society exists. This assumption runs on a continuum from supposing that society exists as an object, external to and coercive of the individuals and groups that make it up, to positing that society exists symbolically through the interpretive moves of individuals in interaction. There are thus two ideal types of theory: Scientific theory assumes that society has an objective existence (structures that operate according to law-like principles) and is therefore defined as a formal and logically sound argument that explains an empirical phenomenon in abstract terms and dynamic relationships. The purpose of scientific theory is to explain, predict, and control. Interpretive theory, on the other hand, assumes that society exists symbolically (through different cultural signs and practices) and is itself an interpretation that uses analogy, types, analytical frameworks, and thick descriptions. The goal of interpretive theory is to understand and convey the cultural meanings and practices of one group to another.

• The second assumption that sociologists make results in a value orientation toward society. This orientation runs on a continuum as well, from believing that sociologists and their theories ought to simply describe how society and/or its different elements work, to holding that sociologists and theories ought to be engaged in making human life better (by being critical about the present configuration of power and pointing the way to improved social arrangements).

• Learning theory is more than memorizing concepts, definitions, and relationships. Intrinsic to theory is the action of theorizing: theory is a particular way of looking at the world that is founded upon active thinking and questioning. Theorists are immersed in their social worlds; they think about them and question them continually.

• Generally speaking, classical theory is concerned with several themes, including but not limited to modernity, the economy and religion, culture, equality and oppression, social cohesion and change, and empiricism.

Building Your Theory Toolbox

Conversations With Perspectives, Assumptions, and Theory—Web Research

• One of the things that we noted in this chapter is that theory is a perspective and that perspectives contain assumptions. Using Google or your favorite search engine, define the three main perspectives in sociology (functionalism, conflict theory, and symbolic interaction) and list at least two assumptions for each perspective.

- Contemporary sociological theory has many different perspectives. Some of these perspectives are rational choice theory, feminist theory, dramaturgy, ethnomethodology, world systems theory, queer theory, and postcolonial theory. Using Google or your favorite search engine, find the definition for at least three of the perspectives listed. Did you find any other perspectives in your search?
- In this chapter, we made distinctions among critical, scientific, and interpretist theory. There is another distinction that is made quite often in sociology, that between sociological theory and social theory. Use Google or your favorite search engine to define sociological and social theories. What are the differences and similarities between the two? Do you define yourself as more of a social or sociological theorist? Why?

Further Explorations—Readings

Scientific Sociology:

Comte, A. (1974). *The positive philosophy.* (H. Martineau, Trans.). New York: AMS Press. (Original translated work published 1855)

Durkheim, E. (1938). *The rules of sociological method* (S. A. Solovay & J. H. Mueller, Trans.; G. E. G. Catlin, Ed.). Glencoe, IL: The Free Press. (Original work published 1895)

Turner, J. H. (1993). *Classical sociological theory: A positivist's perspective.* Chicago: Nelson.

Interpretist Sociology:

Blumer, H. (1969). *Symbolic interactionism: perspective and method.* Englewood Cliffs, NJ: Prentice Hall.

Pressler, C. A., & F. B. Dasilva. (1996). *Sociology and interpretation: From Weber to Habermas.* Albany: State University of New York Press.

Weber, M. (1949). *The methodology of the social sciences.* (E. A. Shils & H. A. Finch, Trans. & Eds.). New York: The Free Press. (Original work published 1904)

Critical Sociology:

Flyvbjerg, B. (2001). *Making social science matter: Why social inquiry fails and how it can succeed again.* Cambridge, MA: Cambridge University Press.

Habermas, J. (1972). *Knowledge and human interests.* New York: Beacon.

Horkheimer, M., & Adorno, T. (1976). *Dialectic of enlightenment.* New York: Continuum. (Original work written in 1944)

Smith, D. (1987). *The everyday world as problematic: A feminist sociology.* Boston: Northeastern University Press.

Theoretical and Sociological Ideas:

Bauman, Z., & May, T. (1990). *Thinking sociologically.* Oxford: Blackwell.

Hurst, C. E. (2000). *Living theory: The application of classical social theory to contemporary life.* Needham Heights, MA: Allyn & Bacon.

Kivisto, P. (2004). *Key ideas in sociology.* Thousand Oaks, CA: Pine Forge Press.

Further Explorations—Web Links

http://ryoung001.homestead.com/Theorists.html (Web page with links to lots of information about the early founders and classical thinkers in sociology)

http://www2.fmg.uva.nl/sociosite/topics/sociologists.html (International site with links to contemporary and classical thinkers)

http://www2.pfeiffer.edu/~lridener/DSS/DEADSOC.HTML (The Dead Sociologists' Society—general information on classical theorists in sociology)

http://www.wsu.edu:8000/~dee/ENLIGHT/PHIL.HTM (This Web page gives a good account of the Enlightenment)

CHAPTER 2

Organic Evolution—Herbert Spencer (British, 1820–1903)

> *Societies, like living bodies, begin as germs—originate from masses which are extremely minute in comparison with the masses some of them eventually reach.* (Spencer, 1876–1896/1975a, p. 451)

It is perhaps fitting that our book begins with Herbert Spencer and ends with Talcott Parsons. Spencer was one of the most influential theorists of the nineteenth century, and Parsons was one of the most significant theorists of the twentieth century. In many ways, Spencer and Parsons have more in common than any other two theorists in this book. Yet Parsons begins his first great theoretical work with the question, "Who now reads Spencer?" and concludes that "Spencer is dead" (Parsons, 1949, p. 3). Nevertheless, we have to acknowledge—and Parsons does—that Spencer must have had an incredible influence at one time for Parsons to begin his tome with such a question. At one point Spencer was fundamental to sociology. Spencer was one of the most widely read writers of his time—his works sold somewhere between 500,000 and 1,000,000 copies in his day. However, Spencer has now been neglected for a good many years. Jonathan Turner (1985), a normally staid theorist, declares at the beginning of *Herbert Spencer: A Renewed Appreciation,* "I have written these pages because I am distressed, even angry, about how contemporary social theorists have treated Herbert Spencer" (p. 7).

Yet while Spencer draws such strong reactions, we cannot make many *direct* connections between Spencer and contemporary sociology, which in itself is curious. Why has such an obviously strong sociological voice been silenced? Turner (1985) points us in the right direction: "At a time when social theorists genuflect at the sacred works of St. Marx, St. Durkheim, and St. Weber, we spit on the grave of Spencer because he held a moral philosophy repugnant to the political biases of many contemporary theorists" (p. 7).

Spencer was a product of his time, as we all are, and believed in social evolution (the notion that there are superior social forms—like the Western European model of technological advancement). It was actually Spencer that coined the phrase **survival of the fittest**, not Darwin. The idea of **social evolution**, or **social Darwinism** as it has been called, is that certain social forms are selected for their ability to better survive in the environment. Natural selection in evolution is kind of like an invisible hand that chooses some forms to live and some to die out. Of course, no real "choice" is made in the way we think of it; some forms (either organic or social) are simply better equipped to survive than others in a particular environment. Natural selection in organic evolution is the reason science gives for why some organisms are only found in certain environments (like the Polar Bear in arctic regions). In the most extreme versions of social selection—eugenics and manifest destiny—it proposes that certain races or groups of people are inferior. However, Spencer did not advocate such extremes, and, as Turner (1985) points out, "Spencer's sociology and his moral philosophy are written in separate places and are worlds apart. One finds far less moralizing in Spencer's sociology than that of either Durkheim or Marx" (p. 7).

Even given Turner's point, why should we care about Spencer today? There are really two reasons we should study Spencer. First, a good deal of what sociologists have been doing since leaving Spencer is "reinventing the wheel." Because Spencer has been neglected, many of his insights have been rediscovered by sociologists ignorant of his work—and there are still many more profound insights in Spencer's work waiting to be unearthed.

Second, more clearly than any other thinker, Spencer gives us a mental space from which to view society in macro-level, functional terms. It was actually Herbert Spencer that formulated the basic tenets of modern functionalist thinking, though his influence has long since been forgotten. Thinking about society in system terms—as responding to brute mechanical forces such as matter, force, and motion—or in structural terms can yield unique and profitable insights, particularly today as the world appears to be moving toward global systems. How are we to understand the relations among nations? For example, is it possible that the institutional arrangements within different nations will influence competition in global capitalism (see Jeffrey A. Hart, 1993)? The United States has different structural configurations among education, economy, and government than do France, Germany, Britain, and Japan. Do those differences matter? Do they make one of these nations more competitive than the others? To answer such questions, we must have a perspective similar to Spencer's.

Spencer in Review

- Spencer was born on April 27, 1820, in Derby, England, the eldest of nine children, the only child to survive to adulthood.
- His father, George (a religious dissenter and Benthamite), was a school teacher and taught Herbert at home until he was 13, after which the boy continued his training with his Uncle Thomas. Part of his schooling was taken up with reading aloud from Harriet Martineau's *Illustrations of Political Economy.*
- At age 17, young Spencer decided he wasn't cut out for a university education and worked as a railway engineer for about four years. Afterward, he supported himself as a journalist and editor (at *The Nonconformist* and the

London *Economist*) until his uncle died and left him money. From that point on, Spencer lived as a private intellectual.

- In his time, Spencer was a radical thinker arguing for technical progress, individual freedom, and liberty; he was a son of the Enlightenment, an evolutionist, and he advocated *laissez-faire* capitalism. Yet, obviously, Spencer was not radical in the same vein as Marx; he lived during the time of mid-Victorian affluence and enjoyed the benefits of the English middle class.
- Though most of the people who intellectually influenced Spencer were his contemporaries (through conversations and readings at various clubs, like the Athenaeum and X Clubs), we can safely say that he was influenced by the works of Thomas Malthus, Adam Smith, and, to a lesser extent, Auguste Comte.
- The title of Spencer's first book indicates his positivistic stand and his Enlightenment concerns: *Social Statistics: The Conditions Essential to Human Happiness Specified, and the First of Them Developed* (1851). He eventually published many defining books in philosophy, psychology, education, and sociology.
- After a lengthy illness, Spencer died on December 8, 1903.

The Perspective: The Evolution of the Universe at a Glance

Okay, let's start off with a bang. How did the universe begin and why does it look the way it does now? We don't want to mess around in this book, so let's go for the big bucks: after you've answered the first question, tell me how society fits in with that explanation of the universe. In other words, is society a natural part of the evolution of the universe, or is it something separate? If it's part of the universe, how does it fit in? Let me make that question a bit more tangible: How are the Milky Way Galaxy (roughly 150,000 light years in diameter, containing about 100 billion stars), the United States, and you related? Those are some big, massive, and maybe silly questions. Personally, I wouldn't know how to begin to answer them, but Herbert Spencer did. In fact, we could actually use his theory to answer each of these questions, because of his particular perspective: Spencer sees things from a positivistic, grand theoretical, evolutionary point of view.

Positivistic grand theory: It is fitting that the first theorist we consider in a book of social and sociological theory is a man who believes in positivism. **Positivism** has been at the center of sociology since its inception. Of course, not all of our theorists adhere to positivism, but most of them orient themselves to sociology through positivism in one way or another. We addressed this issue in our first chapter on theorizing—positivism is one of the philosophies in back of science. Positivistic sociologists assume that the social world is objective.

As a positivist, Spencer seeks to uncover the laws of society through which we can understand, predict, and control our lives. In an interesting move, Spencer (1873/1961) argues for the existence of these laws by appealing to a common belief: "If there is no natural causation throughout the actions of incorporated humanity, government and legislation are absurd . . . social sequences having no ascertainable order, no effect can be counted upon—everything is chaotic" (p. 41). As we will see throughout this book, there are many who do not ascribe to positivism or the idea of causal laws. However, even these people act as if there are discernable causal sequences in social behavior. On a mundane level, every one of us acts as if human behavior is fairly predictable, in elevators, shopping malls, and so on. Spencer's example is more formal: if there are no predictions that can be made about human behavior, why then do we pass legislation geared toward *affecting* human behavior? Obviously, Spencer concludes, society would be a difficult if not impossible enterprise if there were no law-like forces in back of human action. "And if so, it behooves us to use all diligence in ascertaining what the forces are, what are the laws, and what are the ways in which they co-operate" (p. 41).

Spencer was not formally trained in school and never held an academic position. He was not a Ph.D., but he hired Ph.D.'s for his research assistants. He was a recluse who frequented coffee shops and other places of public discourse in order to pick the minds of the brightest intellectuals. As Lewis Coser (2003) puts it in *Masters of Sociological Thought*, "Spencer absorbed his science to a large extent as if through osmosis, through critical discussions and interchanges with his scientific friends and associates" (p. 110).

This kind of background, coupled with the spirit of the age, made Spencer rather unique in the way he saw the world. Because he was not trained in any one discipline, he wasn't restricted by any disciplinary parameters. Spencer wrote freely in biology, philosophy, psychology, sociology, and anthropology. Many of his books were the standard texts of their time—such as *First Principles, The Principles of Biology, The Principles of Psychology, Principles of Sociology.* In addition, at the time that Spencer wrote, the educated world at large held great hopes for progress through science. The Enlightenment and positivism held immense sway; people saw very few boundaries and felt that little was out of their reach.

The result of these two influences is that Spencer formed a grand theory. A **grand theory** is one that explains all phenomena through a single set of concepts. Grand theorists generally assume that everything in the universe operates in system-like ways, and that all systems are built on the same principles and subject to the same dynamics. So biology, psychology, sociology, and physics are all similar systems and can be ultimately explained using the same theory. One of the things that thinking as a grand theorist implies is that you tend to see things working in very abstract and mechanistic terms. The theory absolutely has to be abstract in order to embrace all the phenomena in the universe. For instance, in a grand theory you couldn't use simple psychological terms to explain psychology because they wouldn't apply in sociology or biology—the terms of a grand theory must be more abstract than any one discipline. Thus, the terms of Spencer's theory tend to be very general and as a result perhaps non-intuitive on the surface.

Evolutionary dynamics—matter, force, and motion: According to Spencer, the basic building blocks of the universe (including society and the individual) are **matter, force, and motion.** Keep in mind that Spencer was writing around the middle of the 1800s, so his physics is a bit out of step with contemporary knowledge, but not by much. Physicists today, somewhat like Spencer, argue that the objective universe is made up of material substance (matter) and energy (force). In fact, one of the principles of Einstein's theory of relativity is that matter and energy are equivalent and interchangeable. However, the importance of Spencer's notions of matter, force, and motion doesn't lie in the accuracy of his physics, but, rather, in how he applies it to understanding society. The basic application is that Spencer's chief dynamic in his social theory of evolution is the movement (force and motion) of populations (matter). Populations are set in motion through various levels of force, such as population growth or immigration.

> It is true that Evolution, under its primary aspect, is a change from a less coherent state to a more coherent state, consequent on the dissipation of motion and integration of matter; but this is far from being the whole truth. Along with a passage from the incoherent, there goes on a passage from the uniform to the multiform. Such, at least, is the fact wherever Evolution is compound, which is in the immense majority of cases. (Spencer, 1862/1896, p. 291)

One important inference that Spencer made from his theory of matter, force, and motion is that once matter is put in motion by a force, it will continue in its initial direction until it meets another force or piece of matter. When a moving object meets another object or force, the force behind the initial object will transfer to the second object or force. Think in terms of a pool game. If you hit the cue ball, it will continue moving along its initial path until it hits the edge of the table or another ball. When it hits another ball, both the cue ball and the second ball are affected. Spencer also notes that the force can dissipate due to friction if it doesn't hit another force or piece of matter. Again, the physics here aren't really important for our consideration, but we'll see how these ideas of continuity of motion and persistence of force play themselves out in his social theory, beginning with the dynamics of evolution.

Spencer argues that evolution generally occurs due to three dynamics: the instability of homogeneous units, the multiplication of effects, and segregation. The idea of the **instability of homogeneous units** contends that any group of similar and unconnected entities is intrinsically unstable. Think of our pool game again. Just before the break, all the billiard balls are bunched together and form a group. But because they are all the same and are disconnected, when the cue ball strikes them they will scatter and the group will be gone. However, if the cue ball hits a group of balls of different sizes, weights, and shapes, and if the group is connected with long strings, then the force and movement will be dispersed differently, and the group, while change is inevitable, would still remain a group. Spencer argues that it is like that for social groups. Simple, unconnected, homogeneous groups are unstable. The sameness of the units within the group means that they all have the same resources to respond to changes in the environment. Thus, if one unit doesn't have the necessary resource, then none of them do. Because this kind of uncertainty and instability threatens survival, there is an evolutionary "pressure" for groups to change to more complex forms.

Segregation posits that if a substance or unit is isolated from others, it will be subject to different environmental forces. Here our pool game analogy wears thin, but we can use it to see a little of what Spencer is talking about. After the break, the balls scatter to different parts of the table and some fall in the pockets. Because the balls are in different sections of the table, they will be subject to different forces when you hit the cue ball—some will be hit, others won't, and so on. The problem with the pool game analogy here is that you don't see any changes in the balls or the pool table. The game is too short. However, if the game was played repeatedly over a thousand-year period of time, the balls and the table would exhibit different patterns of wear. Different parts of the table and diverse balls would exhibit unique patterns of wear because they are segregated one from another. Further, as these differences evolve, they also influence one another through increasingly different patterns, so the initial effects would become multiplied and more pronounced (**multiplication of effects**).

Let me give you what may be an easier example to hold onto. In doing this, I'm going to get ahead of myself a bit, but I think it's important for you to see how Spencer is thinking. In early human societies, all the necessities of life came from one social structure: kinship. As societies grew, different structures developed and fulfilled special functions. Thus, family didn't take care of providing food and serviceable goods, the economy did. Because the structures became separate and different from one another, they were subject to different environments and had different relationships with the other structures in society. For example, in modern society, the structure of education has to contend with very different forces than does the economy. As a result of these different forces, these structures will continue to change (evolve) in different ways over time. The effects of differentiation are multiplied.

Evolutionary progress—simple to complex organisms: We've seen a bit of how physics influences Spencer's evolutionary thought; now let's look at some of his intellectual roots in biology. Thomas Malthus, Karl Ernst von Baer, and Charles Darwin were among those in the biological sciences that influenced Spencer. Malthus is most noted for his theory of population growth. He argued that populations tended to grow in size until checked by the four horsemen of the apocalypse: war, pestilence, famine, and disease. Spencer takes Malthus's emphasis on population growth and his use of competition and struggle but argues that these dynamics result in social evolution, not tragedy. From von Baer, Spencer adapts the argument that biological entities evolve from single-celled, undifferentiated organisms into highly differentiated and specialized forms. Spencer argues that the process of social evolution is from simple societies to complex societies, with the complex societies containing those differentiated and specialized structures that have survived and increased survival chances. The relationship between Spencer and Darwin was mutually beneficial: Darwin gleaned a great deal from Spencer; Darwin's ideas in turn encouraged Spencer to view the environment as specifically influential in forming the structure of society.

In brief, then, from the biological thinking of his day, Spencer takes the following ideas: 1) the central place of competition in forming societal and individual differences; 2) the movement from simple to differentiated structures in evolution; and 3) the differential effects of environments produce differences in systems (social or otherwise). Generally, Spencer argues that more complex forms are able to respond in more diverse ways to changes in the environment. They are better able to handle external threats because not every structure is directly threatened, and because there are specialized structures. So, for example, the human body has a specialized immune system that can respond to a variety of threats and is able to adapt to changes in the kinds of danger experienced. It can do a good job because it is not also responsible for digesting food or taking in oxygen from the atmosphere. Because of differentiation and specialization, each subsystem is better able to respond to challenges. According to Spencer, the same is true for society: increasing complexity and specialization lead to enhanced survivability.

An implication of thinking in evolutionary terms is use of the organismic analogy. An analogy is a way of explaining something through comparison. Analogies help us understand new ideas. So, for example, we may not understand the difference between the brain and the mind, but someone might be able to help us by using the analogy of the computer—the brain is like the hardware and the mind like the software. In Spencer's case, he understands society in terms of an organism.

The **organismic analogy** is a way of looking at society that understands the form of society and the way society changes as if it were an organism. For example, in order for you to survive, you need oxygen and you get your oxygen from air. Because of that need and because of that source, your body has a specific structure built inside it—your lungs. It also has other structures and systems (such as the circulatory system) that aid the lungs in fulfilling this need. Fish don't have lungs and neither do plants. They have other structures and systems that are designed to meet their specific needs within their environment.

There are three specific theoretical ideas that come from looking at society like an organism: the relationship between needs and functions, systems thinking, and systemic equilibrium. Societies have specific needs, just like your body. Sociologists call these **requisite needs.** The term underscores the idea that these needs are required for survival. Societies develop certain structures or institutions that meet those needs. Just like your lungs are built differently than your stomach because they fulfill a different function, so, too, different social institutions are built differently from one another because they function to fulfill diverse social needs. Another similarity with your body is that institutions are taken for granted. That is, they work without our usually being aware of them. In fact, the more we are aware of social institutions, the less power they have over our lives.

Thinking in terms of society as an organism also implies that society works like a system. *Systems* are defined in terms of relatively self-contained wholes that are made up of variously interdependent parts. That's a mouthful, but it really isn't as daunting as it might seem. Your body is a system and it's made up of various parts and subsystems that are mutually dependent upon one another (your lungs would have a hard time getting along without your stomach), each part contributing to the good of the whole. Thinking about the body also lets us see that systems are bounded. That is, they exist separate from but dependent upon the environment. There's a definite place where your body ends and the world around you begins. In that sense, your body is relatively self-contained. We have to negotiate the boundary between ourselves and the physical environment, but they are separate. As we will see, that negotiation is part of the needs of the system. So, for Spencer, society acts like a system with mutually dependent parts that are separate from but interacting with the environment.

In addition, systems tend toward **equilibrium.** If you are reading this book, then you are alive. If you are alive, your body is regulating its temperature, among other things. The point is that it is probably either hotter or colder in the environment than your body wants or needs. Your body senses the difference between its goal (98.6 degrees Fahrenheit) and its environment and uses more or less energy in heating itself. Your body keeps itself in balance. Spencer posits that society does the same kind of thing: the social system has internal pressure mechanisms that work to keep society in balance. One of the things that we will notice as we move through Spencer's theory is that change and integration are social responses to system pressures, rather than individual people making decisions. According to Spencer, the mechanisms for integration and change are *in the system itself,* not in the people who live in the system.

Taking these ideas together, to think like Spencer is to see the world in mechanistic terms. Spencer is a positivist; he sees the universe governed by invariant laws and he gives little place to the influence of individual people. Spencer gives us a science of things, not of people per se. Spencer's science of things is cast in grand theoretical terms: matter, motion, and force drive the physical and the social worlds alike. To think like Spencer, then, is to think in abstract, law-like terms. Spencer also sees the universe and everything in it working like a system. Thus, to think like Spencer is to see everything working together as an organism, with each part playing a role in producing the whole. And, finally, to think like Spencer is to see the world in progressive evolutionary terms, in terms of the "survival of the fittest."

The remainder of our chapter is divided into two main sections. In *Social Evolution,* we will find that Spencer gives us three functional requisites—three things that every system needs to survive. In the process of social evolution, which is driven by population growth, the three functions are distributed to different social structures or institutions. The evolution of society, then, is the story of increasing structural complexity and survivability. In *Social Institutions,* we will see what various social institutions (like religion and government) provide for the whole and how they change in response to evolutionary pressures.

Social Evolution

Most social scientists of the eighteenth and nineteenth centuries either assumed an evolutionary point of view or explicated one. Both Marx and Durkheim, for example, assumed an evolutionary-like position. Marx assumed that societies are changing through the material dialectic, though the "evolutionary" change was not seen as progressive until the very last movements when history moves from capitalism to socialism and, finally, to communism. Durkheim argued for a moral–cultural evolution. This evolution moves from particularized to generalized culture and morality. More complex societies demand a more generalized culture and value system. For example, a social group that only has one kind of person in it can afford to have a very specific culture (in fact, it *must* have such a culture). Urban gangs are a good example. But a larger social group that embraces a number of different types of people must have a more general identity and value system. Durkheim argued that religion would facilitate this evolution, particularly in its final phase from nation-state to universal humanity (keep in mind that this kind of religion would be very humanistic and accepting of diversity). In addition, according to Durkheim, improvements in communication and transportation move morals to become increasingly independent of time and space and thus embracing of the "sole ideal of humanity."

Yet both Marx and Durkheim saw problems associated with the evolution of society. Herbert Spencer, on the other hand, saw it more in terms of positive progress coming out of social evolution in general and the Industrial Revolution in particular. Spencer saw evolution as a movement from simple forms to complex forms with greater adaptability. Spencer was born and lived his life in Britain. England, in contrast to France (where Durkheim grew up), had experienced slow and steady growth, politically and economically. There was thus a tendency to see the world in terms of gradual and peaceful change. Spencer, like most British people at the time, saw the Empire as the pinnacle of social evolution. Capitalism and the Industrial Revolution were seen as expressions of this superiority. Spencer therefore saw laissez-faire capitalism, the division of labor, free markets, and social competition as part of the survival of the fittest.

System needs: Just as every living organism has certain needs that must be met for it to survive, society too has specific needs. For example, every society needs some standardized method of providing food, shelter, and clothing to its members. Every society also needs some agreed-upon method of communication, some way of passing down its culture to succeeding generations, some way of achieving a general but morally binding identity and value system, and so on. Each functionalist has her or his own proposed list of needs. As expected, these lists are pretty similar, but they are usually pretty abstract as well, particularly in the case of a grand theorist such as Spencer.

Spencer argues that because all systems basically work in the same manner, all systems should have the same requisite needs. For Spencer, there are three requisite needs or functions: regulatory, operative, and distributive. The **regulatory function** involves those structures that stabilize the relationships between the system and its external environment and between the different internal elements. In short, it regulates boundaries. An example of a regulatory social structure is government or polity. Governments, both local and national, set and enforce laws that control relationships, whether between internal units (such as individuals or corporations) or between our society as a whole and other external systems (such as a foreign country). The printing and regulation of money is another example of how governments manage relationships. When exchanges occur between individuals or companies or nations, it is money and its determined value that standardize those exchanges.

The **operative function** concerns those structures that meet the internal needs of the system. For a society, these needs could be cultural or material. The cultural needs of a society are met through institutions such as education; and the material needs are met through the economy. The operative system also includes such things as values (we can either value material gain or spiritual enlightenment) and communities (the social networks that meet our emotional needs). The **distributive system** involves those structures that carry needed information and substances. It is the transportation (roads, railways, airlines) and communication (telephone, mail, and Internet) networks that move goods and information through society.

These three different functions are not simply a way to categorize and understand what is going on in society, though they certainly do that. Spencer also says that there are at least two kinds of issues associated with these functions. First, Spencer notes that each subsystem will display the same needs. What that means is that every subsystem acts just like a system and can be understood as having the same needs. So, the distributive system, for example, has needs for distributive, operative, and regulatory functions. Spencer doesn't make subsystem needs and analysis a central feature of his theory; but, as we will see, Talcott Parsons does. Another dynamic that Spencer points out is that in social evolution, there is a tendency for the regulatory function to differentiate first. In other words, as we move from simple collectives to more complex societies, the first structure to differentiate and specialize is government.

The organization of every society begins with a contrast between the division which carries on relations habitually hostile with environing societies and the division which is devoted to procuring necessaries of life, and during the earlier stages of development these two divisions constitute the whole. Eventually there arises an intermediate division serving to transfer products and influences from part to part. (Spencer, 1876–1896/1967, p. 214)

Differentiation and specialization: Evolution involves three phases: differentiation, specialization, and integration. The basic premise is that complex organisms have greater chances of survival. Complexity in this case is defined in terms of structure and function: more complex organisms will have a greater number of specialized structures fulfilling the requisite functions of regulation, operation, and distribution. The instability of homogeneous units, segmentation, and multiplication of effects therefore push organisms to differentiate and specialize; once an organism has multiple structures performing specialized tasks, integration becomes a need. Think of single-celled amoebas. Because of their nature, integration isn't an issue: if there is only one cell, there is nothing with which that cell can be integrated. Differentiated and specialized structures by definition perform dissimilar tasks and will tend to move in different directions. Thus multicellular animals must create structural solutions to the problem of integration. The human body, for instance, uses the central nervous system to integrate all its different structures and subsystems.

Obviously, when we are talking about social evolution, we have in mind **social structures.** Before we go any further, I'd like to stop and make sure we have a good definition of social structure—I have found that the idea of social structure is something with which most students have trouble. I want you to think for a moment about how you made your way into the room you are in right now. Unless the room you're in is a single room standing all by itself out in the middle of a field with no roads in or out, then you probably got to the room by driving on roads, walking on paths, and stepping through hallways and doorways. Okay, now imagine that you wanted to get into this room using something besides the door or the window. It would be difficult, wouldn't it? Walking through walls is no easy task. The point of all this is that you used different kinds of structures to get around (roads, walls, floors, doors, and so on). Those structures helped you accomplish your goal of getting to this room (which is itself a structure), but they also restricted and guided your options. Social structures are like material structures: they guide our behavior, but they restrict us as well.

Here's an important point about structures to keep in mind: the *structural* parts of structures are the *connections* among the units. For example, the highway system is considered a societal infrastructure precisely because it contains relationships among all the roads in the system. A single road out in the middle of nowhere would not be considered part of the infrastructure. However, I can stand on the road in front of my house and be connected to every other road on the North American continent precisely because they are structured. Using a different analogy, the substructure of a house isn't created simply by piling 2 × 6's in a haphazard manner. It's putting them together in a specific way that creates the structure. Another way to understand the principle of structure is to compare it to an organizational chart. There are positions in the chart, like vice president or secretary, and there are expectations associated with the positions (like making decisions or typing). The expectations, of course, are based on the relationships among and between the nodes or positions on the chart.

Your status position is like your place in the organizational chart of society. You have several status positions, like student, son or daughter, friend, and so on. Those structured positions tell people what to expect from you—there are roles (behavioral expectations) associated with each position. These roles exist by virtue of the connection to other positions. For example, because you are related differently to other positions, your role as a daughter is different than your role as a worker. Further, you act like a student or you act like a young person not because you *are* a student or young, but because that's what is expected of that status position in this country.

Social structures, then, both restrict and enable human behavior, and are made up of connections among sets of positions that form a network. The interrelated sets of positions in society are generally defined in terms of status positions, roles, and norms. These social and cultural elements create and manage the connections among people, and it is the connections that form the structure. **Structural differentiation** in society, then, is the process through which social networks break off from one another and become functionally specialized. That is, the network of status positions, roles, and norms becomes peculiar to a specific function. Social evolution has involved a movement from simple social forms (like hunter-gatherer societies) to more complex forms (like **postindustrial** societies). Just as the human body has developed specialized structures, such as the heart, to meet certain needs, so society has developed dedicated structures. Initially, all the needs of society were met through a single structure, kinship; but as societies grew, they also became more complex and developed specialized structures to meet the requisite needs. I've pictured this progression in Figure 2.1.

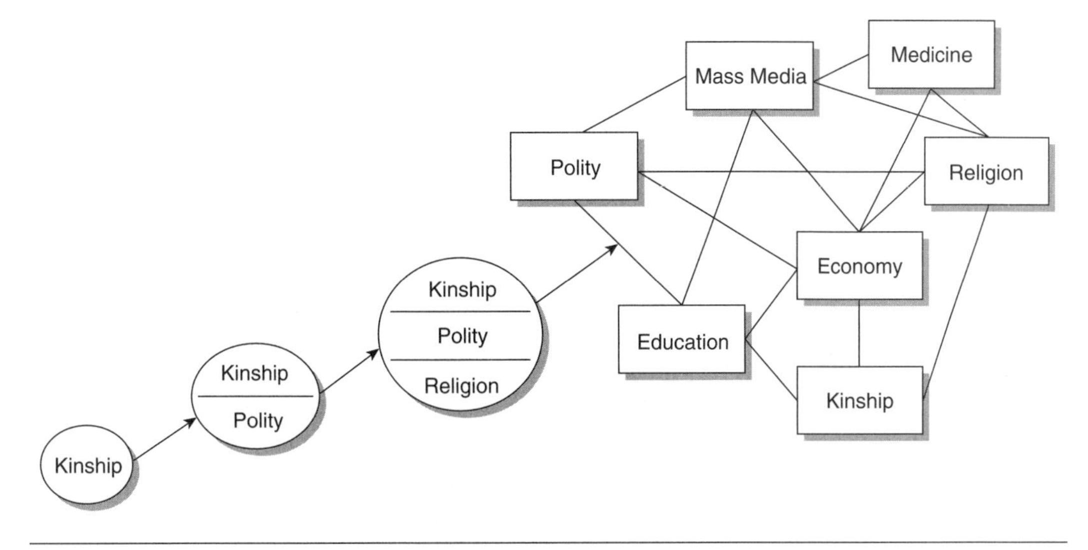

Figure 2.1 Social Evolution—Simple to Complex

Figure 2.1 is not meant to be a dynamic model, nor does it attempt to map out all the steps. It simply presents a picture of the general idea of Spencerian social evolution. Notice that in the first circle, which would represent hunter-gatherer societies, there is only one structure: kinship. What that indicates is that all the needs of society are being met through one social institution. So, for example, the religious needs of the group are met through the same role and status structure as family. That is, the head of the family is also the "priest." The second and third circles indicate that there is some differentiation, but the different institutions are still very closely linked. For example, the first-born son would be expected to go into government and the second-born son into religion. The final picture shows a society that is structurally differentiated and specialized, much like our own. In our society, your family role is generally not associated with the role you will play in religion, school, government, or the economy (to the degree that it is, there is another structure at work—a structure of inequality).

Social evolution is fueled by population growth (social "matter"). As populations grow, they need to expand their structural base in order to meet the needs of the collective. There are two basic ways a population can grow: through a higher birth rate than death rate and/or by **compounding** (the influx of large populations through either military or political conquest). Think back to Spencer's use of physics in his evolutionary theory: compounding increases the force and motion of matter, in this case population. Generally speaking, increased population growth increases the level of structural diversity. According to Spencer, the different ways populations grow and gather—dissimilar levels of force, motion, and matter—produce different types of society.

Spencer gives us two related typologies of society. Typologies are used to categorize and understand some phenomenon. We think in terms of typologies all the time. For example, we use music typologies: Beethoven falls under classical music, Pete Seeger under folk, and Eminem is hip-hop. One of the big differences between our everyday typologies and those used in sociology is that the ones in sociology are more rigorously constructed and defined. We will see that a number of theorists use typologies to understand various aspects of society. Spencer's first societal typology is defined around the processes of compounding that we have been talking about.

There may be **simple, compound, and doubly compound societies.** A simple society is "one which forms a single working whole unsubjected to any other, and of which the parts co-operate, with or without a regulating centre, for certain public ends" (Spencer, 1876–1896/1975a, p. 539). In this definition of simple societies, we see that the defining feature is that the whole must not be subject to political rule from more than one group. I mentioned earlier that the first structure to differentiate is government. The levels of government that form in response to population compounding are, then, the basis of Spencer's first typology. In simple societies, like hunter-gatherer groups, the vertical dimension of government is flat. In other words, if the group has a chief, there is only one chief with no one above him. Compounded societies, on the other hand, not only have larger populations, they also have "several governing heads subordinated to a general head." In doubly compounded societies, there are additional layers of governing bodies. Obviously we are talking about increasing complexity in social structures; this typology notes the importance the political structure plays.

Integration and power: When all the functions were carried out by one social structure, coordinating the way it fulfilled those functions was easy. However, as structures differentiate, they also become segmented and specialized. Each structure develops its own set of status positions, norms, goals, methods of organizing, values, and so on. This process of differentiation continues as each of the subsystems becomes further differentiated through the principle of segmentation and multiplication of effects. For example, airlines, financial markets, automobiles, the Internet, and newspapers are all part of the distributive system, but each has its own language and values. Spencer identifies these issues as problems of **coordination and control:** as societies become more differentiated, it becomes increasingly difficult to coordinate the activities of different institutions and to control their internal and interinstitutional relations. As a result of these problems, pressures arise to centralize the regulatory function (government). Notice the way that is phrased: it is not the case that the individuals or political parties decide that there needs to be increased governmental control; it is the system itself that creates pressures for this centralization.

> The mutual dependence of parts which constitutes organization is thus effectively established. Though discrete instead of concrete, the social aggregate is rendered a living whole. (Spencer, 1876–1896/1975a, p. 448)

Spencer's basic model of evolution is outlined in Figure 2.2. For societies, population growth is the basic matter or mass. The force and motion of the mass is increased dramatically through compounding: the gathering together of large groups of people. Society responds to influxes of population by structurally differentiating. Each of the differentiated structures is functionally specialized; each fulfills one rather than several functions. As a result, structures become mutually dependent. So, if the economic system only produces goods and services necessary

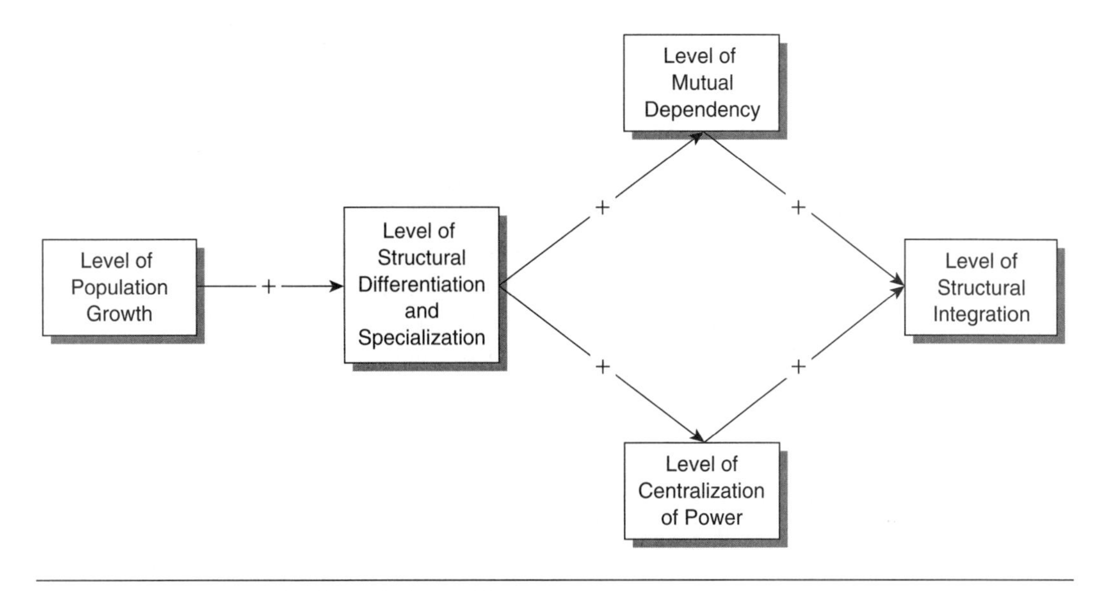

Figure 2.2 Structural Differentiation and Integration

for collective and organic survival, then it must depend upon another structure like the family for socialization. Because diverse structures require organization, the regulatory sector tends to become more powerful and exert greater influence on social units (both organizations and people). Together, mutual dependency and a strong regulatory subsystem facilitate structural integration. The positive feedback arrow from integration to population growth indicates that a well-integrated social system makes possible further population growth and aggregation.

> Organization in excess of need, prevents the attainment of that larger size and accompanying higher type which might else have arisen. (Spencer, 1876–1896/1975b, p. 262)

Centralized authority thus solves the problems of coordination and control, but it in turn can create productive stagnation and resentments over excessive control. Increased authority over people's behavior makes them less likely to innovate. It also hampers the exchange of information, the flow of markets, and so forth. Freedom brings innovation; thus stagnation in the long run creates pressures for deregulation. As you might already be able to tell, this is a kind of cycle that societies go through, but there is a variety of complications in the process.

This cycle of centralization and decentralization brings us to Spencer's second societal typology. In some ways, we can think of his first typology as a rough guide to understanding the structural aspect of social evolution. It focuses on the kind of polity or government a society has defined in terms of its structure. That is, the structure surrounding governmental actions can be measured from simple, with one level of governing structure, to complex, with multiple levels of governing structures. Understanding evolution in terms of the regulation function makes a certain amount of intuitive sense. In terms of evolutionary survival, the most complex and adaptive species is humanity, and it is mainly our brain and mind that give us the evolutionary advantage. Evolution, then, can be seen as the progress toward more and more complex regulatory systems.

Spencer's second typology also focuses on the regulatory system. While with the first he is interested in understanding the structural differentiation of polity, in this one he is looking more at the content of government and its relationship to society—in particular, the economic system. There are two types in this scheme—militaristic and industrial—that are defined around the relative importance of the regulatory and operative functions. The fundamental difference between them is that the militaristic operates according to the principle of compulsory cooperation and the industrial according to voluntary cooperation.

I've listed the defining features of each type in Table 2.1. In **militaristic societies**, the state is dominant and controls the other sectors in society. The religious, economic, and educational structures are all used to further the state's interest. The economy's first mission is to provide for military needs rather than consumer products. In militaristic societies, religion plays a key role in legitimating state activities. The religious structure itself is rather homogeneous, as religion needs to create a singular focus for society. Religion also provides sets of ultimate values and beliefs that reinforce the legitimacy and actions of the state. Information is tightly controlled and the government exercises strong control over the media. Public meetings and individual behaviors are controlled as well. Religion in particular provides ultimate sanctions (death and hell) for social deviance. The status structure also assures individual compliance. The status hierarchy in militaristic societies is much more pronounced and the rituals surrounding status (like bowing or using status titles such as madam, lord, mister, and so forth) are clear, practiced, and enforced.

As you can see from Table 2.1, **industrial societies** are less controlled by the state and are oriented toward economic freedom and innovation. Rather than the collective being supreme, the individual is perceived as the focus of rights, privileges, consumer goods, and so forth. Religion is more diverse and there is a greater emphasis on individualistic reasons for practicing religion than on collective ones. Individual happiness and peace is associated with salvation and religiosity. Education shifts to scientific and inclusive knowledge rather than increasing patriotism through selective histories and ideological practices (like the pledge of allegiance). Information is freed and flows from the bottom up.

Both the typology of compounding and the centralization of power are meant as ways of classifying and understanding social evolution. According to Spencer, the typology of compounding is clear: societies have a tendency to become structurally more complex over time, especially with regards to the regulatory function. This progression is analogously seen in organisms. There has been an overall inclination toward more complex entities because they have a greater chance to survive.

> Of the traits accompanying this reversion towards the militant type, we have first to note the revival of predatory activities. Always a structure assumed for defensive action, available also for offensive action, tends to initiate it. (Spencer, 1876–1896/1975a, p. 569)

Much less clear is the evolutionary path between militaristic and industrial societies. It seems that there is a general evolutionary trend toward industrial societies; Spencer characterizes the movement from industrial to militaristic as regressive. All societies begin as militaristic. Evolutionary routes are characterized by competition and struggles of life and death. Aggressive and protective behaviors became pronounced for humans once we began to use

Table 2.1 Militaristic and Industrial Typology

	Militaristic	*Industrial*
Dominant Institution	The State: the nation (including all social institutions and actions) seen as synonymous with the army; power is centralized	The Economy: freedom of economic exchange and association; profit motivation; regulatory function is diffused
System Equilibrium	Compulsory cooperation	Functional equilibrium maintained through individual choices
Religious Function	Oriented toward legitimating government—direct regulatory function • Government identified as God-ordained • Homogeneous religious types • Legitimating political myths tend to contain ultimate values • Social deviation is defined as sin	Oriented toward the individual—indirect regulatory function • Greater separation of church and state • Heterogeneous religious types • Focus on individual salvation and happiness
Status Structure	Greater status differentiation with distinct categories and clear ceremonial practices	Overlapping and vague status positions with unclear and infrequently practiced ceremonial distinctions
Educational Function	Directly controlled by state with strong ideological socialization	Indirectly monitored by state with emphasis on general and scientific knowledge
Mass Media	Information tightly controlled by state	Information freely flows from the bottom up
Conception of Individual	Individual exists for the benefit of the whole; individual behaviors, beliefs, and sentiments of interest to the state	State exists to protect rights of the individual; duty of individual to resist government intrusion; belief in minority rights

agriculture for survival; it was also at this point that people became aware of themselves as a society and not simply as a family. Land was paramount for survival and something that could be owned and thus taken or protected. The regulatory function, then, grew to be most important and differentiated from the rest. Early society developed through warfare and eventually standing armies and taxation were created. Once established, it is then functional for society in the long run to move toward the industrial type. However, society can "regress." Having set the general tendency, Spencer delineates the reasons why there is likely to be a "revival of the predatory spirit" and the various complications involved in such a regression.

There are three main reasons why a society would, once it has become industrialized, move toward the militaristic type. First, there is pressure in the system toward militaristic society originating from the military complex itself. A military complex is formed by a standing army and the parts of the economy that are oriented toward military production. It's a fact of evolution: once an organism comes into existence, it will fight for its own survival; this is no less the case for social entities than for natural organisms. Thus, a military complex by its very existence is not only available for protection; it is in its best interests to instigate aggression. This is not to deny the existence of legitimate threat, which is the second reason for regression to militarism. Spencer argues clearly that there are times when a system is threatened by other societies, internal pressures, or the natural environment. Under such threat, the system will revert back to militaristic configurations for its own survival. However, a constant pressure toward militaristic society is exerted by the military complex.

The final reason why a society might revert is the presence of territorial subjects. A territorial subject is a social group that lives outside the normal geographic limits of the state, yet is still subject to state oversight. A current example of this kind of relationship for the United States is Puerto Rico. Puerto Rico is a commonwealth of the United States with autonomy in internal affairs, yet its chief officer is the U.S. president and its official currency is the U.S. dollar. The United States not only has a vested interest in Puerto Rico, it is also organizationally and politically involved. As such, the United States is ultimately responsible for Puerto Rico's internal stability and external safety. Issues of investments and responsibilities provide opportunity for military action and the creation of a militaristic society.

However, there are complications on the road to regression. The first complexity is social diversity. When Spencer talks about social diversity and societal types, he generally has race in mind. He, however, also acknowledges that what he has in mind is best "understood as referring to the relative homogeneity or heterogeneity of the units constituting the social aggregate" (1876–1896/1975a, pp. 558–559). Race, in this case, is the most stable social difference because it has the clearest visible cues associated with it. Spencer argues that groups have natural affiliations to band together and to create their own social structures. If such different groups do not affiliate with one another in larger collectives, then society will not be internally integrated. Spencer terms this the "law of incompleteness" and says that it represents a risk to society if an institutional shock were to occur. In other words, segregated societies are unstable because functional unity is difficult to achieve. This situation represents a potential threat: it indicates that the society may not be able to respond quickly enough to changes in the environment (societal or physical). Because of this potential threat, societies with diverse groups that do not associate with one another will tend to be militaristic. However, if the groups *do* affiliate with one another and the differences are perceived to be small, then the society is "relatively well fitted for progress." Such societies tend to produce heterogeneous groups that are functionally linked together, and are thus more flexible and gravitate toward the industrial type.

Another complication of regression is the effects of different institutions and their previous growth. For a society to regress to militarism, it must first have been industrial. Under the conditions of the industrial type, the economy grows and its power can rival or even exceed that of the state; religion diversifies and becomes more focused on the individual; and the culture of individualism grows. Returning to a militaristic posture, then, must counter all of these institutional developments.

I've diagramed these factors and their relationships in Figure 2.3. As I mentioned before, there is a back-and-forth movement from centralized to decentralized state control. The general evolutionary movement is toward industrial forms, but there are also pressures to revert back to militarism that are themselves countered by other social pressures and complications. In the diagram, both the regressive factors (military complex, threat, and territorial subjects) and impediments (social diversity and institutional configurations) are pictured as mitigating forces. In other words, a society that has moved from militaristic to industrial will tend to stay industrial unless the regressive factors mitigate the general evolutionary trend; and once a revival of the predatory spirit begins, it will continue to move society to militarism unless the impediments mitigate the effects. Thus, modern society is generally seen as moving along the continuum from military to industrial forms in response to population pressures, system stagnation, internal or external threats (other nations or changes in physical resources), previous institutional arrangements (such as the military complex or diversified religion), and the level of social integration.

Spencer does allow himself a speculative moment about future types. He says that the change from militaristic to industrial and back again involves a change in belief. In militaristic societies, people believe that the individual exists for the state and common good. Industrial societies represent a reverse of that belief: the state exists for the individual, to protect his or her rights and freedoms. Spencer says that the next type will also require an inverting of belief. In industrial societies, people believe that life is for work. As we'll see when we get to Weber, this is the work ethic of capitalism. The next step in social evolution inverts that belief, from life is for

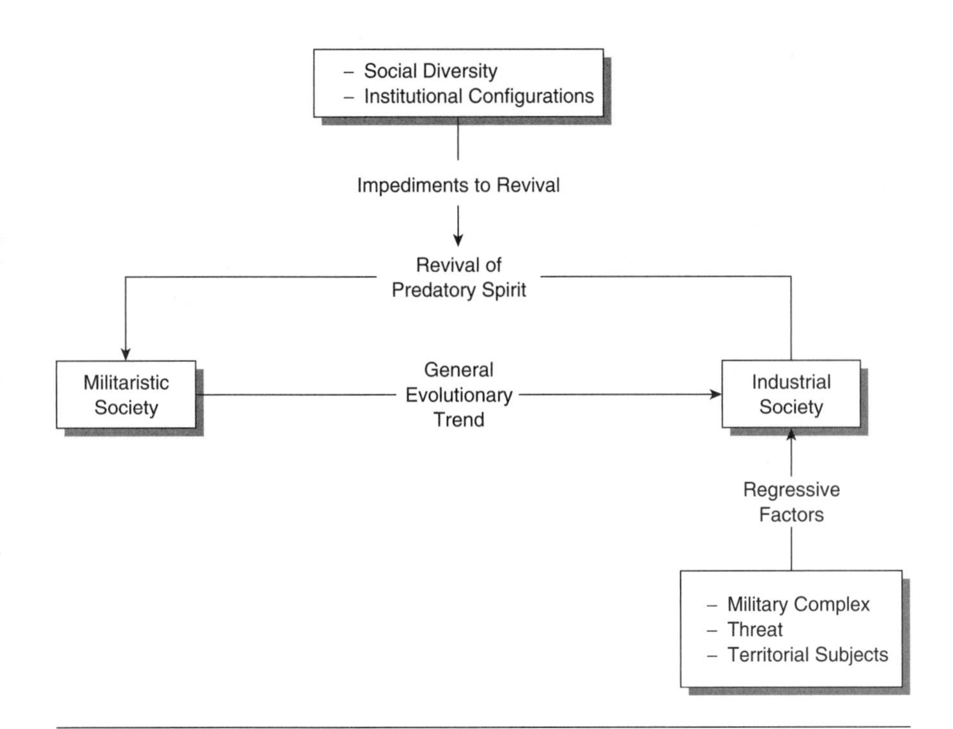

Figure 2.3 Militaristic-Industrial Cycle

work to work is for life. As an illustration, Spencer (1876–1896/1975a) points to "the multiplication of institutions and appliances for intellectual and aesthetic culture and for kindred activities not of a directly life-sustaining kind, but of a kind having gratification for their immediate purpose"; unfortunately, he also says, "I can here say no more" (p. 563).

Social Institutions

Defining institutions: The bulk of Spencer's three-volume set, *The Principles of Sociology,* is dedicated to an analysis of institutions. Spencer, like so many students of sociology, uses the term *institution* but never provides a clear definition. I'm going to give you a definition that is in keeping with Spencer, but it probably goes a bit beyond him as well. **Social institutions** have three interrelated elements. First, institutions have functions: they are collective solutions to survival needs that provide predetermined meanings, legitimations, and scripts for behavior. This notion of function is probably the most important defining feature of an institution. In everyday speech, we tend to use the term **institution** in a variety of ways. We use it to refer to someone or something that is firmly established; for example, "George is an institution around this place"; or, we use it to refer to a significant practice, as in "Baseball is an American institution." Yet, these don't qualify as social institutions because they do not themselves meet a societal need.

The second defining feature of institutions is that they are not reducible to individual actions or agency. Institutional behaviors aren't something that one person does; they are society-wide. Further, institutions resist modification by the individual or even groups of individuals. For example, a number of years ago it was popular among some to see legal marriage as unnecessary and even detrimental. Marriage was seen as part of the Establishment, part of the baggage forced upon us by religion and the state. These feelings of course came out of the hippie and feminist movements in the 1960s. As a result, many people didn't get legally married. Some had private ceremonies in a field surrounded by family and friends; others just exchanged vows between themselves in the privacy of their own home. As a result of all these countercultural behaviors, nothing happened; marriage didn't go away. It's still a booming business, because institutions resist change from agents.

In addition to not being reducible to individual agency and resisting modification, institutions are not subjectively available. Let's take marriage as our example again. There isn't one of us living that was around when marriage was created. We don't have a subjective memory of it or personal orientation toward it. Everything we know about the reasons why marriage was instituted and the functions it fulfills is contained in stories that we tell each other. These stories legitimate our institutions. They give us a justifiable basis for believing in them. Obviously, the stories are the products of generations of politically motivated groups, but they constitute the reasons we believe in our institutions.

The fourth characteristic of a social institution is that it tends to be wrapped in morality; institutions are moral phenomena. In addition to containing their own legitimation, they also provide beliefs and rituals that imbue the institution with rightness and moral energy. One of the functional reasons for this morality is undoubtedly the necessity to protect our solutions to survival needs. Just as any animal will fight to protect its continued existence, so humans must be motivated to shield their cultural institutions. High levels of moral investment also serve to ultimately legitimate our intuitions. We tend to see morality as connected with a higher being. Thus, if our institutional arrangements are moral religious entities, we do not see them as our own creation.

Domestic institutions: Kinship or family is one of the most basic of all social institutions. Its essential function is to facilitate biological reproduction, without which any species will die. In the long run, family functions to provide for patterned relationships between men and women, emotional and physical support, lines of inheritance, and care and socialization for the young.

In the absence of alternatives, kinship is the chief organizing principle of a collective; before there were state bureaucracies or economic organizations, people were organized by kinship. The roles and status positions found in the kinship system informed people how to act toward one another and what their obligations and rights were; and all social functions were met through the status positions and roles of family. Spencer spends a great deal of time recounting the evolution of the family form. As is his custom, he presents large amounts of historical data drawn from several sources, and from the data he draws general conclusions. His basic point is that the monogamous family structure was selected because it is the most fitting in terms of human survival. The basic path of development looks like this: general promiscuity → polyandry and polygyny → monogamy. Monogamy, then, is a relatively recent development in human history. Driving the movement to monogamy is the social need for broader and firmer social networks.

Spencer argues that there is a clear association between the structure of family and the regulation of social action and relationships generally. This idea makes sense when we consider it in the evolutionary model. Humans depend more upon social networks for survival than any other creature. However, these networks aren't created through instincts or sensory experiences (like smell) as they are with most other social animals; they are produced through culture. In fact, it is the inability of humans to generate extensive social networks and behaviors through instinct or the senses that pushes us to move toward monogamy. Maryanski and Turner (1992, p. 13) explain our problem with social structure:

> In proportion to the prevalence of promiscuity, there must be paucity and feebleness of relationships. . . . Family bounds, therefore, are not only weak but cannot spread far; and this implies defect of cohesion among members of the society. (Spencer, 1876–1896/1975a, p. 637)

> Our central finding is that, compared with most Old World monkeys, the reconstructed blueprint of the social structure of the LCA [Last Common Ancestor] of apes and humans reveals the hominoid lineage as predisposed toward low-density networks, low sociality, high mobility, and strong individualism. This pattern of weak tie formation, low sociality, high mobility, and strong individualism was to pose . . . a difficult problem for hominoid species, especially those on the human line, once they moved from a forest to an open-range habitat.

In the beginning of human history, Spencer argues, there was almost complete and universal promiscuity. However, under general promiscuity, social relationships are weak and can only extend as far as the mother and child. Even the siblings themselves are only half-related. As a result, political subordination and control are limited. Only the strongest will lead and leadership will be ever-changing based on contests of strength. The development of religion is also hampered. Spencer argues that early forms of religion involved ancestor worship. When human beings first began to see past the objective line of this life, they saw it in terms of social relationships. The thing that connected us to the next life was kinship. We saw ancestors who had gone before as paving the way for us. It was through them that communication with the spiritual world could occur. Further, some of our first inklings of the idea of the soul came through the idea of ancestral reincarnation. Thus, when kinship only extended as far as mother to child, the ability of people to see past this life was limited.

The link between species survival and sociability pushed humans to begin to control sexual behavior and family relationships. However, the two forms that followed general promiscuity, polyandry and polygyny, were also limited. Polyandry is the practice of a woman having more than one husband (poly = many, andr = men). The major problem with this type is that it produces fewer offspring than other kinship forms. A woman can only birth and care for about one child per year, no matter how many husbands she has. Therefore, a social structure that is built around polyandry is at an evolutionary disadvantage.

In terms of offspring, one man with many wives (polygyny) is a better model, yet it suffers in its ability to produce social stability. Remember that one of humankind's most basic needs is a subsystem that regulates internal and external social relationships. Thus, stabilizing the political structure is one of the first concerns of society. Most initial forms of government were male dominated, due to ever-present war and the use of men to wage war. The reasons why men were used for war probably had less to do with the differences in strength between men and woman, and more to do with the expendability of men. Men are more expendable in terms of biological reproduction than women. Women were needed to nurse the young; and in terms of reproduction, society can do with far fewer men than women (one man can literally inseminate hundreds of women).

There came to be, then, a clear association among men, war, and political governance (war was generally the outcome of relationships with other groups). The problem with polygyny is that the line of succession from chief to son is unclear (initial forms of stable governance came through family lineage). The relationships among mothers, children, and father are clear in polygyny, but which son is preeminent isn't obvious. So, in the ruling families, a pattern began to develop that privileged one wife above all others; and her children were dominant and in line to succeed their father.

Gradually, monogamy was chosen as the primary form of kinship for a number of evolutionary reasons. To begin with, it added to the political stability of the group as competition among wives and offspring was done away with. Spencer also notes that "succession by inheritance" is conducive to stability because it secures the supremacy of the elder; and the use of elders for ruling tends to create what Weber would later call traditional authority. Further, the practice of monogamy was an important component in the evolution of religion, as it favors a single line of ancestral worship. The establishment of religion, of course, also helped to systematize socialization, which led to further social stability. In addition, monogamy tends to increase the overall birthrate and decrease the childhood morbidity rate. We've seen that polyandry results in low birthrates; and when compared to monogamy, so does polygyny. Under polygyny, some men have many wives and some have none; but under monogamy, most men have wives and thus there are more family units capable of producing children. And since monogamy also provides for greater emotional care for everybody concerned, children are better cared for physically and the morbidity rate goes down.

Because of the relationship between family and gender, I want to pause to note that Spencer feels that men and women should have equal rights. He sees them as performing different functions, but the functions themselves are equally needed by society. I must also note that one of the criticisms of Spencer is that he is a social Darwinist, seeing certain societies and races as more advanced—but in his view of equal rights for women, he was in many ways ahead of his time, as the following quotes illustrate.

> Equity knows no difference of sex. In its vocabulary the word man must be understood in a generic and not in a specific sense. The law of equal freedom manifestly applies to the whole race—female as well as male. (1882/1954, p.138)

> Perhaps in no way is the moral progress of mankind more clearly shown, than by contrasting the position of women among savages with their position among the most advanced of the civilized world. (1876–1896/1975a, p. 713)

Ceremonial institutions: Spencer argues that there are three major forms of social control: ceremonial, ecclesiastical, and political. Most sociology students are very familiar with the latter two. Political entities govern human behavior externally through laws, coercive force, and authority. One of the defining characteristics of states is the monopolization of legitimate coercive force. This monopoly serves not only to protect the nation's external boundaries but also to maintain internal peace. Religious, or ecclesiastical, institutions bring about social control internally rather than externally. Religion gives us sets of beliefs about right and wrong behaviors. These moral standards exert their force from inside of the individual. An individual believes that his or her behavior is right and the person wants to please his or her god through proper behavior.

Spencer gives us yet another form of social control in the form of ceremonial institutions. Ceremonies are formal or informal acts or series of acts that link people together hierarchically. Ceremonies may be simple, as in an act of politeness, or quite elaborate, as in prescribed rituals. Obviously there are certain protocols that apply when a head of state visits another country. In the protocol, the relationship between the ruler and the ruled is played out in behaviors. Ceremonies also indicate the hierarchical relationship among nations: some visiting dignitaries are afforded more elaborate ceremonies than others. Ceremonies can also be conventional and everyday.

In this sense, Spencer has in mind forms of address, titles and emblems of honor, forms of dress, and the like. For example, when we use titles like "Professor" or "Doctor" or "Miss," we are reproducing a hierarchical system of authority and control.

Spencer argues that ceremonial institutions are the earliest and most general forms of social control. The other forms of social control, religion and government, both sprang from this primitive type. The earliest forms of ceremony evolved naturally; that is, there was a natural connection between emotional responses and physical behaviors. Emotions cause physical reactions that vary according to the intensity of the feeling. For example, you can judge the strength and type of emotion that someone has for you by the way the person hugs you. When someone is joyful, the person's body reacts by becoming excited and perhaps jumping around; and feeling inferior is expressed bodily by relaxed muscles and lowered gaze and posture.

Ceremonies evolved further by becoming more symbolic, but they initially still maintained a degree of natural sequence. For example, the Dakota's ceremony of burying the tomahawk to symbolize peace is clearly linked to the actual putting away of weapons in order to cease hostilities. Another example that Spencer gives us is the past practice in central South Africa of drinking blood to establish kinship relations. The practice is based on the early belief that all of a person is contained in a part of the person. Spencer traces many of these natural sequences such as the move from taking trophies (such as the head of an enemy) to mutilating some part of an enemy's body to mark conquest.

Ceremonies further changed toward intentional symbolization in societies "which have reached the stage at which social phenomena become subjects of speculation" (Spencer, 1876–1896/1975b, p. 25). Spencer gives us a really interesting idea and relationship here. The idea of social phenomena becoming subjects of speculation is one that would later be important to Weber, and, in a somewhat modified form, important to such contemporary theorists as Anthony Giddens. Something important changes when we go from simply living in society to thinking about society and our social encounters. When we simply live socially, our experience is fairly immediate and all that exists is contained in those kinds of moments. In contrast, when we come to the point in our development when we think more and more about society, when we become reflexive about our place and actions in society, then everything becomes less tied to the moment and more abstract.

In terms of ceremonies, the important thing that happens when society becomes speculative about itself is that the rituals become increasingly arbitrary and intentionally symbolic. An example of this symbolic evolution is the phrase "Mickey Mouse." Mickey Mouse began as a cartoon, but soon evolved to mean something that is not to be taken seriously, to something ineffectual, of little importance, and then worthless. While we can understand perhaps how the phrase Mickey Mouse came to mean something worthless, the beginning and each successive move is symbolically linked rather than naturally connected.

Spencer generally argues that as societies begin to use ecclesiastical and political means of control, ceremonial controls become less used and more inconspicuous. More specifically, this general trend is influenced by the degree of voluntary cooperation versus enforced structures. The higher the level of voluntary cooperation, the less that ceremonial institutions will be used. On the other hand, the higher the levels of inequality in a society and the greater the need for coercive enforcement, the greater will be the use of ceremonial institutions that mark and recreate the differences among social groups.

The power of ceremonial institutions is that they "spontaneously generate afresh" the forms of rule every time human beings interact and use them. One of the interesting things about this section of Spencer's work is that he anticipated work done by contemporary theorists who emphasize the importance of micro-level interaction for building and preserving society. For example, both Erving Goffman and Randall Collins argue that one of the foundations of society is micro-level interaction rituals. One of the things that Goffman (1967) brings out is the unintentional consequences of ritualized behaviors. During encounters with other people, we engage in what he calls "face-work." In face-work we strive to avoid embarrassing ourselves or others by maintaining role-specific behavior and not noticing when action doesn't meet expectations. Our intent is to save face; the effect is that we save the interaction and thus the patterned interactions upon which society is built. Collins (1988) argues that behavior such as deference and demeanor rituals produces and reproduces the status hierarchy of society. Our demeanor tells others the level of social honor that we expect in a given situation; and deference refers to the behaviors through which we demonstrate respect to others (e.g., proffering titles of respect, exhibiting downcast eyes, and so forth).

Political institutions: Society is defined through cooperation; and Spencer tells us that there are two principal ways through which cooperation occurs: spontaneous cooperation due to individual motives of exchange; and consciously devised cooperation. When groups consciously create organization, they become aware of public ends or goods for the first time. In other words, it is through political organization that a group becomes reflexively aware of itself. Up until that point, the primary awareness is individual. So, political organization is necessary for society. Through it we become aware of ourselves as a whole, and it provides direction for cooperative efforts and restraint on individual behaviors.

> The mere gathering of individuals into a group does not constitute them a society. A society, in the sociological sense, is formed only when, besides juxtaposition there is cooperation. . . . Cooperation, then, is at once that which cannot exist without a society, and that for which a society exists. (Spencer, 1876–1896/1975b, p. 244)

Spencer sets out to understand the fundamental forms of political organization and how they evolved. He argues that undergirding every type of government is a primitive form. The basic political division in society is between strength and age. With age comes experience, but it is usually accompanied by reduction in strength. The most distinguished individuals, then, incorporate both strength and wisdom; and, among those with strength and wisdom, some are more distinguished than others. Therefore, according to Spencer, there are three divisions to the basic political structure: the masses of young and weak, those who are strong and/or experienced, and those elite few who are the best among the strong and experienced.

Spencer argues that every political structure is but a derivation of this one. What began as distinctions among the predominant man, the superior few, the inferior many, became in successive forms the head chief, subordinate chiefs, and warriors; kings, nobles, and people; and in our own society, the chief executive officer, elected officials, and citizens. Further, Spencer (1876–1896/1975b) proposes that every type of political structure is simply differing combinations in these three areas: "a despotism, an oligarchy, or a democracy, is a type of government in which one of the original components has greatly developed at the expense of the other two" and "the various mixed types are to be arranged according to the degrees in which one or other of the original components has the greater influence" (p. 317). Thus, in a democracy the citizens are most important; in an oligarchy the power is vested in the middle section of small groups of leaders; and despotism is defined by power exercised exclusively by the chief person.

We have already talked about most of the dynamics through which political forms evolve. The first mechanism is population growth. As populations grow, especially through compounding, the levels of governing structure become more complex. In terms of Spencer's triune political structure, that means that additional levels of strong and/or experienced and elite positions are created (or what we might refer to as upper and middle management). The second influencing factor is the environment wherein the society is located, and this influences the degree of centralization of power. The environment is composed of both other societies and physical resources. These two elements are interrelated for societies. The presence of war is more likely when physical resources are few. Nations with few natural resources must obtain them from others, either through military conquest or political trade affiliations. In earlier times, military conquest was used, since political economic agreements are based on established and rather complex political entities. Thus, the complexity of the political structure is narrowed during times of war or threat (military type) and widened during times of peace and development (industrial type).

In addition to the basic structure of government, Spencer also explains the foundation of power. He notes that we have a tendency to think of political power as a feature of the governing structure itself. But that's incorrect. The political structure was built upon something more fundamental: the public sentiment of community. Spencer is arguing that underneath and prior to the political and judiciary systems that everywhere appear to guide our lives, is a feeling of being part of something larger than ourselves; sociologists would later refer to this as social solidarity. This collective sentiment is stronger in more primitive societies and is most clearly expressed through religious sanctions. This argument of Spencer's is strikingly similar to the one Durkheim would make some 25 years later: the source of society is an emotional feeling of attachment that provides moral basis for social identity and action.

According to Spencer, political authority finds its legitimacy in emotion derived from ancestors and its power in socialization. Weber, writing almost 50 years later, would call this kind of legitimation traditional authority. It's an authority based on time-honored customs; it has the weight of time and history in back of it. Again anticipating Weber, Spencer argues that political power that is ascribed to a single individual is unstable. The leader will eventually die and the governing body be recreated. However, linking government to family lineage and traditional authority creates a stable form that can survive from one generation to the next (Spencer, 1876–1896/1975b, pp. 321–344).

Political power is also stabilized through socialization. Spencer doesn't use the word socialization, but the process he describes is exactly what contemporary sociologists mean by the term. Through proper training during youth, people come to share similar ideas, sentiments, ideas of right and wrong, language, and so forth. The trick of socialization is that when it is successful, these socially constructed items appear as subjective and natural to the individual.

> As fast as custom passes into law, the political head becomes still more clearly an agent through whom the feelings of the dead control the actions of the living. (Spencer, 1876–1896/1975b, p. 323)

As a side note, I want to mention that Lloyd Warner theoretically elaborates this Spencerian notion of ideas and emotions from the past influencing our behaviors today. Warner (1959) defines culture as "a symbolic organization of the remembered experiences of the dead past as newly felt and understood by the living members of the collectivity" (p. 5). The important element of this definition of culture is the emotional attachment that symbols from the past can bring. The people, ideas, and experiences of the past tend to take on sacred qualities. Think about what happens after a person dies. There is a tendency to speak of the individual in positive and reverent terms. The failures and idiosyncrasies of the individual are forgotten and her or his virtues are extolled and linked to timeless values. Warner tells us the same things happen to collective memories. We can recast meanings because signs and symbols become freed from their immediate controls (their social reality isn't immediately present) and at the same time they acquire a sacred power. The reason I mention Warner, besides the fact that he is an interesting theorist, is that he gives us yet another example of how Spencer's ideas prefigure quite a bit of sociological work. Warner himself attributes his ideas to Durkheim, rather than Spencer.

Ecclesiastical institutions: We have seen religion referenced in our discussion of other institutions. Particularly, we have seen that religion functions as an agent of social control. Religion infuses values and morals with supernatural power. According to Spencer, religion also functions to reinforce and justify existing social structures, particularly those built around inequality. If social structures are seen as an extension of a god's spiritual order, and if behaviors are seen as holy directives, then questioning social arrangements becomes a matter of questioning the god, and inappropriate behavior becomes sin.

As with all social institutions, Spencer is primarily concerned with tracing its evolutionary development. Any student of religion is faced with one inarguable fact: through the course of human history, religion has changed. Our initial forms of religion were not monotheistic or ethical. They were, in fact, marked by the presence of many gods displaying quite human behaviors of greed, lust, and so on. The question, then, with which we are faced, is why did our conceptions of deity change over time? Did God reveal her- or himself gradually, or was the idea of God linked to developments in society and human personality? Spencer (1876–1896/1975c) argues for the second alternative: "Among social phenomena, those presented by Ecclesiastical Institutions illustrate very clearly the general law of evolution" (p. 150).

I'd like to take a moment to point out that we are not necessarily faced with an either/or proposition. It isn't the case that God either revealed him/herself or the idea of God evolved. If God had to progressively reveal the truth to humans, it was because we weren't ready either socially or psychologically. That would imply that there would be corresponding changes in society (and/or personality) and in our idea of God. If God works through history, then we would expect to find similarities between social and personality structures and the type of religion practiced. On the other hand, if there is no God, then we would expect to find similarities between social and personality structures and the type of religion practiced. Either way, the results are the same.

> In their normal forms, as in their abnormal forms, all gods arise by apotheosis. (Spencer, 1876–1896/1975c, p. 19)

Spencer argues that all religions share a common genesis. Religion began, according to Spencer, as "ghost-propitiation." Spencer argues that human beings became aware of themselves as double. We have the waking, walking self, and we also have a self that can be absent from the body. We became aware of this other-self through dreams and visions. During dreams, we have out-of-body experiences. We travel to faroff places and engage in some pretty amazing behaviors. In thinking about these experiences, primitive humans concluded that we are double. There's part of us that is tied to the earth and time, but there is also part of us that can transcend both.

Moving from conceptualizing the other-self from dream states to thinking that the other self survives death is a short and quite intuitive step. Death seems a lot like permanent sleep; it's a sleep from which we don't awake. If we soul-travel when we are temporarily asleep, then we must also travel when we permanently sleep. This idea led early humans to minister to the dead, especially dead relatives. These dead relatives could cause mischief if not taken care of, but they could also prove to be valuable allies in times of need. Further, these disembodied spirits also came to be seen as the source for life on this earth through reincarnation. Early humans were quite in tune with the cycles of nature. Trees that appear to die in winter came back to life in the spring. Thus it was not a very big step, once souls were seen as surviving death, to see them also as the source for the other-self in life. In this way, early humans began their ancestor worship.

Obviously, ancestor worship is linked to family, and the head of the family was also the head of religious observance. However, two factors that came from military conquest brought about significant social and structural changes. First, the religious function differentiated from family. As human beings moved from nomadic existence to settled farming, war became more and more common. People were tied to the land in a way never before realized and land became a limited resource that had to be protected from other groups who had outgrown the ability of their land to produce enough food. The increased involvement in war by the clan chief meant that he had to shift his priestly responsibilities to another family member. Eventually, these religious functions were separated structurally as well.

The other factor that influenced the development of religion, which was also the result of military conquest, was the incorporation of more and more conquered people into the collective. Each of these groups of people had its own family god and ancestral worship. Polytheism was therefore the natural result of war and conquest. Further, the presence of this competitive chaos of gods created pressures for there to be a professional class of priests. The priests began to impose order on this array of deities, which of course reflected the social order of society: the gods that became most important were the gods of the most important families.

As societies continued to evolve to more complex forms of organization, so did the structure of religion. Organized pantheons developed as the political structure became more centralized and class inequalities increased. The hierarchy of higher and lower gods thus reflected the organized divisions of society. Because polity relies a great deal on religion for legitimation, the more warlike, and thus more successful, societies developed images of vengeful and jealous gods. And as the political structure became more centralized and bureaucratized, the religious structure did as well. When governments rely on religion for legitimation, it is in the best interest of the professional priestly class to organize and unite. As religious leaders organize and unite, they have to simplify the religious belief system, which in turn leads to monotheism. Max Weber gives a more detailed account of the evolution from magic to ethical monotheism, but the basic outline of Weber's thought is here in Spencer.

Thinking About Modernity and Postmodernity

While many of the theorists we will look at give us specific ideas to use to think about postmodernity, I want to employ Spencer to form the general problem. Spencer gives us a clear theory of modernity. His overall interest is to explain the evolution of society from primitive, simple structures to modern, complex structures. His argument is that as populations grow and compound, structures differentiate. Differentiated structures are held together by mutual need and a powerful regulatory system.

Defining postmodernity: A postmodern society is one wherein social institutions are not simply differentiated, they are fragmented. There are at least five important reasons for this fragmentation and de-institutionalization: the level of structural differentiation, the level of transportation and communication technologies and infrastructures, the level of market velocity and expansiveness, the level and rate of commodification, and the level of institutional doubt (Allan & Turner, 2000). I don't expect you to completely understand each of these variables right now (we will be looking at them in subsequent chapters), but I do want you to have a basic idea of what we mean by modernity and postmodernity.

Modernity and *postmodernity* are terms that are used in a number of disciplines. In each, the meaning is somewhat different, but the basic elements are the same: modernity is characterized by unity and postmodernity is distinguished by disunity. In modern literature, for example, the unity of narrative is important. Novels move along according to their plot, and while there might be twists and turns, most readers are fairly confident in how time is moving and where the story is going. Postmodern literature, on the other hand, doesn't move in predictable patterns of plot and character. Stories will typically jump around and are filled with indirect and reflexive references to past styles of writing or stories. The goal of a modern novel is to convey a sense of continuity in the story; the goal of a postmodern work is to create a feeling of disorientation and ironic humor.

In the social world, modernity is associated with the Enlightenment and defined by progress; by grand narratives and beliefs; and through the structures of capitalism, science, technology, and the nation-state. To state the obvious, postmodernism is the opposite or the critique of all that. According to most postmodern thought, things have changed. Society is no longer marked by a sense of hope in progress. People seem more discouraged than encouraged, more filled with a blasé attitude than optimism. We'll explore some of the specifics of the postmodern critique of grand narratives, science and technology, capitalism, and the nation-state as we work our way through the book.

However, right now I want to give you a couple of handles on postmodernism. First let me say that any attempt to capture postmodern ideas is doomed to fail on one level or another. Postmodern theory is based on the idea of the inability of categories to express human reality. If I say, then, that I'm going to reduce postmodernism to a few concepts or categories, I'm already outside the postmodern project. Yet, it is good to begin somewhere, and here is where I choose to begin.

The various postmodern theories seem to me to boil down to two main issues or effects: culture and the individual (Allan & Turner, 2000). Because of certain social changes, culture has become more important in our lives than social structure, and we are more individually oriented than group oriented. Thus, culture and the individual are more important now than they ever have been. Nevertheless, both culture and the individual are less effectual than ever before as well. Culture is important, but it has become fragmented and unable to aid in the production of a unified society. Further, a decisive break between reality and culture has occurred, such that all we have are texts and subjective meanings. The individual is more important, too, but has become de-centered and unable to emotionally connect in any stable or real fashion.

Religion—a postmodern case in point: We will explore these themes as we move through our theorists, but let me give you an example of what we're talking about. Spencer argues that there is a symbiotic connection between religion and the state. Religion serves to legitimate the state, and the state in turn provides religion with structure and protection. In this relationship between the state and religion, Spencer also explores the evolution of religion from polytheism to monotheism. Let's ask ourselves a rather natural question: If religion is evolving, what's next? Spencer actually has a section dedicated to the idea of the future of religion, but he doesn't get very far. To see what might be going on with religion in a postmodern society, let's look at what Thomas Luckmann, a contemporary social theorist, has to say.

According to Luckmann (1991), society has always used religion to create ultimate meanings. Like Spencer, Luckmann acknowledges that these meanings are used to legitimate social institutions, as well as to answer impossible questions (such as, what is the meaning of life? Why did my mother die?). Religion also serves to stabilize our meaning systems. We can intentionally change our meanings—the changes we brought about in the meaning of race and gender in the United States are good examples. However, because meaning is socially constructed, it will tend to shift and change on its own. Let's look at it in terms of what we do to the natural environment. Most of us have lived in houses with yards. "Yards" don't naturally happen. We have to kill the weeds; rototill the soil or bring topsoil in; we have to plant all the trees, grass, and bushes in the exact places we want them. I remember as a child, my parents had a plan drawn up to help them build the yard. Once we put in the yard, we have to work to keep it up. If we don't, the yard will revert back to nature and be gone. The same is true with symbolic meanings. Once in place, they have to be kept up and stabilized in order to continue to exist. Religion does an amazing job of stabilizing our meaning by equating our constructed meanings with eternal truths and by conceptualizing deviance and disorder with sin.

However, according to Luckmann, modern structural differentiation and specialization has made the great transcendences (the ultimate meanings of life) that religion provides structurally unstable. This structural instability has resulted in the "privatization of religion" (1991, p. 176), a new form of religion.

Other parts of society have filled in the gap left by the loss of highly structured religion and turned it to profitable business. As a result, the individual is faced with a de-monopolized market created by mass media, churches and sects, residual nineteenth-century secular ideologies, and substitute religious communities. The products of this market form a more or less systematically arranged meaning set that refers to minimal, intermediate, but rarely great transcendences. In other words, the meanings provided through mass media and commodities are small and fleeting. Under these conditions, a set of meanings may be taken up by an individual for a long or short period of time and may be combined with elements from other meaning sets.

What Luckmann is describing is an environment where meanings—particularly ultimate meanings—are held in doubt and where individual people can pick and choose which meanings they want to hold to and which they don't. Most of us like to think that we ought to be free to choose what we believe and what we don't believe. I think this level of freedom is certainly preferable to the enforced meaning of more traditional societies (think of the Salem witch trials), but there are other effects of this level of freedom as well. If every person can pick and choose what meaning suits them, then meaning by definition is unstable and borders on meaninglessness.

Not too long ago, I met someone that exemplifies what Luckmann is talking about. When she introduced herself to me, she told me she was a "Christian-Pagan." That threw me for a loop. The word pagan originally meant "country dweller," but it got its current meaning of "heathen" or "irreligious hedonist and materialist" from Christians who used it to refer to non-believers and people who practiced magic. So, tell me, what is a Christian-Non-Believer? Or, better yet, what is a Christian-Non-Christian? You might respond and say, "Well, that's her choice and she should be able to create her own meanings for her own life." I don't disagree, but this kind of situation begs a larger, more important question: How will this loss of predictable meaning influence society?

We will return to these issues surrounding modernity and postmodernity in most of our chapters. Throughout the book, we are going to explore the main issues of postmodernism, culture and self, using some of the ideas that our theorists will provide. My hope for this exercise is to better attune you to the possibilities of the historical moment in which we live.

Summary

- Spencer's perspective is in many ways foundational for sociology. He explicates the basic premises with which most sociologists will either agree or disagree. Spencer's point of view is founded on positivism, the search for the invariant laws that govern the universe. He approaches society as if it were an object in the environment, and his purpose as a theorist is to simply describe how this object works. Notice that, for his time, Spencer had a fairly radical point of view; yet, his is not a critical theory. The reason for this blend of radical yet descriptive perspective is that Spencer sees the dynamics of progress existing within the evolutionary process. It is less important, then, to try and change the way people think about society. For Spencer, society works much like an organic system, with various parts functioning for the welfare of the whole. He extends the analogy to explicate the evolutionary changes in society, from simple to complex structural differentiation. The evolution of society occurs in much the same way that evolution occurs universally: through the motion and force of matter driven by the instability of homogeneous units, segmentation, and the multiplication of effects.

- All systems have universal needs: regulation, operation, and distribution. Every system, including society, meets these needs through various kinds of structures. The ways in which the needs are met produce the unique features of each system and society. Generally speaking, there is an evolutionary trend toward greater structural differentiation and specialization, with system needs being met through separate and distinct structures. For society, this trend is driven by increases in population size. As the population increases, especially through compounding, structures differentiate and specialize in function. This process produces interstructural dependency but also problems in coordination and control. In response, society tends to centralize the regulatory subsystem, which, along with dependency, facilitates system integration. However, too much regulation can cause stagnation and pressures for deregulation. In reaction to these system pressures, societies in the long run move back and forth on a continuum between militaristic and industrial.

- Social institutions are special kinds of structures. They are organized around societal needs, resist change, and are morally infused. All institutions, however, change slowly through evolutionary pressures, and because they are functionally related, change is mutual and generally in the same direction. Generally speaking, family evolves from universal promiscuity to monogamy (monogamy is evolutionarily chosen because it provides more explicit and stable social relations); religion moves from polytheism to monotheism (monotheism has the evolutionary advantage of providing a single legitimating and enforcing mechanism for diverse populations and bureaucratic states); the state generally moves from militaristic with simpler political structures to industrial with more complex structures; and ceremonial institutions generally recede in importance as religion, family, and polity become successful structures. In spite of this, the political and ceremonial structures tend to move back and forth on a continuum in response to vested interests, perceived threat, and the level of social inequality.

- Modernity for Spencer is characterized by high levels of structural differentiation. This increased complexity gives modern societies greater adaptability and increased chances of survival. Postmodernity is characterized by institutional and cultural fragmentation. The main difference between differentiation and fragmentation revolves around unity—in differentiated systems, the different elements are linked together, whereas in a fragmented system, the parts are not as organized and have greater freedom. The effects of postmodernity are themselves contradictory: culture and self (subjective experience) are simultaneously becoming more important and less real and meaningful. Religion is a case in point: religion in postmodernity is less centrally organized and thus open to increasingly idiosyncratic interpretations.

Building Your Theory Toolbox

Conversations With Herbert Spencer—Web Research

Almost anytime we talk with people we learn something about them as individuals. Spencer isn't here with us now, but we can still learn about him as a thinking, feeling person. What kinds of personal information would you likely pick up if you talked with Herbert Spencer? For example, did you know that he invented a type of paperclip? Perhaps not earth-shattering information, but it's still interesting. To find out some intriguing information about Spencer, use Google or your favorite search engine to answer the following questions. You may have to read biographical statements or combine terms (like "Spencer and knowledge") to find your answers.

- In Spencer's time, it was customary for gentlemen to belong to clubs or associations. It appears that for Spencer his club memberships were particularly important. What were the Athenaeum and X Clubs? What was Spencer's X Club name? What is the Royal Society and what was Spencer's association with it?
- As we will see, most of our theorists had interesting relationships with religion. What was Spencer's religious background? What does it mean to be a Benthamite (follower of Jeremy Bentham)? How do you think these factors influenced the way Spencer saw the world?
- In view of the fact that Spencer was a founding thinker in many disciplines, such as psychology, sociology, physics, biology, and so on, the way he was educated is interesting. How much formal education did Spencer have? How much formal education is required today to do the work that Spencer did? What social changes do you think have contributed to these differences?

- Spencer was very typical of how Enlightenment thinkers viewed knowledge. In fact, his way of thinking about knowledge continues to be dominant today. What kind of knowledge did Spencer think is "of most worth"? How does this fit with Spencer's perspective? What do you think might be some problems with defining this type of knowledge as most worthy?
- Many times the social circumstances in our theorists' lives give us insights into the social and historical differences between their time and ours. Spencer's death may have been associated with what we would today call "drug abuse." What drug was it and why did he use it? What do you think this issue tells us about social control in Spencer's modernity and our time?
- Sometimes there's more to our theorists than meets the eye. Who was George Eliot and how was Spencer related to George?

Passionate Curiosity

Seeing the World (using the perspective)

Each of our theorists has a worldview. This worldview is informed by assumptions and values concerning the universe, the social world, and/or what it means to be human. This perspective forms the base upon which each theorist does his or her work; it is fundamental to what makes their work unique. Most of their concepts and theories flow out of their worldview. Given Spencer's perspective, answer the following questions (Another way of stating this issue is, what would a Spencerian analysis of __________ look like?):

- Evaluate the idea of grand theory. What benefits come from having one universal theory to explain all phenomena? Spencer's grand theory is based on the ideas of evolution and systems. Do you think that everything in the universe is part of one system or another? If not within systems, how then do different particulars relate to one another (for example, how are institutions related)?
- Using Google or your favorite search engine, find a definition of globalization. How would Spencer view globalization? (Hint: think in terms of systems and complexity)
- Consult a recent newspaper and read articles about social institutions like the state and the economy. How would Spencer interpret these current events?

Engaging the World (using the theory)

Thinking now of Spencer's theory (his concepts and relationships), answer the following questions:

- Spencer argues that monogamy developed because of its functional use in creating social relationships. Many people feel that monogamy is currently threatened. If monogamy is in danger, how do you think Spencer would theorize about it? (Remember, he would think about it in progressive, evolutionary terms.)

- Spencer argues that ceremonial institutions are at the heart of social control. Do you think the use of ceremonial institutions has gone up or down since Spencer's time? Theoretically, what do you think this change would indicate?
- Recalling our definitions of structures and institutions, do you think there are more social institutions today than one hundred years ago? If so, what does this imply about society's survivability? What does it imply about the state? Given what Spencer says about the state and structural differentiation, would you say that society is more modern or postmodern?
- Using Google or your favorite search engine, type in "censored news stories." Read through a few of the sites. Based on your reading, do you think that the news is being censored in this country? If so, or if not, how would Spencer's theory explain it?

Weaving the Threads (synthesizing theory)

There are central themes about which most of our theorists speak. These themes include modernity; social institutions such as the state, economy, and religion; culture; diversity, equality, and oppression; social cohesion and change; and empiricism. A good approach to theory is to pay attention to how these themes are developed. As I noted in the beginning chapter, one of the ways we can build theory is through synthesis: compare, contrast, and bring together elements from different theorists. The following questions are based on Spencer's theorizing and are meant to begin your thinking about these themes:

- What are the functions of kinship in society? How did the structure of kinship aid in the evolution of society? If the structure of kinship has changed since Spencer's time, what do you think that means (remember to think like Spencer)?
- What is religion's function in society? How has religion evolved over time? In other words, what are the social factors that contributed to changes in religion?
- What is the basis of political power in society? What factors bring about changes in the type of state and its structure?
- How do societies differentiate? Once societies differentiate, how do they integrate? Thinking like Spencer, would you conclude that this country has higher or lower structural complexity and differentiation as compared to even 50 years ago?
- What are the basic themes of modernity and postmodernity?

Further Explorations—Readings

Paxton, N. L. (1991). *George Eliot and Herbert Spencer: Feminism, evolutionism, and the reconstruction of gender.* Princeton: Princeton University Press. (Interesting analysis of Spencer's relationship to Eliot and feminist thought)

Spencer, H. (1904). *An autobiography.* New York: Appleton. (Good source for background to Spencer's thought and life)

Turner, J. H. (1985). *Herbert Spencer: A renewed appreciation.* Beverly Hills, CA: Sage. (Clear introduction to Spencer's thought)

Further Explorations—Web Links: Spencer

http://www.iep.utm.edu/s/spencer.htm (Site maintained by The Internet Encyclopedia of Philosophy; good overview of Spencer's life and work)

http://cepa.newschool.edu/het/profiles/spencer.htm (Site maintained by the Department of Economics at the New School; good source for further Web links about Spencer)

http://plato.stanford.edu/entries/spencer/ (Site maintained by Stanford Encyclopedia of Philosophy; solid review of Spencer's philosophy)

Further Explorations—Web Links: Intellectual Influences on Spencer

Thomas Malthus: http://cepa.newschool.edu/het/profiles/malthus.htm (Site maintained by the Department of Economics at the New School; good source for information and further Web links about Malthus)

Karl Ernst von Baer: http://www.zbi.ee/baer/ (Site maintained by the von Baer museum, Estonia; background, ideas, further readings)

Charles Darwin: http://www.aboutdarwin.com/index.html (Site maintained by AboutDarwin.com; lots of interesting historical facts about Darwin and his work)

CHAPTER 3

Engines of Change—Karl Marx (German, 1818–1883)

A knowledge of the writings of Marx and Engels is virtually indispensable to an educated person in our time.... For classical Marxism... has profoundly affected ideas about history, society, economics, ideology, culture, and politics; indeed, about the nature of social inquiry itself.... Not to be well grounded in the writings of Marx and Engels is to be insufficiently attuned to modern thought, and self-excluded to a degree from the continuing debate by which most contemporary societies live insofar as their members are free and able to discuss the vital issues. (Tucker, 1978, p. ix)

Karl Marx forms the foundation for much sociological as well as social thinking (even for people unaware of his influence). Marx is the one that gave us the initial insight to see patterns of conflict evolving and revolving around systems of inequality. So, anytime we notice, understand, and care about inequalities, we are seeing the world like Marx did. Marx also taught us to pay attention to the economy and the state and how the elite in those institutions use power and ideology. If we see class as an important determinant in life, then we are thinking like Marx. If we think that the government and other social institutions may actually be used by the upper class to facilitate inequality, then we are thinking like Marx. If we feel that some of the things that people believe may in the long run keep them oppressed or prevent them from reaching their full potential, then we are thinking like Marx.

> History itself is a *real* part of *natural history*—of nature's coming to be man. Natural science will in time subsume under itself the science of man, just as the science of man will subsume under itself natural science: there will be *one* science. (Marx, 1932/1978a, pp. 90–91)

More than his influence on social thinking in general, Marx has been a dominant figure that theorists have either argued against, like Weber, or used in unique ways to understand society and the way inequality works. One such school of thought is the Frankfurt School, the birthplace of critical theory. We've been talking about the Enlightenment, positivism, and the idea of progress. Marx's theory generally falls under this perspective. Marx feels that history and society can and should be studied scientifically; and he is decidedly an empiricist. He argues that societies change and history moves forward in response to dialectic structural forces in the economy; thus, for Marx, society functions much like a machine, according to law-like principles that can be discovered and used. Most importantly for our present discussion, Marx posits that human consciousness is materially based as well. He argues that consciousness is directly related to economic, material production; thus he argues against idealism, religion, and most philosophy. While Marx is critical of society, he also holds that a scientific approach is the path to true knowledge that would liberate the oppressed.

The Frankfurt School, formed in 1922 at the University of Frankfurt in Germany, inverts Marx's emphasis on the empirical world and science. The events in Germany during the 1920s and 1930s created an intellectual atmosphere where the state's use of ideology apart from class relations became an important focus of research. Nationalism—pride in one's national identity at any cost—took root in prewar Germany and came to fruition under the Nazis. As a result of not only WWII but also WWI, the burning question for many social and behavioral scientists became, how is it possible for people to believe in such a destructive national ideology?

Some of the answers, such as those from Erich Fromm, focused on psychological issues. Others, such as the Frankfurt School, focused on the social production of knowledge and its relationship to human consciousness. This kind of Marxism focuses on Marx's Hegelian roots. We will talk more about Georg Wilhelm Hegel in a bit, but for now the salient point is that this refocus on Hegel produced a major shift from material explanations of consciousness to idealistic and cultural ones. Thus, like Marx, the Frankfurt School focuses on ideology; but, unlike Marx, critical theory sees idological production as linked to culture and knowledge rather than class and material relations of production. Ideology, then, is more broadly based and insidious than Marx supposed.

Max Horkheimer became the director of the Frankfurt School in 1930 and continued in that position until 1958. Horkheimer criticized the contemporary Western belief that positivistic science was the instrument that would bring about necessary changes, positing instead that the questions that occupy the social sciences simply reflect and reinforce the existing social and political order. Horkheimer believed that the kind of instrumental reasoning or rationality that is associated with science is oriented only toward control and exploitation, whether the subject is the atom or human beings. Science is thus intrinsically oppressive, and a different kind of perspective is needed to create knowledge about people.

Jürgen Habermas, the current director since 1963, picked up Horkheimer's theme and argues that there are three kinds of knowledge and interests: empirical, analytic knowledge that is interested in the technical control of the environment (science); hermeneutic or interpretive knowledge that is interested in understanding one another and working together; and critical knowledge that is interested in emancipation. Because scientific knowledge seeks to explain the dynamic processes found within a given phenomenon, science is historically bound. That is, it only sees things as they exist. That being the case, scientific knowledge of human institutions and behaviors can only describe and thus reinforce existing political arrangements (since society is taken "as is"). As such, science in sociology is ideological. Critical knowledge, on the other hand, situates itself outside the historical normative social relations and thus isn't susceptible to the same limitations as science. Thus, science is limited, and truly important social questions must be addressed from outside the historical confines of present-day experience. The intent of critical knowledge is to get rid of the distortions, misrepresentations, and political values found in our knowledge and speech. Critical theory is based in praxis and sees an inseparable relationship between knowledge and interests.

Antonio Gramsci (1928/1971) gives us another important cultural extension of Marx. For Gramsci, the condition of the West is past the point where the materially based Marxist revolution can occur. Revolution for Gramsci is not the simple product of external economic forces as in Marx, but revolutions come out of and are preceded by intense cultural work. This cultural work is based upon critical knowledge. Marx implies, and Gramsci makes specific, that the only way to truly see the whole social system and its problems is to stand outside of it. So, in order to understand the capitalist system with its ideology you must be outside of it and grasp it as a whole. This obviously presents a problem for the elite: the system is created by and works for the benefit of the capitalist ruling class. Standing outside the system is particularly difficult for capitalists. Workers, on the other hand, while subject to the ideology of the system, are also by definition outside of it. The working class by its very position of alienation is capable of seeing the true whole, the knowledge of class relations from the standpoint of the entire society and its system of production and social relations.

Another school of Marxist thought that is particularly noteworthy is world systems theory, initiated by Immanuel Wallerstein. According to Wallerstein (1974/1980), individual states are currently linked together in a world economic system. There are four types of states in this world economy: the core, which contains the great military and economic powers of the day; the semi-periphery—states that are between the core and the periphery in terms of economic and military power; the periphery, which is the colonial or undeveloped countries; and the external areas available for conquest. The periphery is the core's exploited labor force, thus living conditions are poor and labor is forced. The core, as a result of having the periphery to exploit, enjoys light taxation, relatively free labor, and a high standard of living. The key to the world economic system is the presence of external areas to exploit. When there are no more external areas, world capitalism will dominate. And once that happens, the internal contradictions of capitalism will play themselves out and the end result will be a world socialist government. Thus, rather than emphasizing the critical consciousness that is implicit in Marx, as does the Frankfurt School, world systems theory simply lifts the empirical class dynamics out from the national level and situates them globally.

I won't spend this much time talking about the contemporary influences of all of our theorists. I've gone into a bit of detail with Marx because I think it is important for us to see that Marx's influence is stronger today than it has ever been. Above I talked about two main paths through which Marxism has come to us today: critical and world systems theory. These two paths have, in turn, informed countless other contemporary perspectives and theories. Among them are feminist standpoint theory; critical race theory; globalization studies; the Birmingham School of Cultural Studies; literary criticism; the postmodern theories of Douglas Kellner, Ben Agger, Fredric Jameson, and others; critical media studies; and the list goes on. Of course, this is one of the defining features of a classic: it continues to influence contemporary thought. However, I would hazard a guess that Marx's thought has specifically influenced our social, political, and sociological thinking more than any other thinker in this book.

I end this introduction with a quote from Friedrich Engels (1978b), spoken at Marx's graveside.

> On the 14th of March, at a quarter to three in the afternoon, the greatest living thinker ceased to think. He had been left alone for scarcely two minutes, and when we came back we found him in his armchair, peacefully gone to sleep—but for ever.
>
> An immeasurable loss has been sustained. . . . The gap that has been left by the departure of this mighty spirit will soon enough make itself felt.
>
> Just as Darwin discovered the law of development of organic nature, so Marx discovered the law of development of human history: the simple fact, hitherto concealed by an overgrowth of ideology, that mankind must first of all eat, drink, have shelter and clothing, before it can pursue politics, science, art, religion, etc.; that therefore the production of the immediate material means, and consequently the degree of economic development attained by a given people or during a given epoch, form the foundation upon which the state institutions, the legal conceptions, art, and even the ideas on religion, of the people concerned have been evolved, and in the light of which they must, therefore, be explained, instead of vice versa, as had hitherto been the case.
>
> But that is not all. Marx also discovered the special law of motion governing the present-day capitalist mode of production, and the bourgeois society that this mode of production has created . . .
>
> Such was the man of science. But this was not even half the man. Science was for Marx a historically dynamic, revolutionary force . . .
>
> For Marx was before all else a revolutionist. His real mission in life was to contribute, in one way or another, to the overthrow of capitalist society and of the state institutions which it had brought into being, to contribute to the liberation of the modern proletariat, which he was the first to make conscious of its own position and its needs, conscious of the conditions of its emancipation. Fighting was his element. And he fought with a passion, a tenacity and a success such as few could rival . . .
>
> And, consequently, Marx was the best hated and most calumniated man of his time. Governments, both absolutist and republican, deported him from their territories. Bourgeois, whether conservative or ultra-democratic, vied with one another in heaping slanders upon him. All this he brushed aside as though it were a cobweb, ignoring it, answering only when extreme necessity compelled him. And he died beloved, revered and mourned by millions of revolutionary fellow workers—from the mines of Siberia to California, in all parts of Europe and America—and I make bold to say that, though he may have had many opponents, he had hardly one personal enemy.
>
> His name will endure through the ages, and so also will his work. (p. 681)

Marx in Review

- Born on May 5, 1818, in Trier, one of the oldest cities in Germany, to Heinrich and Henrietta Marx. Both parents came from a long line of rabbis. His father was the first in his family to receive a secular education (he could recite numerous passages from Enlightenment thinkers)—Heinrich was a lawyer who allowed himself to be baptized Protestant in order to avoid anti-Semitism; a move that was not entirely successful.
- At seventeen, Karl Marx enrolled in the University of Bonn to study law. It was there that he came in contract with and joined the Young Hegelians, who were critical of Prussian society (specifically, because it contained poverty, government censorship, and religious discrimination). The Young Hegelians were particularly critical of the Prussian state's use of religious legitimation. Bonn was also a party school and young Marx spent a good deal of his time in beer halls. His father thus moved him to a more academically oriented university (Friedrich-Wilhelms-Universität in Berlin) where, much to his father's chagrin, Marx's interests turned to philosophy. Because of his political affiliations, Marx was denied a university position by the government. Marx turned to writing and editing, but had to battle government censorship continually.
- In 1843, Marx moved to Paris with his new wife, Jenny von Westphalen. In Paris, he read the works of reformist thinkers who had been suppressed in Germany and began his association with Friedrich Engels. During his time in Paris, Marx wrote several documents that were intended for self clarification (they were never published in his lifetime) but have since become important Marxian texts (*Economic and Philosophic Manuscripts of 1844* and *The German Ideology,* which was finished in Brussels).
- Over the next several years, Marx moved from Brussels, back to Paris, and then to Germany. Much of his movement was associated with revolutions that broke out in Paris and Germany in 1848. That year also marks the publication of *The Communist Manifesto.* Finally, in 1849, Marx moved to London, where he remained. He spent the early years of the 1850s writing several historical and political pamphlets.
- In 1852, Marx began his studies at the British Museum. There he would sit daily from 10AM to 7PM, studying the reports of factory inspectors and other documents that described the abuses of early capitalism. This research formed the basis of *Das Kapital,* his largest work. During this time, three of his children died of malnutrition.
- The workers' movements were quiet after 1848, until the founding of the *First International.* Founded by French and British labor leaders at the opening of the London Exhibition of Modern Industry, the union soon had members from most industrialized countries. Its goal was to replace capitalism with collective ownership. Marx spent the next decade of his life working with the *International.* The movement continued to gain strength worldwide until the Paris Commune of 1871. The Commune was the first

worker revolution and government. Three months after its formation, Paris was attacked by the French government. Thirty thousand unarmed workers were massacred.

- Marx continued to study but never produced another major writing. His wife died in 1881 and his remaining daughter a year later. Marx died in his home on March 14, 1883.

The Perspective: Human Nature, History, and Reality

The nucleus of Marx's thought contains two issues, both of which come from the world of philosophy. Three interrelated questions have dominated philosophy since its beginning, some 2500 years ago: What is reality? How do we know what we know? And, how are humans uniquely aware of their world and themselves? The first question asks what kinds of things exist and that field of study is called ontology (the study of being or existence). The second question asks how knowledge is created and its field of study is called epistemology (the study of knowledge). The third issue looks at how people are aware of themselves and their surroundings; it is simply referred to as the philosophy of consciousness.

Species-being: Marx actually builds his sociology from his answers to these questions. Let's consider the issues of consciousness and knowledge first. It seems humans are aware in a way that other animals are not. First, we are not simply aware of the environment; we are also conscious of our own awareness of it. And second, we can be conscious of our own existence and give it meaning. There are many philosophical and some sociological speculations about how this came about. Marx proposes a rather unique answer to the problem of human consciousness: **species-being.**

> By means of it [species-being] nature appears as *his* work and his reality. The object of labor is, therefore, the *objectification of man's species life*; for he no longer reproduces himself merely intellectually, as in consciousness, but actively and in a real sense, and he sees his own reflection in a world which he has constructed. (Marx, 1932/1995, p. 102)

Marx argues that the unique thing about being human is that we create our world. All other animals live in a kind of symbiotic relationship with the physical environment that surrounds them. Zebras feed on the grass and lions feed on the zebras, and in the end the grass feeds on both the zebras and the lions. The world of the lion, zebra, and grass is a naturally occurring world, but not so for the human world. Humans must create a world in which to live. They must in effect alter or destroy the natural setting and construct something new. The human survival mechanism is the ability to change the environment in a creative fashion in order to produce the necessities of life. Thus, when humans plow a field or build a skyscraper, there is something new in the environment that in turn acts as a mirror through which humans can come to see their own nature. Self-consciousness as a species that is distinct from all others comes as the human being observes the created human world. There is, then, an intimate connection between producer and product: *the very existence of the product defines the nature of the producer.*

Here's an illustration to help us think about this: Have you ever made anything by hand, like clothing or a woodworking project, or perhaps built a car from the ground up? Remember how important that thing was to you? It was more meaningful than something you buy at the store simply because you had made it. You had invested a piece of yourself in it; it was a reflection of you in a way that a purchased commodity could never be. But this is a poor illustration because it falls short of what Marx truly has in mind. Marx implies that human beings in their natural state lived in a kind of immediate consciousness. Initially, human beings created everything in their world by hand. There weren't supermarkets or malls. If they had a tool or a shirt, they had made it or they knew the person who did. Their entire world was intimately connected. They saw themselves purely in every product. Or, if they had bartered for something, then they saw an immediate social relationship with the person who had made the thing. When they looked into the world they had produced, they saw themselves, they saw a clear picture of themselves as being human (creative producers), and they also saw intimate and immediate social relations with other people. The world that surrounded them was immediately and intimately human. They created and controlled and understood themselves through the world that they had made.

Notice a very important implication of Marx's species-being: human beings by their nature are social and altruistic. Marx's vision of human is based on the importance of society in our species survival. We survive collectively and individually because of society. Through society we create what is needed for survival; if it were not for society, the human animal would become extinct. We are not equipped to survive in any other manner. What this means, of course, is that we have a social nature—we are not individuals by nature. Species-being also implies that we are altruistic. Altruism is defined as uncalculated commitment to others' interests. Not only are we not individualistic by nature, we are not naturally selfish. If human survival is based on collective cooperation, then it would stand to reason that our most natural inclination would be to serve the group and not the self. The idea of species-being is why Marx believes in communism—it's the closest economic system to our nature state. Further, Marx would argue that, under conditions of modernity/capitalism, we don't see these attributes in humans because we exist under compromising structures. It is capitalism that teaches us to be self-centered and self-serving. This effect of capitalistic structures is why Marx argues that society will go through a transition stage of socialism on its way from capitalism to communism.

Marx's theory of species-being also has implications for knowledge and consciousness. In the primitive society that we've been talking about, humans' knowledge about the world was objective and real; they held ideas that were in perfect harmony with their own nature. According to Marx, human ideas and thought come about in the moment of solving the problem of survival. Humans survive because we creatively produce, and our clearest and most true ideas are grounded in this creative act. In species-being, people become truly conscious of themselves and their ideas. Material production, then, is supposed to be the conduit through which human nature is expressed, and the product ought to act as a mirror that reflects back our own nature.

Let's try an analogy to get at this extremely important issue. There are a limited number of ways you can know how you physically look (video, pictures, portraits, mirrors, and so forth). The function of each of these methods is to represent or reproduce our image with as little distortion as possible. But what if accurate representation was impossible? What if every medium changed your image in some way? We would have no true idea how we physically look. All of our ideas would be false in some way. We would think we see ourselves but we wouldn't. Marx is making this kind of argument; but not about our physical appearance; he's concerned with something much more important and fundamental—our nature as humans. We think we see it, but we don't.

We need to take this analogy one step further: notice that with mirrors, pictures, and videos, there is a kind of correspondence between the representation and its reality. What I mean is that each of these media presents a visual image, and in the case of our physical appearance, that's what we want. Imagine if you asked someone how you looked and the person played an audio cassette tape for you. That wouldn't make any sense, would it? There would be no correspondence between the mode of representation and the initial presentation. This, too, is what Marx is telling us. If we want to know something about our human nature, if we want to see it represented to us, where should we look? What kind of medium would correspond to our nature? Marx is arguing that every species is defined by its method of survival or existence. Why are whales, lions, and hummingbirds all different? They are different because they have different ways of existing in the world. What makes human beings different from whales, lions, and hummingbirds? Humans have a different mode of existence. We creatively produce what we need—we make products, and we are the only species that does.

So, where should we look to understand our nature? What is the medium that corresponds to the question? If we want to know how we look physically, we look toward visual images. But if we want to know about our nature, we must look to production and everything associated with it. Thus, according to Marx, production is the vehicle through which we can know human nature. However, Marx says that there is something wrong with the medium. Under present conditions (capitalism), it gives a distorted picture of who and what we are.

If we understand this notion of species-being, then almost everything else Marx says falls into place. To understand species-being is to understand alienation (being cut off from our true nature), ideology (ideas not grounded in creative production), false consciousness (self-awareness that is grounded in anything other than creative production), and we can also understand why Marx placed such emphasis on the need for class consciousness in social change. This understanding of human nature is also why Marx is considered an economic determinist. The economy is the **substructure** from which all other structures (**superstructure**) of human existence come into being and have relevance. We will explore these ideas more later on, but for now let's move on to the second part of the core of Marx's thought—his consideration of reality/ontology.

Material dialectic: While ontology doesn't become part of Marx's sociology, as consciousness does, his ontological work results in the concept of the **material dialectic,** or what is generally called dialectical materialism, which is essential for his theory. In Marx's time, there were two important ways of understanding the issue of reality: idealism and materialism. **Idealism** posits that reality only exists in our idea of it. While there may indeed be a material world that exists in and of itself, that world exists for humans only as it appears. The world around us is perceived through the senses, but this sense data is structured by innate cognitive categories. Thus, what appears to humans is not the world itself but our idea of it.

On the other hand, **materialism** argues that all reality may be reduced to physical properties. In materialism, our ideas about the world are simple reflections; those ideas are structured by the innate physical characteristics of the universe. Marx feels that both of these extremes do not correctly consider humans as social animals. He proposes another way of understanding reality and consciousness; he terms this way of thinking naturalism, or humanism.

Marx rejects brute materialism out of hand, but he has to consider idealism more carefully. One of idealism's strongest supporters, Georg Wilhelm Friedrich Hegel, died just four years prior to Marx enrolling at the University of Bonn, and at the time, Hegel still held a significant place in the thinking of German philosophers (in fact, Hegel is still a powerful figure in philosophy). Hegel was an idealist and argued that material objects (like a chair or a rock) truly and completely only exist in our concept of them. But Hegel took idealism to another level, using it to argue for the existence of God (the ultimate concept); he argued that the ideal took priority over the material world. According to Hegel, human history is a dialectical unfolding of the Truth that reality consists of ideas and that the material world is nothing more than shadow. This dialectical unfolding ends in the revelation of God.

A dialectic contains different elements that are naturally antagonistic to one another; Hegel called them the thesis and antithesis. The dialectic is like an argument or a dialog between elements that are locked together (The word *dialectic* comes from the Greek word *dialektikos,* meaning discourse or discussion.). For example, to understand "good," you must at the same time understand "bad." To comprehend one, you must understand the other: good and bad are locked in a continual dialog. Hegel argued that these kinds of conflicts would resolve themselves into a new element or synthesis, which in turn sets up a new dialectic: every synthesis contains a thesis that by definition has conflicting elements. Hegel's ideational dialectic would look something like I have diagrammed in Figure 3.1.

Marx liked the historical process implied in Hegel's dialectic, but he disagreed with its ideational base. Marx, as we have seen, argues that human beings are unique because they creatively produce materials to fill their own material needs. Since the defining feature of humanity is *production,* not ideas and concepts, then Hegel's notion of idealism is false, and the dialectic is oriented around material production and not ideas—the material dialectic. Thus, the dynamics of the historical dialectic are to be found in the economic system, with each economic system inherently containing antagonistic elements (see Figure 3.2). As the antagonistic elements work themselves out, they form a new economic system.

Notice that the engine of progress according to the dialectic is conflict. For Hegel, ideas resolve themselves and humanity comes closer to the Truth because there is a natural antagonism within the idea itself. Marx of course sees this occurring in

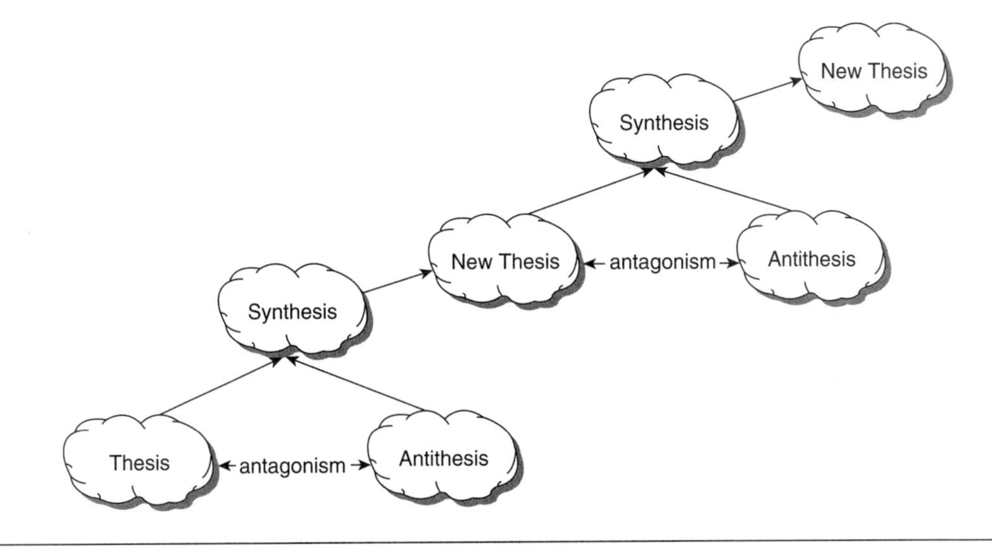

Figure 3.1 Hegel's Dialectic

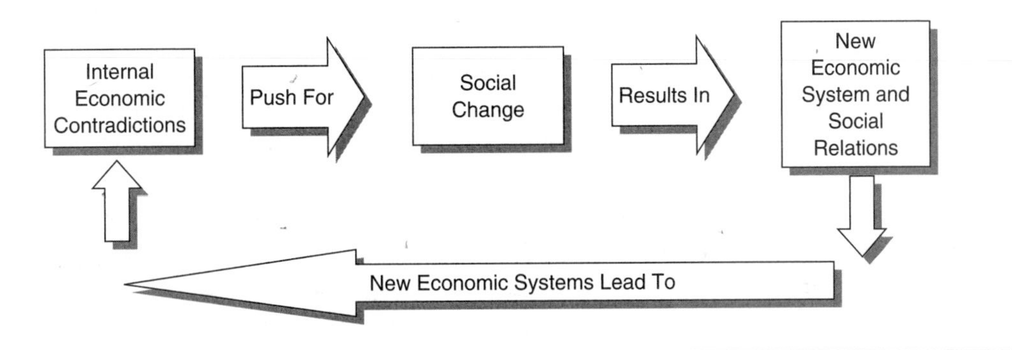

Figure 3.2 Marx's Material Dialectic

material or social relations, so the way societies change and progress is through conflict—the engine of social change is dialectical conflict. Here we see a general point about the conflict/critical model of society, and of course another element in Marx's thinking.

We can think of this way of seeing society as an upheaval model. According to this perspective, society is not like an organism that gradually and peacefully becomes more complex in order to increase its survival chances, as in functionalism. Rather, society is filled with human beings who exercise power to oppress and coerce others. Periods of apparent peace are simply times when the powerful are able to dominate the populace in an efficient manner. But, according to this model, the suppressed will become enabled and will eventually overthrow and change the system. Social change, then, occurs episodically and through social upheaval.

There's an important point here: for Marx, revolution is unavoidable. It is certainly the case that Marx was critical about capitalism because of a personal point of view: he hated the abuses he saw. Remember, at that time governmental controls such as

OSHA did not exist and capitalists required their workers to work in abject conditions. But here we can also see a more logical, philosophical reason behind the critical point of view. If history is the result of structural forces locked in a dialectic, then it makes sense and is philosophically consistent to look at society using a conflict perspective.

In mapping out the past of the historical dialectic, Marx categorizes five different economic systems (means of production along with their relations of production): preclass societies, Asiatic societies, ancient societies, feudal societies, and capitalist societies. Preclass societies are like hunter–gatherer groups. These were small groups of people with a minimal division of labor (one that Marx termed the **natural division of labor**) and communal ownership of property (termed **primitive communism**). Asiatic societies were a special form in that they had particular problems to overcome due to their large populations. There was thus a tendency to form "oriental despotism" to solve these problems. Ancient societies developed around large urban centers, such as Rome. Private property and slave labor came into existence, as well as significant class inequality. Ancient societies were replaced by feudal systems wherein the primary economic form was serf labor tied to the land of the aristocracy. Feudal systems were replaced by capitalist systems. Eventually, the capitalist system will be replaced by socialism and that by communism. The specific dynamics that Marx says caused these shifts in economic systems are not important in our consideration right now. What is important to see is that for Marx, social change comes about because of inherent contradictions in the economic structure. What this implies is that to think like Marx is to think as a structuralist.

There are a number of dualisms in sociology. One of them concerns the tension between agency and structure: how free (agency) are people to be and act apart from social constraints (structure)? As a structuralist, Marx feels that social structures profoundly influence human thought, feelings, and action. Social change comes about not simply because of the free actions of the people, but because of changes in the social structure. These changes are prompted by the dialectical elements within every economic system.

To think like Marx, then, is to be driven by two main ideas. The first is the notion of species-being. Seeing society through this lens means to understand basic human nature as defined through production. It also implies that all true ideas are materially based. If human nature is founded on a unique way of existing in the world through creative production, then our most human (humane and humanistic) ideas must spring from the economy. This means two things. It first implies that true ideas are not abstract concepts with no basis in material reality. Human ideas are grounded and real, if they spring from creative production.

The other implication of this way of understanding consciousness and knowledge concerns false ways of knowing and existing. If ideas come from any other source than creative production, then they are simply counterfeit realities that lead to false consciousness and alienation. We will talk at length about these issues shortly, but for now I want us to see an important repercussion of this concept: to think like Marx means to be concerned about the inner, subjective world that human beings experience.

Marx has often been seen as anti-spiritual. This notion is far from the truth. Marx is deeply concerned with lifting human experience out of the quagmire and placing our feet on higher ground. Species-being implies that humans are altruistic, social beings, but our sociability has been cut off and each person stands alone and naked due to cold capitalistic considerations. Species-being also implies that the individual person is filled with a creative capacity that has been disconnected and denied in the search for profit. Rather than being a cold materialist, Marx gives us a "spiritual existentialism in secular language" (Fromm, 1961, p. 5). To think like Marx, then, is to be critical of humanity's inhumanity. "Marx's philosophy is one of protest; it is a protest imbued with faith in man, in his capacity to liberate himself, and to realize his potentialities" (Fromm, 1961, p. vi).

The second idea that drives Marx's thought is the material dialectic. Thus, to think like Marx also means to think in historical, structural terms. While Marx is overwhelmingly concerned with the authenticity of human experience, he sees that it is the economic structure that moves history and influences our inner person. Society, then, is objective and causative, through the economy and class relations. To think like Marx also means to have an historical perspective. It's easy for us to be weighed down by the demands and problems of our lives. And it is thus easy for us to be concerned only with small segments of time and society. Yet Marxian thought is different; it is bigger. C. Wright Mills (1959) made this distinction clear when he spoke of personal troubles and public issues. He called this point of view the sociological imagination:

> The first fruit of this imagination—and the first lesson of the social science that embodies it—is the idea that the individual can understand his own experience and gauge his own fate only by locating himself within his period, that he can know his own chances in life only by becoming aware of those of all individuals in his circumstances. (p. 5)

The balance of this chapter is divided into two smaller parts. In *The Basic Features of Capitalism,* we will be looking at class and class structure; value and exploitation; and industrialization, markets, and commodities. Throughout this section, we will see that these elements of capitalism are locked in dialectic, one that provides the engine of historical change and will eventually lead to the demise of capitalism, according to Marx. In the second section, we will talk about *The Ramifications of Capitalism.* In Marx's understanding, economic systems are not sterile creatures; they create particular kinds of consciousnesses and relations with the world at large. Thus, capitalism brings with it alienation, private property, commodity fetish, false consciousness, and ideology. All of these issues exist and are understood because of species-being. But Marx holds out a hope for us through the dialectic: class consciousness will be produced as the structural conflicts resolve themselves, and capitalism will pass, not peacefully, into socialism.

The Basic Features of Capitalism

Class and class structure: As we have already seen, Marx is an economic determinist, which means he views the dynamic behind history as the process of production.

When Marx speaks about production, he is concerned with three intertwined issues: the means or actual process of production, the social relationships that form because of production, and the end result of production—the product. The **means of production** refers to the methods and materials that we use to bring into being those things that we need to survive. On a small scale, we might think of the air-hammers, nails, wood, concrete mixers, and so forth that we use to produce a house. Inherent within any means of production are the **relations of production:** in this case, the contractor, subcontractor, carpenter, financier, buyer, and so forth. In the U.S. economy, we organize the work of building a house through a contracting system. The person who wants the house built has to contract with a licensed builder who in turn hires different kinds of workers (day laborers, carpenters, foremen, etc.). The actual social connections that are created through particular methods of production are what Marx wants us to see in the concept of the relations of production.

Of course, Marx has something much bigger in mind than our example. In classic feudalism, for instance, people formed communities around a designated piece of land and a central manor for provision and security. At the heart of this local arrangement was a noble who had been granted the land from the king in return for political support and military service. At the bottom of the community was the serf. The serf lived on and from the land and was granted protection by the noble in return for service. Feudalism was a political and economic system that centered on land—land ownership was the primary means of production. People were related to the land through oaths of homage and fealty (the fidelity of a feudal tenant to his lord). These relations functioned somewhat like family roles and spelled out normative obligations and rights. The point here, of course, is that the way people related to each other under feudalism was determined by that economic system and was quite different than the way we relate to one another under capitalism: most of us don't think of our boss as family.

For Marx, human history is the history of class struggles. Marx identifies several different types of classes, such as the feudal nobility, the bourgeoisie, the petite bourgeoisie, the proletariat, the peasantry, the subproletariat, and so on. As long as these classes have existed, they have been antagonistic toward each other. Under capitalism, however, two factors create a unique class system. The first thing capitalism does is lift economic work out of all other institutional forms. Under capitalism, the relationships we have with people in the economy are seen as distinctly different from religious, familial, or political relations. For most of human history, all these relationships overlapped. For example, in agriculturally based societies, family and work coincided. Fathers worked at home and all family members contributed to the work that was done. Capitalism lifted this work away from the farm, where work and workers were embedded in family, and placed it in urban-based factories. Capitalist industrialization thus disembedded work from family and social relations. Contemporary gender theorists point out that this movement created dual spheres of home and work, each controlled by a specific gender.

Marx, on the other hand, sees gender inequality reaching further back than capitalism, though capitalism has certainly accentuated gender problems. Marx, and more specifically Engels, was actually among the first to write on the issue of gender.

In 1877, an American anthropologist, Lewis H. Morgan, pu argued for the matrilineal origins of society. Both Marx and Eng a significant discovery. Marx planned on writing a thesis based c and made extensive notes along those lines. Marx never finished th Engels, however, using Marx's notes, did publish *The Origin of the Property, and the State* in 1884. The argument is fairly simple, yet i of the basic ways in which we understand how gender inequality an sion of women came about.

As with all of Marx and Engels' work, Engels begins with a conceptio tive communist beginnings. In this setting, people lived communally, sharing everything, with monogamy rarely, if ever, practiced. Under such conditions, family is a social concern rather than a private issue, with children being raised by the community at large, rather than by only two parents. As a matter of fact, because identifying the father with any certainty was impossible in pre-modern society, paternity itself wasn't much of an issue. Thus, in Marx's way of thinking, "the communistic household implies the supremacy of women" (Engels, 1884/1978a, p. 735).

The key in the transition from matrilineal households to patriarchy is wealth. Primitive societies generally lived "from hand to mouth," but as surplus began to be available, it became possible to accumulate. As men began to control this wealth, it was in their best interest to control inheritance, which meant controlling lineage. Engels (1884/1978a) says that the way this happened is lost in prehistory, but the effect "was the *world-historic defeat of the female sex*" (p. 736). In order to control the inheritance of wealth, men had to control fertilization and birth, which meant that men had to have power over women. Control over women's sexuality and childbirth is why, according to Engels, we have developed a dual morality around gender—women are considered sluts but men are studs if they sleep around. Marriage, monogamy, and the paired family (husband and wife) were never intended to control men's sexuality. They were created to control women's sexuality in order to assure paternity. This control, of course, implied the control of the woman's entire life. She became the property of the man so the man could control his property (wealth). Quoting Marx, Engels (1884/1978a) concludes, "The modern family contains in embryo not only slavery . . . . It contains within itself in *miniature* all the antagonisms which later develop on a wide scale within society and its state" (p. 737).

The second unique feature of class under capitalism is its bipolarization. That is, under capitalism, class tends to be structured around two positions—the **bourgeoisie** (owners) and the **proletariat** (workers). Marx does talk about other classes in capitalism, but they have declining importance. The petite bourgeoisie is the class of small land and business owners and the *lumpenproletariat* is the underclass (like the homeless). While the lumpenproletariat played a class-like position in early French history, Marx argues that because they have no relationship to economic production at all, they will become less and less important in the dynamics of capitalism. The petite bourgeoisie, on the other hand, does constitute a legitimate class in capitalism. However, this class shrinks in number and becomes less and less important, as they are bought out and pushed aside by powerful capitalists. It's important to note that while most people in capitalist countries see business size as the result

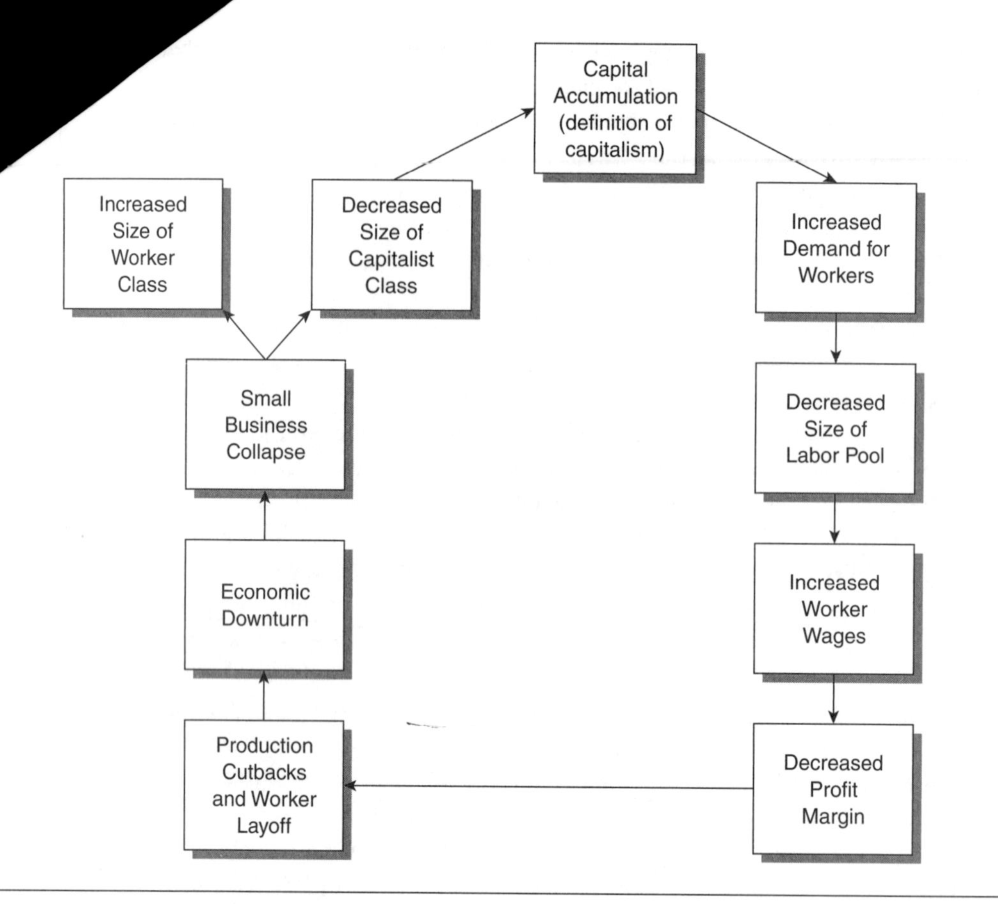

Figure 3.3 Marx's Business Cycle

of competition, Marx sees it as a result of structural dialectic processes. In other words, for Marx there is a hidden structure underneath the surface of these behaviors we generally think of as individualistic and competitive.

One of the principal dialectical processes that helps to polarize the classes is capitalism's business cycle (see Figure 3.3). Capitalism is defined by the reinvestment of profit to make more profit. This intrinsic part of capitalism brings with it certain elements that are destructive to capitalism. As capitalists reinvest **capital,** the demand for labor goes up. The increased demand for labor causes the labor pool, the number of unemployed, to shrink. As with any commodity, when demand is greater than the supply, the price goes up—in this case wages. The increase in wages causes profits to go down. As profits go down, capitalists cut back production, which precipitates a crisis in the economy. The crisis causes more workers to be laid off and small businesses (the petite bourgeoisie) to fold. These small businesses are bought out by the larger capitalists and the once small-scale capitalists become part of the working class. The result of this process being repeated over time is that the class of dependent workers increases and capital is centralized into fewer and fewer hands. Thus, the existing capitalists accumulate additional capital and the entire cycle starts again.

Figures that measure wealth and ownership over time are difficult to produce. Wealth in the United States is usually well hidden and ferreting out ownership lines is thorny at best. Nevertheless, we can find numbers that seem to substantiate the effects that Marx is talking about. While these are indirect measures at best, it is interesting to note that according to *Forbes* magazine (see Kennickell, 2003), the average wealth of the top 400 richest people in the United States rose from $921 million in 1989 to $2.1 billion in 2002. The distribution of income over time has likewise shifted upwards in the United States, according to Census data. In 1967, the top fifth of the population received 43.8% of the distributed income and the bottom fifth received just 4%. In 2001, the top fifth was paid 50.1% of the available income and the bottom fifth received 3.5%. The big drop in income between 1967 and 2001 actually was in the middle fifth: they dropped from 17.3% to 14.6%. Like I said, these figures don't prove Marx's theory, but they do provide some indication that the control of capital is centralizing into fewer and fewer hands.

> But not only has the bourgeoisie forged the weapons that bring death to itself; it has also called into existence the men who are to wield those weapons—the modern working class—the proletariat. (Marx & Engels, 1848/1978, p. 478)

As a result of this business cycle, three things occur. First, the size of the capitalist class shrinks. For Marx that means that when the revolution does come about, it will be easy to take over power from fewer capitalists. Of course, the second thing that occurs is that the size of the working class army increases, which will give them more power. And finally, the gap between the owners and the workers becomes more and more apparent, leading to a **bi-polarization of conflict** (the reduction of conflict to two parties).

The bi-polarization of conflict is a necessary step for Marx in the process of social change, and an important one for conflict theory as a whole. Overt and intense conflict are both dependent upon bi-polarization; crosscutting interests, that is, having more than one issue over which groups are in conflict, tend to pull resources (emotional and material) away from conflict. For example, during WWII it was necessary for the various nations to coalesce into only two factions, the Axis and the Allied powers. So the United States became strange bedfellows with the Soviet Union in order to create large-scale, intense conflict. Note that each time the United States has engaged in a police action or war, attempts are made to align resources in as few camps as possible. This move isn't simply a matter of world opinion; it is a structural necessity for violent and overt conflict. The lack of bi-polarized conflict creates an arena of crosscutting interests that can drain resources and prevent the conflict from escalating and potentially resolving.

Value and exploitation: Marx forms much of this theory of capitalism in direct contrast to the political economists of his day. For most political economists in Marx's time, commodities, value, profit, private property, and the division of labor were seen as natural effects of social evolution. However, Marx takes a critical perspective and sees these same processes as instruments of oppression that dramatically affect people's life chances.

One of the important issues confronting early political economists concerned the problem of value. We may have a product, such as a car, and that product has value, but from where does its value come? More importantly, why would anybody pay more for the car than it is worth? Well, you say, only a sucker would pay more for a car than it is worth. Yet, as you'll see, there is a way in which we all pay more for every commodity or economic good than it is worth. That was what struck the early economists as a strange problem to be solved.

In order to understand this issue, Adam Smith, an early economist and author of *Inquiry Into the Nature and Causes of the Wealth of Nations* (1776), came up with some useful concepts. He argues that every commodity has at least two different kinds of values: use-value and exchange-value. **Use-value** refers to the actual function that a product contains. This function gets used up as the product is used. Take a bottle of beer, for example. The use-value of a bottle of beer is its taste and alcoholic effect. As we drink the bottle, those functions are expended. Beer also has exchange-value that is distinct from use-value. **Exchange-value** refers to the rate of exchange one commodity bears when compared to other commodities. Let's say I make a pair of shoes. Those shoes could be exchanged for 1 leatherbound book or 5 pounds of fish or 10 pounds of potatoes or 1 cord of oak wood, and so on.

This notion of exchange-value poses a question for us: what do the shoes, books, fish, and potatoes have in common that allow them to be exchanged? I could exchange my pair of shoes for the leatherbound book and then exchange the book for a keg of beer. The keg of beer might have a use-value for me where the leather book does not; nonetheless, they both have exchange-value. This train of exchange could be extended indefinitely with me never extracting any use-value from the products at all, which implies that exchange- and use-value are distinct. So, what is the common denominator that allows these different items to be exchanged? What is the source of exchange-value?

Smith argues, and Marx agrees, that the substance of all value is human labor: "Labour, therefore, is the real measure of the exchangeable value of all commodities" (Smith, 1776/1937, p. 30). There is labor involved in the book, the fish, the shoes, the potatoes, and in fact everything that people deem worthy of being exchanged. It is labor, then, that creates exchange-value. If we stop and think for a moment about Marx's idea of species-being, we can see why Smith's notion appealed to him: the value of a product is the "humanness" it contains.

This explanation is termed the *labor theory of value,* and it is the reason why Marx thinks **money** is so insidious. The book, the shoes, and the potatoes can have exchange-value because of their common feature—human labor. If we then make all those commodities equal to money—the universal value system—then labor is equated with money: "This physical object, gold [or money] . . . becomes . . . the direct incarnation of all human labour" (Marx, 1867/1977, p. 187), which of course adds to the experience of alienation from species-being.

In employing the two terms, Smith tends to collapse them, focusing mainly on exchange-value. Nevertheless, Marx maintains the distinction and argues that the difference between use-value and exchange-value is where profit is found (we pay more for a product than its use-value would indicate). Smith eventually argued that profit is simply added by the capitalist. Profit in Smith's theory thus becomes arbitrary and controlled by the "invisible hand of the market." However, in analyzing value, Marx discovers a particular kind of labor—surplus labor—and argues that profit is better understood as a measurable entity that he calls **exploitation.**

Like Smith, Marx distinguishes between the use-value of a product and its exchange-value in the market. Capital, then, is created by using existing commodities to create a new commodity whose exchange-value is higher than the sum of the original resources used. This situation was odd for Marx. From where did the added value come? His answer, like Smith's original one, is human labor. But he took Smith's argument further. Human labor is a commodity that is purchased for *less than its total worth.* According to the value theory of labor, the value of any commodity is determined by the labor time necessary for the production or replacement of that commodity. So, what does it cost to produce human labor? The cost is reckoned in terms of the necessities of life: food, shelter, clothing, and so on. Marx also recognizes that a comparative social value has to be added to that list as well. What constitutes a "living wage" will thus be different in different societies. Marx calls the labor needed to pay for the worker's cost of living **necessary labor.** The issue for Marx is that the cost of necessary labor is less than that of what the worker actually produces.

The capitalist pays less for a day's work than its value. I may receive $75.00 per day to work (determined by the cost to bare sustenance the worker plus any social amenities deemed necessary), but I will produce $200.00 worth of goods or services. The necessary labor in this case is $75.00. The amount of labor left over is the **surplus labor** (in this case, $125.00). The difference between necessary labor and surplus labor is the rate of exploitation. Different societies can have different levels of exploitation. For example, if we compare the situation of automobile workers in the United States with those in Mexico, we will see that the level of exploitation is higher in Mexico (which is why U.S. companies are moving so many jobs out of the country). Surplus labor and exploitation are the places from which profit comes: "The rate of surplus-value is . . . an exact expression for the degree of exploitation . . . of the work by the capitalist" (Marx, 1867/1977, p. 326).

By definition, capitalists are pushed to increase their profit margin and thus the level of surplus labor and the rate of exploitation. There are two main ways in which this can be done: through absolute and relative surplus labor. The capitalist can directly increase the amount of time work is performed by either lengthening the workday, say from 10 to 12 hours; or he or she can remove the barriers between "work" and "home," as is happening as a result of increases in communication (computers) and transportation technologies. The product of this lengthening is called **absolute surplus labor.** The other way a capitalist can increase the rate of exploitation is to reduce the amount of necessary labor time. The result of this move is called **relative surplus labor.** The most effective way in which this is done is through industrialization. With industrialization, the worker works the same number of hours but her or his output is increased through the use of machinery. These different kinds of surplus labors can get a bit confusing, so I've compared them in Figure 3.4.

The figure starts off with the type of surplus labor employed. Since there is always exploitation (you can't have capitalism without it), I've included a "base rate" for the purpose of comparison. Under this scheme, the worker has a total

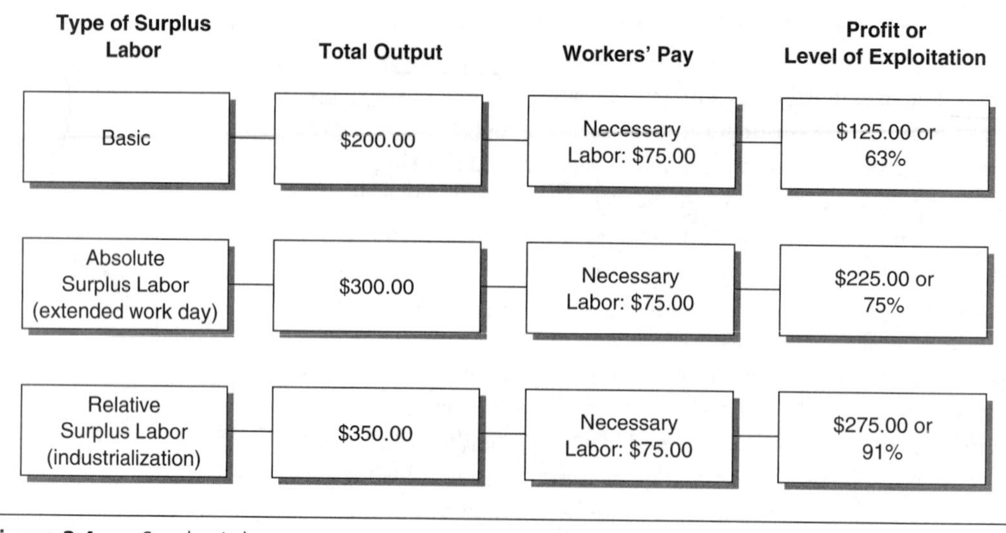

Figure 3.4 Surplus Labor

output of $200.00; he or she gets paid $50.00 of the $200.00 produced, which leaves a rate of exploitation of about 63%, or $125.00. The simplest way to increase profit, or the rate of exploitation, is to make the worker work longer hours or take on added responsibilities without raising pay (as a result of downsizing, for example). In our hypothetical case, the wage of $75.00 remains, but the profit margin (rate of exploitation) goes up to 75%. By automating production, the capitalist is able to extract more work from the worker, thus increasing the total output and the level of exploitation (91%). If this seems natural to us (capitalists have a right to make a profit), Marx would say that it is because we have bought into the capitalist ideology. We should also keep in mind that in their search for maintaining or increasing the rate of exploitation, capitalists in industrialized nations export their exploitation—they move jobs to less developed countries.

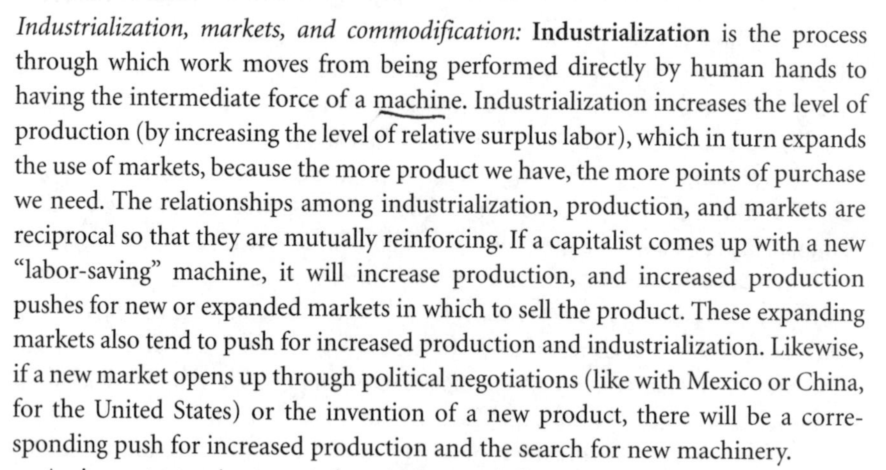

Industrialization, markets, and commodification: **Industrialization** is the process through which work moves from being performed directly by human hands to having the intermediate force of a machine. Industrialization increases the level of production (by increasing the level of relative surplus labor), which in turn expands the use of markets, because the more product we have, the more points of purchase we need. The relationships among industrialization, production, and markets are reciprocal so that they are mutually reinforcing. If a capitalist comes up with a new "labor-saving" machine, it will increase production, and increased production pushes for new or expanded markets in which to sell the product. These expanding markets also tend to push for increased production and industrialization. Likewise, if a new market opens up through political negotiations (like with Mexico or China, for the United States) or the invention of a new product, there will be a corresponding push for increased production and the search for new machinery.

An important point to note here is that capitalism requires expanding markets, which implies that markets and their effects may be seen as part of the dialectic of capitalism. In capitalism, there is always a push to increase profit margins. To increase profits, capitalists can expand their markets horizontally and vertically, in addition to increasing the level of surplus labor. In fact, profit margins would slip if capitalists did not expand their markets. For example, one of the main reasons that you probably have a CD player is that the market for cassette tape players bottomed out (and now there is a push for MP3 and newer technologies). Most people who were going to buy a cassette player had already done so, and the only time another would be purchased is for replacement. So, capitalists invented something new for you to buy so that their profit margin would be maintained.

In general, **markets** refer to an arena in which commodities are exchanged between buyers and sellers. For example, we talk about the grocery market and the money market. These markets are defined by the products they offer and the social network involved. Markets in general have certain characteristics that have consequences for both commodities and people. They are inherently susceptible to expansion (particularly when driven by the capitalist need for profit), abstraction (so we can have markets on markets, like stock market futures or the buying and selling of home mortgage contracts), trade cycles (due to the previous two issues), and undesirable outputs (such as pollution); and they are amoral (so they may be used to sell weapons, religion, or to grant access to health care).

The speed at which goods and services move through markets is largely dependent upon a generalized medium of exchange, something that can act as universal value. Barter is characterized by the exchange of products for one another. The problem with bartering is that it slows down the exchange process because there is no general value system. For example, how much is a keg of beer worth in a barter system? We can't really answer that question because the answer depends on what it is being exchanged for, who is doing the exchanging, what their needs are, where the exchange takes place, and so on. Because of the slowness of bartering, markets tend to push for more generalized means of exchange—such as money. Using money, we can give an answer to the keg question, and having such an answer speeds up the exchange process quite a bit. Marx argues that as markets expand and become more important in a society, and the use of money for equivalency becomes more universal, money becomes more and more the common denominator of *all* human relations. As Marx (1932/1978b) says,

> By possessing the *property* of buying everything, by possessing the property of appropriating all objects, *money* is thus the *object* of eminent possession. The universality of its *property* is the omnipotence of its being. It therefore functions as the almighty being. Money is the *pimp* between man's need and the object, between his life and his means of life. But that which mediates *my* life for me, also *mediates* the existence of other people *for me*. For me it is the *other* person. (p. 102)

The expansive effect of markets on production is called the process of **commodification.** The concept of commodification describes the process through which more and more of the human life-world is turned into something that can be bought or sold. So, instead of creatively producing the world as in species-being, people increasingly buy (and sell) the world in which they live. This process of commodification becomes more and more a feature of human life because it continually expands. Think of a farming family living in the United States around 1850 or so. That family bought some of what they needed, and they bartered for other things, but the family itself produced much of the necessities of life. Today the average American family buys almost everything they want or need. The level of commodification is therefore much higher today.

These mutually reinforcing relationships are pictured in Figure 3.5. As you can see, I've placed profit motivation as the driving force. This indicates, as Marx would argue, that these effects are a result of an intrinsic feature of capitalism, and thus part of the dialectic. Marx sees these relationships as mutually reinforcing and multiplying—that is, they continue to expand at ever-increasing rates. Notice also the feedback loop from the level of commodification. As more and more of our life-world becomes fair game for commodification, the possibilities for new markets expand.

As you can tell, once in place, these elements of capitalism are self-reinforcing. Money facilitates exchanges in markets; money and the drive for profit push the size and exchange rate of markets, which in turn pushes for increased production and commodification. One of the things that Marx points out is that human beings have the unusual ability to create their own needs. Animal needs are basically tied to instinct and survival, but once humans begin to create commodities, we create our own needs (I really do *need* a sub-woofer for my stereo). The potential, then, for the production of commodities is endless; commodification once begun takes on a life of its own. As the ideas of capitalism take root and people begin to see the human life-world in terms of things that can be bought or sold, there is a constant urge to find what else can be used to make a profit.

The capitalist drive for expanding profits and the endless potential for commodification is one of the prime factors in back of what is now called globalization. While

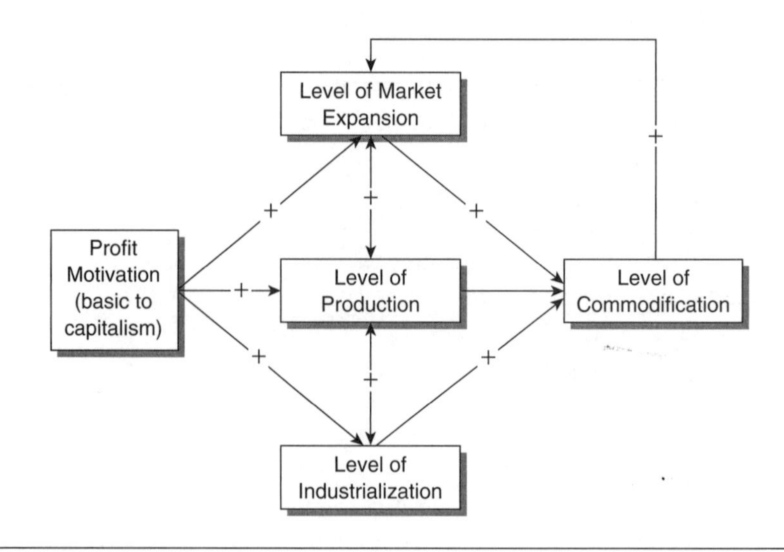

Figure 3.5 Commodification

there are other factors at work (such as global information and communication systems), expansive markets and commodification are particularly salient. Marx foresaw some of this. He writes in *The Communist Manifesto,* "The need of a constantly expanding market for its products chases the bourgeoisie over the whole face of the globe. It must nestle everywhere, settle everywhere, establish connexions everywhere. The bourgeoisie has through its exploitation of the world-market given a cosmopolitan character to production and consumption in every country" (Marx & Engels, 1848/1978, p. 476). Marx indicates that global markets and commodification lead to an interesting result—a cosmopolitan character to production and consumption. As commodities are fitted for the global market and as we become globalized consumers, the purchased good (and the consumer) becomes emptied of its local meanings and idiosyncrasies. For example, I have a friend from China who, when we go out for Chinese food, says, "This isn't Chinese food." In order to make commodities palatable to everyone, we have to make them bland (and this doesn't simply apply to food). Recently, George Ritzer (2004b) in a provocative book refers to this kind of phenomenon as "the globalization of nothing within the realm of consumption, itself proliferating throughout the globe at a breathtaking rate" (p. x). We will talk more about the trivializing effects of commodification when we get to Simmel; but it begins here with Marx.

The Ramifications of Capitalism

Alienation, private property, and commodity fetish: We often think of alienation as the subjective experience of a worker on an assembly line or at McDonald's. This perception is partly true. To get a little better handle on it, however, let's think about the differences between a gunsmith and a worker in a Remington plant. The gunsmith is a craftsperson and would make every part of the gun by hand—all the metal work, all the woodwork, everything. If you knew guns, you could tell the craftsperson just by looking at the gun. It would bare the person's mark because part of the craftsperson was in the work, and the gunsmith would take justifiable pride in the piece.

> The development of Modern Industry, therefore, cuts from under its feet the very foundation on which the bourgeoisie produces and appropriates products. What the bourgeoisie, therefore, produces, above all, is its own grave-diggers. (Marx & Engels, 1848/1978, p. 483)

Compare that experience to the Remington plant worker. Perhaps she is the person who bolts the plastic end piece on the butt of the rifle, and she performs the task on rifle after rifle, day after day. It won't take long until the work becomes mind numbing. It is repetitive and non-creative. Chances are, this worker won't feel the pride of the craftsperson but will instead experience disassociation and depression. We can see in this example some of the problems associated with a severe division of labor and over-control of the worker. There's no ownership of the product or pride as there is in craftsmanship. We can also see how the same issues would apply to the worker at McDonald's. All of it is mind-numbing, depressive work, and much of it is the result of the work of scientific management.

Frederick Taylor was the man who applied the scientific method to labor. He was interested in finding the most efficient way to do a job. Efficiency here is defined in terms of the least amount of work for the greatest amount of output. Under this system, the worker becomes an object that is directly manipulated for efficiency and profit. Taylor would go out into the field and find the best worker. He and his team would then time the worker (this is the origin of time-management studies) and break the job down into its smallest parts. In the end, Taylorism created a high division of labor, assembly lines, and extremely large factories.

The problem with understanding our Remington and McDonald's examples as alienation is that it focuses on the subjective experience of the worker. It implies that if we change the way we control workers—say from Taylorism to Japanese management, as many American companies have—we've solved the problem of alienation. For Marx, that wouldn't be the case. While alienation implies the subjective experience of the worker (depression and disassociation), it is more accurate to think of it as an objective state. So workers under the Japanese system are not structurally *less* alienated because they are more broadly trained, are able to rotate jobs, work holistically, and have creative input. Workers *are* alienated under all forms of capitalism, whether they feel it or not. Alienation is a structural condition, not a personal one. Workers are of course more likely to revolt if they experience alienation, but that is a different issue.

Alienation always exists when someone other than the worker owns the means of production and the product itself. Having said that, we can note that Marx actually talked about four different kinds of alienation. **Alienation** in its most basic sense is separation from one's own awareness of being human. We know we are human; Marx doesn't mean to imply that we don't have the idea of being human. What he means is that our idea is wrong or inaccurate. Recall Marx's argument of species-being: that which makes us distinctly human is creative production, and we become aware of our humanity as our nature is clearly reflected back to us by the mirror of the produced world. At best, the image is distorted if we look to see our nature in the things we own. It's the right place to look, but according to Marx, products should be the natural expression of species-being and production should result in true consciousness. The commodity itself does not reflect humanity, and thus, everything about our commodities is wrong. At worst, the image is simply false. Marx argues that if we look elsewhere for our definition and knowledge of human nature (such as using language; having emotions; possessing a soul, religion, rationality, free-will, and so on), it is not founded on the essential human characteristic—free and creative production. Thus, the human we see is rooted in false consciousness.

> We arrive at the result that man (the worker) feels himself to be freely active only in his animal functions—eating, drinking, and procreating, or at most also in his dwelling and in personal adornment—while in his human functions he is reduced to an animal. The animal becomes human and the human becomes animal. (Marx, 1932/1995, p. 99)

From this basic alienation, three other forms are born: alienation from the work process, alienation from the product itself, and alienation from other people. In species-being, there is not only the idea of the relationship between consciousness and creative production, there is also the notion that human beings are related to one another directly and intimately through species-being. We don't create a human world individually; it is created collectively. Therefore, under conditions of species-being, humans are intimately and immediately connected to one another. The reflected world around them is the social, human world that they created. Imagine, if you can, a world of products that are all directly connected to human beings (not market forces, advertising, the drive for profit, and so forth). Either you made everything in that world or you know who did. When you see a product, you see yourself or you see your neighbor or your neighbor's friend. Thus, when we are alienated from our own species-being, because someone else controls the means and ends of production, we are estranged from other humans as well.

Finally, of course, we are alienated from the process of work and from the product itself. There is something essential in the work process, according to Marx. The kind of work we perform and how we perform that work determines the kind of person we are. Obviously, doctors are different from garbage collectors, but that isn't what Marx has in mind. An individual's humanity is rooted in the work process. When humans are cut off from controlling the means of production (the way in which work is performed and for what reason), then the labor process itself becomes alienated. There are three reasons for this alienated labor, which end in the alienation of the product. First, when someone else owns the means of production, the work is external to the worker; that is, it is not a direct expression of his or her nature. So, rather than being an extension of the person's inner being, work becomes something external and foreign. Second, work is forced. People don't work because they want to; they work because they have to—under capitalism, if you don't work you die. Concerning labor under capitalism Marx (1932/1995) says, "Its alien character is clearly shown by the fact that as soon as there is no physical or other compulsion it is avoided like the plague" (pp. 98–99). And third, when we do perform the work, the thing that we produce is not our own; it belongs to another person.

Further, alienation, according to Marx, is the origin of **private property**—it exists solely because we are cut off from our species-being; someone else owns the means and ends of production. Underlying markets and commodification—and capitalism itself—is the institution of private property. Marx felt that the political economists of his day assumed the fact of private property without offering any explanation for it. These economists believed that private property was simply a natural part of the economic process. But for Marx, the source of private property was the crux of the problem. Based on species-being, Marx argues that private property emerged out of the alienation of labor.

Marx also argues that there is a reciprocal influence of private property on the experience of alienation. As we have seen, Marx claims that private property is the result of alienated labor. Once private property exists, it can then exert its own influence on the worker and it becomes "the realization of this alienation" (Marx, 1932/1995, p. 106). Workers then become controlled by private property. This notion is most clearly seen in what Marx describes as *commodity fetish:* workers become infatuated with their own product as if it were an alien thing. It confronts them not as the work of their hands, but as a commodity, something alien to them that they must buy and appropriate.

Commodity fetish is a difficult notion and it is hard to come up with an illustration. Workers create and produce the product, yet we don't recognize our work or ourselves in the product. So we see it outside of us and we fall in love with it. We have to possess it, not realizing that it is ours already by its very nature. We think *it,* the object, will satisfy our needs, when what we need is to find ourselves in creative production and a socially connected world. We go from sterile object to sterile object, seeking satisfaction, because they all leave us empty. To use a science fiction example, it's like a male scientist who creates a female robot, but then forgets he created it, falls in love with it, and tries to buy its affection through money. As I said, it is a difficult concept to illustrate because in this late stage of capitalism, *our entire world and way of living are the examples.* In commodity fetish, our perception of self-worth is linked with money and objects in a vicious cycle.

In addition, in commodity fetish we fail to recognize—or in Marxian terms, we **misrecognize**—that there are sets of oppressive social relations in back of both the perceived need for and the simple exchange of money for a commodity. We think that the value of the commodity is simply its intrinsic worth—that's just how much a Calvin Klein jacket is worth—when, in fact, hidden labor relations and exploitation produce its value. We come to need these products in an alienated way: we think that owning the product will fulfill our needs. The need is produced through the commodification process, and in back of the exchange itself are relations of oppression. In this sense, the commodity becomes reified: it takes on a sense of reality that is not materially real at all.

Contemporary Marxists argue that the process of commodification affects every sphere of human existence and is the "central, structural problem of capitalist society in all its aspects" (Lukács, 1922/1971, p. 83). Commodification translates *all human activity and relations* into objects that can be bought or sold. In this process, value is determined not by any intrinsic feature of the activity or the relations, but by the impersonal forces of markets, over which individuals have no control. In this expanded view of commodification, the objects and relations that will truly gratify human needs are hidden, and the commodified object is internalized and accepted as reality. So, for example, a young college woman may internalize the commodified image of thinness and create an eating disorder such as anorexia nervosa that rules her life and becomes unquestionably real. Commodification, then, results in a consciousness based on reified, false objects. It is difficult to think outside this commodified box. There is, in fact, a tendency to justify and rationalize our commodified selves and behaviors.

False consciousness and religion: Alienation, false consciousness, and ideology go hand in hand. In place of a true awareness of species-being comes **false consciousness,** consciousness built on any foundation other than free and creative production. Humans in false consciousness thus come to think of themselves as defined through the ability to have ideas, concepts, and abstract thought, rather than production. When these ideas are brought together in some kind of system, Marx considers them to be ideology. Ideas function as **ideology** when they are perceived as independent entities that transcend historical, economic relations: ideologies contain beliefs that we hold to be true and right, regardless of the time or place (like the value of hard work and just reward). In the main, ideologies serve to either justify current power arrangements (like patriarchy) or to legitimate social movements (like feminism) that seek to change the structure.

Though Marx sometimes appears to use the terms interchangeably, in some ways I think it is important to keep the distinction between false consciousness and ideology clear. Ideologies can change and vary. For example, the ideology of consumerism is quite different than the previous ideologies of the work ethic and frugality, yet they are all capitalist ideologies. The ideologies behind feminism are different than the beliefs behind the racial equality movement, yet from Marx's position, both are ideologies that blind us to the true structure of inequality: class. Yet false consciousness doesn't vary. It is a state of being, somewhat like alienation in this aspect. We are by definition in a state of false consciousness because we are living outside of species-being. The very way through which we are aware of ourselves and the world around us is false or dysfunctional. The very *method of our consciousness* is fictitious.

> [I]t is clear that the more the worker spends himself, the more powerful the alien objective world becomes which he creates over-against himself, the poorer he himself—his inner world—becomes, the less belongs to him as his own. The more man puts into God, the less he retains in himself. (Marx, 1932/1978b, p. 72)

Generally speaking, false consciousness and ideology are structurally connected to two social factors: religion and the division of labor. For Marx, *religion* is the archetypal form of ideology. Religion is based on an abstract idea, like God, and religion takes this abstract idea to be the way through which humans can come to know their true nature. So, for example, in the evangelical Christian faith, believers are exhorted to repent from not only their sinful ways but also their sinful nature and to be born again with a new nature—the true nature of humankind. Christians are thus encouraged to become Christ-like because they have been created in the image of God. Religion, then, reifies thought, according to Marx. It takes an abstract (God), treating it as if it is materially real, and it then replaces species-being with non-materially based ideas (becoming Christ-like). It is, for Marx, a never-ending reflexive loop of abstraction, with no basis in material reality whatsoever. Religion, like the commodity fetish, erroneously attributes reality and causation. We pour ourselves out, this time into a religious idea, and we misrecognize our own nature as that of god or devil. Religion is ideological because it is based in and reifies ideas (**reification**).

There is also a second sense in which religion functions as ideology. Marx uses the term ideology as *apologia*, or a defense of one's own ideas, opinions, or action. In this kind of ideology, the orientation and beliefs of a single class, the elite, become generalized and seem to be applicable to all classes. Here the issue is not so much reification as class consciousness. The problem in reification is that we accept something as real that isn't. With ideology, the problem is that we are blinded to the oppression of the class system. This is in part what Marx means when he claims that religion is "the opium of the people." As we have seen, Marx argues that because most people are cut off from the material means of production, they misrecognize their true class position and the actual class-based relationships, as well as the effects of class position. In the place of class consciousness, people accept an ideology. For Marx, religion is the handmaiden of the elite; it is a principal vehicle for transmitting and reproducing the capitalist ideology.

Thus, in the United States we tend to find a stronger belief in American ideological concepts (such as meritocracy, equal opportunity, work ethic, poverty as the result of laziness, free enterprise, and so on) among the religious (particularly among the traditional American denominations). We would also expect to see religious people being less concerned with the social foundations of inequality and more concerned with patience in this life and rewards in the next. These kinds of beliefs, according to Marx, dull the workers' ability to institute social change and bring about real equality.

It is important to note that this is a function of religion in general, not just American religion. The Hindu caste system in India is another good example. There are five different castes in the system: Brahmin (priests and teachers), Kshatriya (rulers), Vaishya (merchants and farmers), Shudra (laborers and servants), and Harijans (polluted laborers, the outcastes). Position in these different castes is a result of birth; birth position is based on karma (action); and karma is based on dharma (duty). There is virtually no social mobility among the castes. People are taught to accept their position in life and perform the duty (dharma) that their caste dictates so that their actions (karma) will be morally good. This ideological structure generally prevents social change, as does the Christian ideology of seeking rewards in heavenly places.

For Marx, then, religion simultaneously represents the furthest reach of humanity's misguided reification and functions to blind people to the underlying class conditions that produce their suffering. Yet Marx (1844/1979) also recognizes that those sufferings are articulated in religion as the sigh of the oppressed: "*Religious* suffering is at the same time an *expression* of real suffering and a *protest* against real suffering. Religion is the sigh of the oppressed creature, the sentiment of a heartless world, and the soul of soulless conditions. It is the *opium* of the people" (p. 54). Marx thus recognizes that religion also gives an outlet to suffering. He feels that religion places a "halo" around the "veil of tears" that is present in the human world. The tears are there because of the suffering that humans experience when they don't live communally and cooperatively. Marx's antagonism toward religion, then, is not directed at religion and God *per se,* but at "the *illusory* happiness of men" that religion promises.

Most if not all of Marx's writings on religion were in response to already existing critiques (mostly from Georg Hegel and Ludwig Feuerbach). Marx, then, takes the point of view that "the criticism of religion has largely been completed." What Marx is doing is critiquing the criticism. He is responding to already established ideas about religion, not religion itself necessarily. What Marx wants to do is to push us to see that the real issue isn't religion; it is the material, class-based life of human beings that leads to actual human suffering.

In keeping with that notion, a school of contemporary Marxism, Humanistic Marxism, most notably the area of Liberation Theology, sees religion as an agent of social change. Liberation theologians argue that there are two poles of Christian expression. The one pole is the classic ideological version, where religion serves to maintain the establishment. The other pole emphasizes compassion for the human condition and leadership in social change. The first gives importance to the "meek and mild Jesus" who taught that the proper response to oppression was to turn the other cheek. The second stresses the Christ in the temple who in rage overturned the tables of the moneychangers and drove them out with a whip. So, there are some Marxists that argue that religion can function as an agent of social change, but that kind of religion is very specific and is not encountered very often.

Marx also sees ideology and alienation as structurally facilitated by the *division of labor* (how the duties are assigned in any society). Marx talks about several different kinds of divisions of labor. The most primitive form of separation of work is the "natural division of labor." The natural division was based upon the individual's natural abilities and desires. People did not work at something for which they were ill suited, nor did they have to work as individuals in order to survive. Within the natural division of labor, survival is a group matter, not an individual concern. Marx claims that the only time this ever existed was in preclass societies. When individuals within a society began to accumulate goods and exercise power, the "forced division of labor" replaced the natural division. With the forced division individual people must work in order to survive (sell their labor) and they are forced to work at jobs they neither enjoy nor have the natural gifts to perform. The forced division of labor and the **commodification of labor** characterize capitalism.

This primary division of labor historically becomes extended when mental labor (such as that performed by professors, priests, philosophers) is divided from material labor (workers). When this happens, reification, ideology, and alienation reach new heights. As we've seen, Marx argues that people have true consciousness only under conditions of species-being. Anytime people are removed from controlling the product or the production process, there will be some level of false consciousness and ideology. Even so, workers who actually produce a material good are in some way connected to the production process. However, with the separation of mental from material labor, even this tenuous relationship to species-being is cut off. Thus, the thought of those involved with mental labor is radically cut off from what makes us human (species-being). As a result, everything produced by professors, priests, philosophers, and so on has some reified ideological component and is generally controlled by the elite.

Class consciousness: So far we have seen that capitalism increases the levels of industrialization, exploitation, market-driven forces like commodification, false consciousness, ideology, and reification, and it tends to bifurcate the class structure. On the other hand, Marx also argues that these factors have dialectical effects and will thus push capitalism inexorably toward social change. Conflict and social change begin with a change in the way we are aware of our world. It begins with class consciousness.

Marx notes that classes exist objectively, whether we are aware of them or not. He refers to this as a "class in itself," that is, an aggregate of people who have a common relationship to the means of production. But classes can also exist subjectively as a "class *for* itself." It is the latter that is produced through class consciousness. **Class consciousness** has two parts: the subjective awareness that experiences of deprivation are determined by structured class relations and not individual talent and effort; and the group identity that comes from such awareness.

Industrialization has two main lines of effects when it comes to class consciousness. First, it tends to increase exploitation and alienation. We've talked about both of these already, but remember that these are primarily objective states for Marx. In other words, these aren't necessarily subjectively felt—alienation isn't chiefly a

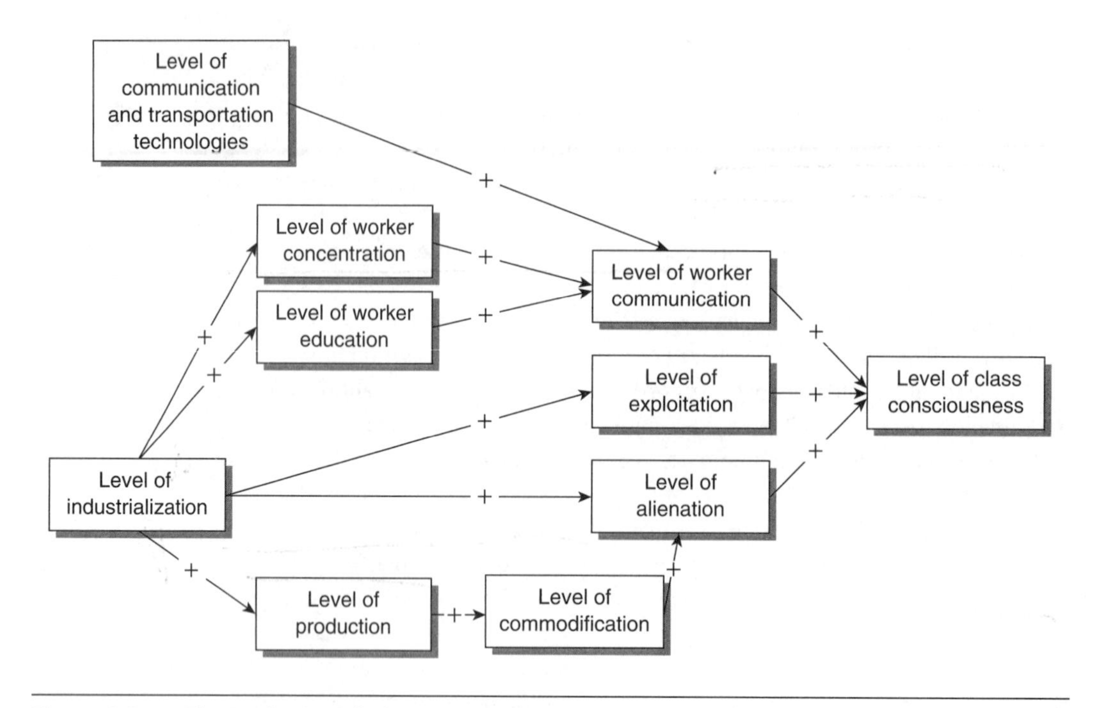

Figure 3.6 The Production of Class Consciousness

feeling of being psychologically disenfranchised; it is the state of being cut off from species-being. Humans can be further alienated and exploited, and machines do a good job of that. What happens at this point in Marx's scheme is that these objective states can produce a sense (or feeling) of belonging to a group that is disenfranchised, that is, class consciousness.

As you can see from Figure 3.6, industrialization has positive relationships with both exploitation and alienation. As capitalists employ machinery to aid in labor, the objective levels of alienation and exploitation increase. As the objective levels increase, so does the probability that workers will subjectively experience them, thus aiding in the production of class consciousness. Keep in mind that industrialization is a variable, which means that it can increase (as in when robots do the work that humans once did on the assembly line—there is a human controlling that robot, but far, far removed from the labor of production) or decrease (as when "cottage industries" spring up in an economy).

The second area of effect is an increase in the level of worker communication. Worker communication is a positive function of education and ecological concentration. Using machines—and then more *complex* machines—requires increasing levels of technical knowledge. A crude but clear example is the different kinds of knowledge needed to use a horse and plow compared to a modern tractor. Increasing the use of technology in general requires an increase in the education level of the worker (this relationship is clearly seen in today's computer-driven U.S. labor market).

In addition, higher levels of industrialization generally increase the level of worker concentration. Moving workers from small guild shops to large-scale machine shops or assembly lines made interaction between these workers possible in a way never before achievable, particularly during break and lunch periods when hundreds of workers can gather in a single room. Economies of scale tend to increase this concentration of the workforce as well. So for a long time in the United States, we saw ever-bigger factories being built and larger and larger office buildings (like the Sears Tower and World Trade Center—and it is significant that terrorists saw the World Trade Center as representative of American society). These two processes, education and ecological concentration, work together to increase the level of communication among workers. These processes are supplemented through greater levels of communication and transportation technologies. Marx argues that communication and transportation would help the worker movement spread from city to city.

So, in general, class consciousness comes about as workers communicate with each other about the problems associated with being a member of the working class (like not being able to afford medication). The key to Marx's thinking here is to keep in mind that these things come about due to structural changes brought about simply because of the way capitalism works. Capitalists are driven to increase profits. As a result, they use industrialization, which sets in motion a whole series of processes that tend to increase the class consciousness of the workers. Class consciousness increases the probability of social change. As workers share their grievances with one another, they begin to doubt the legitimacy of the distribution of scarce resources, which in turn increases the level of overt conflict. As class inequality and the level of bi-polarization increase, the violence of the conflict will tend to increase, which in turn brings about deeper levels of social change.

However, class consciousness has been difficult to achieve. There are a number of reasons given as to why this is true, but many Marxist approaches focus on the relationships among a triad of actors: the state, the elite, and workers. Marx felt that as a result of the rise of class consciousness, and other factors such as the business cycle, workers would unite and act through labor unions to bring about change. Some of the work of the unions would be violent, but it would eventually lead to a successful social movement. In the end, the labor movement would bring about socialism.

Marx sees two of the actors in the triad working in collusion. He argues that under capitalism, the state is basically an arm of the elite. It is controlled by capitalists and functions with capitalist interests in mind. Many of the top governing officials come from the same social background as the bourgeoisie. A good example of this in the United States is the Bush family. Of course, electing a member of the capitalist elite to high public office not only prejudices the state toward capitalist interests, it is also an indicator of how ideologically bound a populace is.

In addition, as C. W. Mills argues, the elite tend to cross over, with military men serving on corporate boards and as high-placed political appointees, and CEOs functioning as political advisees and cabinet members, and so on. A notable example of these kinds of interconnections is Charles Erwin Wilson, president of General Motors from 1941 to 1953. He was appointed Secretary of Defense under President Dwight D. Eisenhower. At his senate confirmation hearing, Wilson spoke the words that epitomize the power elite: "For years, I thought what was good for our country was good for General Motors—and vice versa."

In response to the demands of labor unions, certain concessions were ultimately granted to the working and middle classes. Work hours were reduced and healthcare provided, and so forth. Capitalists working together with a capitalist-privileging state granted those concessions. Thus, even when allowances are granted, they may function in the long run to keep the system intact by maintaining the capitalists' position, silencing the workers, and preventing class consciousness from adequately forming.

The state is also active in the production of ideology. Marx sees the state as somewhat ill-defined. In other words, where the state begins and ends under capitalist democracy is hard to say. The state functions through many other institutions, such as public schools, and its ideology is propagated through such institutions, not only through such direct means as the forced pledge of allegiance in the United States but also through indirect control measures like specific funding initiatives. A good deal of the state's ideology is, of course, capitalist ideology. As a result, the worker is faced with a fairly cohesive ideology coming from various sources. This dominant ideology creates a backdrop of taken-for-grantedness about the way the world works, against which it is difficult to create class consciousness.

The movement of capitalist exploitation across national boundaries, which we mentioned before, accentuates the "trickle-down" effect of capitalism in such countries as the United States. Because workers in other countries are being exploited, the workers in the United States can be paid an inflated wage (inflated from the capitalists' point of view). This is functional for capitalism in that it provides a collection of buyers for the world's goods and services. Moving work out from the United States and making it the world's marketplace also changes the kind of ideology or culture that is needed. There is a movement from worker identities to consumer identities in such economies.

In addition, capitalists use this world labor market to pit workers against one another. While wages are certainly higher due to exported exploitation, workers are also aware that their jobs are in jeopardy as work is moved out of the country. Capitalism always requires a certain level of unemployment. The business cycle teaches us this—that zero unemployment means higher wages and lower profits. The world labor market makes available an extremely large pool of unemployed workers. So, workers in an advanced industrialized economy see themselves in competition with much cheaper labor. This global competition also hinders class consciousness by shifting the worker's focus of attention away from the owners and onto the foreign labor market. In other words, a globalized division of labor pits worker against worker in competition for scarce jobs. This competition is particularly threatening for workers in advanced capitalist countries, like the United States, because the foreign workers' wage is so much lower. These threatened workers, then, will be inclined to see their economic problems in terms of global, political issues rather than class issues.

Further, the workers divide themselves over issues other than class. We tend to see ourselves not through class-based identities, which Marx would argue is the identity that determines our life chances; instead, we see ourselves through racial, ethnic, gender, and sexual preference identities. Marxists would argue that the culture of diversity and victimhood is part of the ideology that blinds our eyes to true social inequality, thus preventing class consciousness.

Thinking About Modernity and Postmodernity

Machines of production and consciousness: Marx has a very clear notion of modernity, though he doesn't use the concept itself. As we've seen, Marx argues that society in general evolves and changes due to variations in the economic system. The means of production determines the relations of production (or the social relations). Marx named five different economic societies: preclass societies, Asiatic societies, ancient societies, feudal societies, and capitalist societies. Societies that contain the capitalist dynamics that we have been talking about are, of course, modern. We would expect, then, for a Marxian view of postmodernity to argue that something has changed the basic configuration of capitalism. Interestingly, we will find as we move through each of our theorists that many of the postmodern thinkers we will consider have some relation to Marxian theory. For now, though, I want us to focus on the central issue for Marx: the means of production.

Fredric Jameson (1984) argues that modern capitalism has gone through three distinct phases, each linked to a particular kind of technology. Early-market capitalism was distinguished by steam-driven machinery; mid-monopoly capitalism was characterized by steam and combustion engines; and late-multinational capitalism is associated with nuclear power and electronic machines. The important Marxian issue here is the relationship between the mode of production and consciousness. As we have seen, Marx argues that consciousness is explicitly tied to production. Human production is supposed to act as a mirror that reflects our nature. Thus, the way in which production is performed is critically important.

This idea is not as extreme as it might seem. Think about digging in your garden. If you use a shovel and hoe to dig the earth, and you bend down and plant the seeds by hand, you will find that you'll notice things about the earth, yourself, and your work. You will smell the loam of the earth; you'll feel the consistency of the soil in your hands and on your knees as you kneel; you'll sense the soil, water, and seed as they mix together; and when the plants sprout, you will see your sweat and toil reflected in the ground and new life. But if you use a tractor, it won't be the same; and if you use a computer to guide the tractor as it plows, plants, and harvests, you will be utterly removed from the smells, tastes, and feel of earth, sun, water, and the human body.

Let me give you another illustrative example. The aesthetics of airplane mechanics are generally different than those of symphony conductors. Part of the difference is undoubtedly due to psychological dispositions. But the majority of the differences are due to the kind of work they do. Constantly working on real machines, getting your hands dirty and banged up, and having to exert physical strength and mechanical ingenuity all day, gives a person a particular perspective and cultural disposition. Likewise for a symphony conductor, interpreting a musical score and working to present that interpretation through an orchestra gives the conductor a certain perspective and disposition, as well. We become what we do. Consequently, there are significant social and individual effects when the general means of production change in a society.

Machines of reproduction and schizophrenic culture: Jameson argues that human reality and consciousness was non-problematically represented by the aesthetic of the machine in earlier phases of capitalism. In other words, workers could touch and see the machines that were used to create products. While alienated, they were still connected to the production process and could, most importantly, also experience the alienating process through their senses. But in postindustrial societies, through multinational capitalism, workers are using machines less and less. Even if they are in a manufacturing sector, the machines are more and more controlled through computers. Those people not in manufacturing today, like the rising service class, use machines of reproduction (movie cameras, video, tape recorders, computers, and so forth) rather than production. Because the machines of late capitalism reproduce knowledge rather than produce it, and because reproduction is always focused more on the medium than the message, Jameson argues that the link from production to signification (culture and meaning) has broken down. Jameson characterizes this breakdown as the schizophrenia of culture and argues that our culture is filled with "free-floating signifiers."

Let's back up a minute and try and understand what Jameson is saying. What he and Marx are telling us is that there is, or was, a chain of real relationships from the material world and the human world of ideas. At one time, our ideas and the language we used were embedded in the physical world. Let's use our garden example again, but place it in a horticultural society (one that lives by planting and growing food). If you are working in that group, your knowledge of farming is firsthand and real. You know how to grow food and the food is there for you when it is done growing. When you talk about growing, the language you use is firsthand, objective, and real. But if we put an owner between the product and the person, there is a break in that continuity. If we insert a tractor and even more equipment in the chain, there are still more breaks in the signification chain. Further, if instead of *production* equipment, we use machines of *reproduction,* then there isn't any objective reference for the signification chain at all. Machines of reproduction don't produce anything; they only reproduce images or text about something else.

Jameson argues that the schizophrenic nature of culture and the immense size and complexity of the networks of power and control in multinational capitalism produce three distinct features of the postmodern era. One, the cultural system is now dominated by image rather than actual signed reality. These media images, or simulacrum (identical copies of something that never existed), have no depth of meaning; meaning is fleeting and fragmented. Two, there is a consequent weakening of any sense of either social or personal biographic history. In other words, it is becoming increasingly difficult for us to create and maintain consistent narratives about our national or personal identities. And three, postmodern culture is emotionally flat. The further culture is removed from actual social and physical reality, the harder it is to invest with strong emotion.

Besides offering us the provocative idea of postmodernity, these ventures into postmodern theory also give us an opportunity to see how a theorist's perspective can be used. In other words, it is very possible that a theory can be Marxian but not exactly what Marx himself might say. The theorist's perspective can be used apart from his or her explicit theory. That's what Jameson does with Marx. Jameson ends up saying that culture is set loose from its economic base and can have independent effects. This isn't something that Marx himself said. Nevertheless, Jameson's theory is Marxian because it originates in a Marxian perspective.

Summary

- Marx's perspective is created through two central ideas: species-being and the material dialectic. Species-being refers to the unique way in which humans survive as a species—we creatively produce all that we need. The material dialectic is the primary mechanism through which history progresses. There are internal contradictions within every economic system that push society to form new economic systems. The dialectic continues until communism is reached, a system that is in harmony with species-being.

- Every economic system is characterized by the means and relations of production. The means of production in capitalism is owned by the bourgeoisie and generally consists of commodification, industrial production, private property, markets, and money. One of the unique features of capitalism is that it will swallow up all other classes save two: the bourgeoisie and proletariat. This bifurcation of class structure will, in turn, set the stage for class consciousness and economic revolution.

- Capitalism affects every area of human existence. Through it, individuals are alienated from each aspect of species-being and creative production. The work process, the product, other people, and even their own inner being confront the worker as alien objects. As a result, humankind misrecognizes the truth and falls victim to commodity fetish, ideology, and false consciousness. However, because capitalism contains dialectical elements, it will also produce the necessary ingredient for economic revolution: class consciousness. Class consciousness is the result of workers becoming aware that their fate in life is determined primarily by class position. This awareness comes as alienation and exploitation reach high levels and as workers communicate with one another through increasing levels of education, worker concentration in the factory and city, and communication and transportation technologies.

- Jameson's argument concerning postmodernity is based on Marx's theory of consciousness: human consciousness is materially based; that is, because of species-being, there is a direct relationship between the method of production and the ideas we have about ourselves and the way we perceive the world. In primitive communism, we had direct and objective knowledge of the world and our selves. Under early capitalism, we suffered from false consciousness and ideology, but this knowledge was still materially grounded in that we were connected to machines of production. In that state, we could have come to a real sense of alienation and class consciousness. However, in postmodernity, the economy is shifting from machines of production to machines of *re*production. With machines of production, there was a connection between culture and the material world. With machines of reproduction, that connection is broken. As a result, cultural signs and symbols are cut loose from their mooring and lose any grounded sense of meaning or reality. From a Marxian point of view, this shift from machines of production to reproduction, from materially grounded ideas to free-floating signifiers, contributes to the prevention of class consciousness.

Building Your Theory Toolbox

Conversations With Karl Marx—Web Research

With Marx, we are lucky in that we have an account of someone who did have a conversation with him. He was interviewed for the *Chicago Tribune* and the article appeared on January 5, 1879. The article contains some personal observations about how Marx looked and acted, and what his home was like. Using your search engine, find the *Chicago Tribune*'s article on Marx and answer the following questions. As you are reading, try to imagine what it would be like to talk with the man who probably had the single greatest impact on world history in the 20th century. What would you ask him if you were there?

- How was Marx treated in Germany and France?
- What personal characteristics of Marx stand out to you?
- What kinds of books did Marx read? What was his breadth of knowledge?
- What was the International Society? What were its beliefs and purposes? Which of the elements in the platform of the International Society have been fulfilled? Are there any that you think ought to be implemented but haven't yet been? Which do you think are bad ideas? Why?
- What does Marx say about religion?
- What does Marx say about violent revolution?

Passionate Curiosity

Seeing the World (using the perspective)

Each of our theorists has a unique perspective. If we can understand the perspective, then almost everything else in the theory falls into place. Marx's perspective is one of the most profound, well-thought-out viewpoints we will come across.

- Marx founds his theory on an assumption about human nature. Why is everything we think about human nature more of an assumption than fact? (Notice that Marx builds an argument to convince us of this assumption, but it still isn't a fact.)
- Evaluate the idea of species-being. Do you agree with Marx's argument concerning human nature? Are we by nature social and altruistic? If you disagree with Marx, what do you assume about human nature? Also—and this is extremely important—upon what do you base your assumption? Recall that Marx based his assumption on the way humans uniquely exist (creative production). So, if you disagree, what is the basis of your assumptions concerning human nature?

Engaging the World (using the theory)

- Using Google or your favorite search engine, type in "job loss." Look for sites that give statistics for the number of jobs lost in the past five years or so (this can be a national or regional number). How would Marx's theory explain this?
- Using Google or your favorite search engine, type in "class structure in U.S." Read through a few of the sites. Based on your research, what is happening to the class structure in the United States? How would Marx explain this?
- Get a sense about the frequency and uses of plastic surgery in this country. You can do this by watching "makeover" programs on TV, or by doing Web searches, or reading through popular magazines, or any number of ways. How can we understand the popularity and uses of plastic surgery using Marx's theory? (Hint: think of commodification and commodity fetish.) Can you think of other areas of our life for which we can use the same analysis?

Weaving the Threads (synthesizing theory)

There are central themes in classical theory. As we proceed through our discussions of the different theorists, it is a good idea to pay attention to these themes and see how the theorists add to or modify them. In the end, we should have a clear and well-informed idea of the central dynamics in sociological theory. Thinking about Marx's theory, answer the following questions:

- What is the fundamental structure of inequality in society? How is inequality perpetuated (in other words, how is it structured)?
- How does social change occur? Under what conditions is conflict likely to take place?
- What is religion? What function does religion have in society?
- How did gender inequality begin?
- What is the purpose of the state? How is it related to other social structures? Where does power reside?

Further Explorations—Readings

Appelbaum, R. P. (1988). *Karl Marx.* Newbury Park, CA: Sage. (Part of the *Masters of Social Theory* series; short, book-length introduction to Marx's life and work)

Bottomore, T., Harris, L., & Miliband, R. (Eds.). (1992). *The dictionary of Marxist thought* (2nd ed.). Cambridge, MA: Blackwell. (Excellent dictionary reference to Marxist thought)

Fromm, E. (1961). *Marx's concept of man.* New York: Continuum. (Insightful explanation of Marx's idea of human nature)

McClellan, D. (1973). *Karl Marx: His life and thought.* New York: Harper & Row. (Good intellectual biography)

Further Explorations—Web Links: Marx and Marxists

http://www.marxists.org/archive/marx/works/index.htm (Site maintained by Marxists Internet Archive; good online source for Marx and Engels' writings)

http://www.uta.edu/huma/illuminations/ (Site maintained by Douglas Brown and Douglas Kellner; critical theory project)

http://en.wikipedia.org/wiki/Karl_Marx (Site maintained by Wikipedia Internet Encyclopedia; Marx's life, theory, and influence; additional links also provided)

http://cepa.newschool.edu/het/profiles/marx.htm (Site maintained by Department of Economics, New School; Web sources for most of Marx's major writings and other Web resources)

http://www.marxists.org/archive/index.htm (Site maintained by Marxists Internet Archive; searchable library of Marx and Marxist writers)

Further Explorations—Web Links: Intellectual Influences on Marx

Georg Wilhelm Friedrich Hegel: http://plato.stanford.edu/entries/hegel/ (Site maintained by Stanford University; excellent source for Hegel's life, philosophy, influences, and published works)

Ludwig Feuerbach: http://plato.stanford.edu/entries/ludwig-feuerbach/ (Site maintained by Stanford University; excellent source for Feuerbach's life, philosophy, influences, and published works)

Adam Smith: http://econ161.berkeley.edu/Economists/smith.html (Site maintained by Brad DeLong, Professor of Economics at the University of California, Berkeley; brief biography and extensive quotes)

CHAPTER 4

Cultural Consensus— Émile Durkheim (French, 1885–1917)

[A] compelling case can be made that, more than any other classical figure, it is to Durkheim that the contemporary cultural revival . . . is most deeply in debt. (Alexander, 1988b, p. 4)

The work of Émile Durkheim lies at the heart of sociology. Spencer was the first to make us aware of the questions surrounding integration in the face of structural diversity, but it is Durkheim who understood the moral basis of integration. He helped us see that structural differentiation and social diversity go hand in hand. Structures may indeed be interdependent, but structures don't act in and of themselves. For social integration to work, then, we also need social solidarity—something that brings people together and gives them a sense of belonging and a willingness to sacrifice. Thus, Durkheim set the stage for the modern sociological analysis of social and cultural integration, one of the defining questions for sociology during most of the 20th century.

One of the reasons that Durkheim is concerned with moral integration is due to his assumptions about human nature. Where Marx assumes that humans are social and naturally altruistic, Durkheim assumes that people apart from society are self-centered and driven by insatiable desires. So, Durkheim's issue with social integration is actually more basic than Spencer's. Certainly, Durkheim gives us an answer to the question of integration in the face of structural differentiation, but he also gives us an answer to the deeper problem of human egoism. If we assume, as Durkheim does, that individuals tend to go off each in his or her own direction, then how can this thing called society work? Durkheim came up with an ingenious answer: the collective consciousness. Today, sociologists usually talk about norms, values, and beliefs, but in back of those terms lies what Durkheim meant by the collective consciousness.

Not only did Durkheim set the agenda for much of sociology, he also asked the primary questions about the nature of society: What is society and how does it exist? People like Montesquieu and Rousseau had thought about society previous to Durkheim, but he wasn't satisfied with the answers they provided. In response, Durkheim wrote a book called *The Rules of Sociological Method* in 1895. The book not only set out a sociological methodology that is still used today, it also presented a full argument about how society exists as sets of social facts.

Recall from our discussion in Chapter 1 about interpretive theory that the issue concerning the facticity (the quality of existing as a fact) of social institutions is tricky. Science assumes that the universe is empirical; that is, the universe is only made up of things that we can touch, see, smell, hear, or taste—things that are objectively real and exist in time and space. I'd like to pose a problem for you: Show me a social structure. Any structure will do, but let's take education for example. Where is the structure of education? Is it in the books, or in the buildings, or in the rules and regulations? Can you touch, taste, see, or in any way sense the actual structure? The problem is that social structures are not really *there* in the same sense that an ocean is there. How do you make a science out of something that isn't quite empirical? This is the fundamental question of scientific sociology, and Durkheim provides a wonderful answer in his idea of social facts, an answer that continues to lie at the very foundation of our concept of institutions.

So far we have considered two of Durkheim's primary issues (social integration and social facts) and have seen the general way in which both the problems and answers have formed the basis of sociological thought. There is one other Durkheimian issue that I'd like for us to contemplate. Generally, the question is about culture, but the specific issue is about signs, symbols, and categories. One of the concepts that the idea of social science is based upon is the notion of *independent effects.* In other words, if society can be studied scientifically, then it must contain some form of its own laws of action apart from the people who make it up and it must be able to independently act upon people. Durkheim poses this kind of question about culture and argues that culture exists independently and operates autonomously. Another way to put this is to say that culture is structured—signs, symbols, and categories are related to one another in a way that influences how people think, feel, and act. Durkheim was really one of the first to consider such a thing, but this idea came to form an entire school of research called linguistic structuralism, which later influenced semiotics and post-structuralism (two influential contemporary schools of cultural analysis).

Durkheim's thinking thus forms the framework for many, many areas of social and sociological inquiry. If you are going to do work in the sociology of religion, the sociology of knowledge, the sociology of emotions, or the sociology of education, or if you are going to think about the place that rituals have in our lives (like Erving Goffman or Randall Collins), or if you want to understand culture, then you are going to have to base at least part of your work on Durkheim. Most importantly, Durkheim is a key founding thinker in structural-functionalism. He argues that society emerged in an evolutionary manner from more simple forms. The pinnacle of evolution is not the individual human; it is, rather, society. Further, society is like any other entity: after a more complex form comes into existence, it takes on a life of its own and can only be understood in its own terms. Just as hydrogen and oxygen when brought together form a whole (water) that exists in a way that is greater than its individual parts, so society is a whole that is greater than the individual people that form it. Durkheim also argues that organic and organic-like systems, such as the human body or society, create structures that operate for the benefit of the whole. Just as you cannot understand what a human lung does apart from the body as a whole, you cannot understand what the structure of religion is all about without seeing it as part of the whole of society.

Finally, Durkheim was one of the most politically active and influential sociologists that has ever lived, despite the fact that Durkheim's sociology does not appear political in the least. Durkheim was an active sociologist and a child of the Enlightenment. He believed that sociology should be used to improve the human condition, that human beings could and should think logically about their condition and apply the laws of society that science can discover. Durkheim was the first actual professor of sociology in France, and he became that country's Minister of Education. As Minister, he led a reformation of the French education system: he created a new kind of curriculum for public schools and he started a revolutionary teacher-training program. Durkheim's desire was to create a civil morality that would lead students to become committed to the institutions of society. Durkheim provides us with an excellent model of how to use sociology in bringing about social change from within the system.

Durkheim in Review

- Born in Epinal, France, on April 15, 1858. His mother, Melanie, was a merchant's daughter, and his father, Moïse, was a rabbi, descended from generations of rabbis.
- Durkheim did well in high school and attended the prestigious *École Normale Supérieure* in Paris, the training ground for the new French intellectual elite.
- The first years (1882–1887) after finishing school, Durkheim taught philosophy in Paris, but felt philosophy was a poor approach to solving the social ills he was surrounded by. During this time, he also traveled to Germany and observed the work of Wilhelm Wundt. He published several articles on the relationship between philosophy, social science, and public education in Germany.
- In 1887, Durkheim was appointed as *Chargé d'un Cours de Science Sociale et de Pédagogie* at the University of Bordeaux. Durkheim thus became the first teacher of sociology in the French system. Durkheim and his desire for a science of morality proved to be a thorn in the side of the predominantly humanist faculty. During this year, Durkheim also married Louise Dreyfus; they later had two children, Marie and André.
- Between 1893 and 1897, Durkheim completed and published his French dissertation (*The Division of Labor*), *The Rules of Sociological Method,* and *Suicide,* in addition to a Latin thesis on Montesquieu. Having thus established the basis for French sociology, in 1898 he founded a sociological journal for his work, *L'Année Sociologique.* The journal flourished for many years and became the leading journal of social thought in France.
- In 1902, Durkheim took a post at the Sorbonne and by 1906 was appointed Professor of the Science of Education, a title later changed to Professor of Science of Education and Sociology. In this position, Durkheim was responsible for training the future teachers of France and served as chief advisor to the Ministry of Education.

- In December 1915, Durkheim received word that his son, André, had been declared missing in action (WWI). André had followed in his father's footsteps to *École Normale* and was seen as an exceptionally promising social linguist. Durkheim had hoped his son would complete the research he had begun in linguistic classifications. The following April, Durkheim received official notification that his son was dead. Durkheim withdrew into a "ferocious silence." After only a few months following his son's death, Durkheim suffered a stroke; he died at the age of 59 on November 15, 1917.

The Perspective: Social Facts and the Law of Culture

> Even if [social expectations] conform to my own sentiments and I feel their reality subjectively, such reality is still objective, for I did not create them; I merely inherited them through my education. (Durkheim, 1895/1938, p. 1)

Social facts: Let's begin with Durkheim's consideration of society. For the longest time, people weren't much concerned with society. It was something that was simply taken for granted—kind of like atoms: they form the basis of most everything we deal with but we didn't know they existed for most of our history. However, beginning in the 1700s, many intellectuals became concerned with defining society and understanding its origins and effects. This new concern arose mostly because of massive changes in society. The sleeping beast could no longer be taken for granted. People had to understand how society was changing and how it might influence human behavior and their quality of life.

Most of the speculations of the time were formed in philosophical terms and began with certain assumptions about the natural state of human beings apart from society. Thomas Hobbes, for example, argued that human nature is basically warlike; on the other hand, John Locke believed that humans are naturally peaceful.

> Here, then, is a category of facts with very distinctive characteristics: it consists of ways of acting, thinking and feeling, external to the individual, and endowed with a power of coercion, by reason of which they control him. (Durkheim, 1895/1938, p. 3)

In contrast to philosophical beginnings, Charles Montesquieu argued that the study of society should begin empirically: with what we see, not with what we think. In some ways, Montesquieu's book *The Spirit of Laws* may be considered the first empirical work in sociology. In it, Montesquieu argues that society must be viewed as an empirical object. As an independent object, its properties and processes could be discovered through observation. Interestingly enough, Montesquieu also argues that the human being is really the product of society, not the other way around. It follows, then, that any attempt to understand society that begins with human nature would have to be false, since human nature is itself a creation of society. From Durkheim's position, one must begin with society to understand human beings.

Durkheim draws from Montesquieu in his thinking about society. He argues that society exists as an empirical object, almost like a physical object in the environment. Durkheim uses the concept "social fact" to argue for the objectivity of society and scientific sociology. According to Durkheim, **social facts** gain their facticity because they are external to and coercive of the individual. We can't smell, see, taste, or touch them, but we can feel their objective influence. *Webster's* defines *empirical* as "based on the results of experiment, observation, or experience, and not from mathematical or scientific reasoning; having reference to actual facts." Things that are empirical, then, aren't simply in your head; you can experience them in some way. The experience of empirical facts seems to imply that things are unavoidable and unchangeable (or at least difficult to avoid or change). Here's an analogy to illustrate. If you and I see a rock in our path, one of us would have to be crazy to say it isn't there. If I step on it, I would feel it. If I throw it at you, you would feel the rock, too, whether you wanted to or not. You wouldn't have to think about it to make it hurt; and it would be difficult to change the effect. The rock is simply there, and its effects do not depend upon our interpreting or reasoning or our individual perspective.

Durkheim argues that society is like the rock in these ways: its effects are unavoidable and difficult to change and are not dependent upon you or me. Society is a fact because it doesn't depend upon us for its effects. In this way, we *feel* society every time we think of doing something outside the norm. The social pressure of moral conformity is simply there, like the rock—external to and coercive of the individual.

Durkheim distinguishes between material and nonmaterial elements in social facts. Material elements are like cultural artifacts: they are what would survive if the present society no longer existed. For example, a wheelbarrow is a material social fact. Nonmaterial elements consist of symbolic meanings and collective sentiments. Often such features are attached to material objects, like the meanings behind statues and flags, but many times our most important meanings and feelings have a more abstract existence, like love and freedom.

In addition, society exists as a social fact because it exists ***sui generis***—a Latin term meaning "of its own kind." Durkheim uses the term to say that society exists in and of itself, not as a "mere epiphenomenon of its morphological base." Society is more to Durkheim than simply the sum of all the individuals within it. Society exists as its own kind of entity, obeying its own rules and creating its own effects: "*The determining cause of a social fact should be sought among the social facts preceding it and not among the states of the individual consciousness*" (Durkheim, 1895/1938, p. 110, emphasis Durkheim's).

Let me give you an example of what Durkheim means when he says that the cause of social facts is other social facts. One of Durkheim's most famous studies is *Suicide.* In it, he studied suicide rates, not individual suicides. The suicide rate ("the proportion between the total number of voluntary deaths and the population of every age and sex") is a social fact. The suicide rate measures a collective's "definite aptitude for suicide" at any given historical moment and "is itself a new fact *sui generis,* with its own unity, individuality and consequently its own nature" (Durkheim, 1897/1951, p. 46). The suicide rate of any given society can be understood through social types—**egoistic, altruistic, anomic,** and **fatalistic suicides**—and each of these types is caused by its relationship to two other social facts: group attachment and behavior regulation. In other words, the suicide rate (a social fact) is caused by other social facts (group attachment and behavior regulation).

Let's push this idea of society *sui generis* a bit further. What are some implications of something existing independently? It may sound simple, but if something exists independently, it can have independent effects. When we are talking about society, this is an extremely important point. It means that society can act on its own. I'm going to give you what I think is an astonishing quote from Durkheim. It's something he said in *The Elementary Forms of the Religious Life.* When I first discovered this statement, I immediately ran to one of my undergraduate professors and showed him. He said, "It's interesting, but Durkheim didn't really mean that." After all these years, I'm still not so sure. But take a read and see what you think. Pay particular attention to the way he says the collective consciousness (society) acts.

> If collective consciousness is to appear, a *sui generis* synthesis of individual consciousnesses must occur. The product of this synthesis is a whole world of feelings, ideas, and images that follow their own laws once they are born. They mutually attract one another, repel one another, fuse together, subdivide, and proliferate; and none of these combinations is directly commanded and necessitated by the state of the underlying reality. Indeed, the life thus unleashed enjoys such great independence that it sometimes plays about in forms that have no aim or utility of any kind, but only for the pleasure of affirming itself. (Durkheim, 1912/1995, p. 426)

Collective consciousness: Durkheim lived in an important age (as do we, I think). Massive social changes that had begun about 100 years previously, began to come to fruition. France, Durkheim's homeland, had gone through a series of upheavals—the Enlightenment, the French Revolution, the Napoleonic Era, the Franco-Prussian War, and so on. The French Revolution in particular had profound consequences for monarchies throughout Europe. New ways of governing were tested, and the common man became increasingly important. The "masses" were born as a political entity. The Industrial Revolution also happened during this time, which brought about large-scale migration of populations from the countryside to the city, helped to radically change the class structure, and removed work from the home, thus creating the dual spheres of gender.

At this time, people were also beginning to come in contact with other peoples and cultures in ways never before experienced. Colonization, mass media such as newspapers, and improvements in transportation (e.g., trains) made the world smaller. People were challenged with new ideas; they found out their beliefs weren't the only ones—new worlds were possible. They also had speedy news of events from around the world. Thus, far-off events that would previously have been unknown began to be seen as having local significance.

The world was in upheaval, especially when compared to the relatively changeless times prior to modernity. These constant changes led many to speculate about humanity and society. Jean Jacques Rousseau, for example, in describing the natural state of man, stated, "In all the universe the only desirable things he knows are food, a female, and rest." Rousseau argued that man fell from this idyllic state through the invention of society. He made the case that the notion of private property came through society and with it came competition, jealousy, and conflict. Society created new passions in humanity and there came to be no limit to our desires, thus self-interest and exploitation now ruled. Rousseau's solution to the negative influences of social institutions lay in the production of a "general will" to which all could commit. Rousseau argued that this general will could be produced through a civil religion, rather than a religion oriented toward the spiritual world, and through a common socialization, rather than through family socialization.

> The totality of beliefs and sentiments common to the average members of a society forms a determinate system with a life of its own. It can be termed the collective or common consciousness. (Durkheim, 1893/1984, pp. 38–39).

Durkheim was caught up with the questions of his age as well. Like Rousseau, he wondered about the stability of society, but, in contrast to Rousseau, he felt that human nature is not idyllic. Durkheim assumes that humans apart from society are filled with "individual appetites" that are "by nature boundless and insatiable." The locus of the problem, then, is different for Durkheim than for Rousseau, but the solution is basically the same. Society must have a "general will" or "collective consciousness" to which people can commit. This collective consciousness is Durkheim's single, overriding concern. He sees it as society's requisite function. The collective consciousness is the one thing that society needs in order to survive. Without it, society will become ill and eventually perish.

For Durkheim, the *collective consciousness* is the totality of ideas, representations, beliefs, and feelings that are common to the average members of society. There does exist, of course, the individual consciousness. However, whatever unique ideas, feelings, beliefs, impressions, and so forth that an individual might have are by definition idiosyncratic. In other words, "individual consciousnesses are actually closed to one another" (Durkheim, 1995/1912, p. 231). The collective consciousness, on the other hand, does allow us a basis for sharing our awareness of the world. Yet the function of this body of culture is not simply to express our inner states to one another; the collective consciousness contributes to the making of our individual subjective states. It is through the collective consciousness that society becomes aware of itself and we become aware of ourselves as social beings.

Durkheim divides the collective consciousness into two basic features: cognitive and emotional. Durkheim argues that the collective consciousness contains primary symbolic categories (time, space, number, cause, substance, and personality). These categories are primary because we can't think without using them. They form our basic cognitions or consciousness of the world around us. These categories are, of course, of social origin for Durkheim, originating with the physical features of society. (The way the population is dispersed in space, for example, influences the way we conceive of space.) Durkheim did some work (*Primitive Classification*) in this area of cognitive categories, especially with his pupil and nephew Marcel Mauss. His work in this area also influenced Ferdinand De Saussure, the founder of French Structuralism (which led to poststructuralism and influenced postmodernism). However, I think for Durkheim the more important aspect of the collective consciousness is emotional. Social emotions or sentiments "dominate us, they possess, so to speak, something superhuman about them. At the same time they bind us to objects that lie outside our existence in time" (Durkheim, 1893/1984, p. 56).

If, as Durkheim supposes, humans are naturally self-serving, then why will self-centered human beings act collectively and selflessly? Durkheim argues that rational exchange principles are not enough. Because our entire being is involved in action, we need to be emotionally bound to our culture. We have to have an emotional sense of something greater than ourselves. This feeling of something greater is what underlies morality. We act socially because it is moral to do so. While we can always give reasons for our actions, many of our actions—especially social actions—generally come about because of *feelings* of responsibility: "Whence, then, the feeling of obligation? It is because in fact we are not purely rational beings; we are also emotional creatures" (Durkheim, 1903/1961, p. 112). So to think like Durkheim is to always be concerned with the emotional foundations of social life.

Social interaction and Durkheim's Law: Durkheim's concern with emotion brings us to the final element in his triad of ideas. I call it "Durkheim's Law." It's a law of interaction: Individuals come together and interact as there is an initial attraction. What the attraction is isn't important to Durkheim. What *is* important is what happens afterward. Attraction leads to interaction, which leads to formation of a "limited group," and the individuals will "be taken up with it and reckon it in their conduct." In other words, every time we interact with other people, we form a temporary sense of a social group. Durkheim (1887/1993) argues that, "Once a group is formed, nothing can hinder an appropriate moral life from evolving" (pp. 23–24).

> It is impossible for men to live together and be in regular contact with one another without their acquiring some feeling for the group which they constitute through having united together, without their becoming attached to it, concerning themselves with its interests and taking it into account in their behaviour. (Durkheim, 1893/1984, p. xliii)

Let me rephrase all this in terms of Durkheim's Law: when people interact, they use and form culture. It is inevitable. We cannot interact with someone without using culture—the culture of speech, the culture of value and morality, and so on—and every time we use culture, we affirm and modify it. Further, the more we interact with someone or with some group, the more we will affirm and form culture. Here's an example: I have a friend, Steve, with whom I have been getting together for over six years. When we first started hanging out, most everything we said had reference to other groups with which we associated. In other words, everything we said referenced other groups and cultures. Now, after six years of interacting, quite a bit of what we talk about and even some of the words we use are particular to us. We have our own code, we have our own morality, and we have our own sense of group boundaries and identity.

I'm sure my example isn't new to you; you've had the same experience yourself many times. Let's think about our common experiences in light of Durkheim's Law. If you interact with one person just one time, how much particular culture do you end up sharing? None. But if you interact with that same person over a long period of time, you will end up sharing quite a bit of particularized culture. So, what would happen if you have two friends that you interact with separately over a six-year period of time? You'll end up with two sets of particularized cultures. Obviously, since all of you are still part of a broader culture, the cultures won't be drastically different, but they will be diverse nonetheless. Now, what would happen if you interacted with those same two people over the same six-year period, but you interacted more often with one than the other? You will have more particularized culture and emotional investment with the one you interacted with more. Now expand our examples to larger groups over longer periods of time, and we can see how the collective consciousness is formed.

The collective consciousness is founded upon Durkheim's Law. When lots of people interact with one another over long periods of time, then a distinct group culture emerges, one that has its own set of norms, values, beliefs, linguistic codes, and so on. More than that, when large groups of people interact, they can create powerful emotions. These emotions get attached to the group culture, and as a result, the group culture can take on a life of its own. It becomes an object, a social fact, something that is external to and coercive of the individual.

To think like Durkheim, then, is to place primary importance on culture and to be concerned with how social solidarity is created, maintained, bridged, and destroyed. To think like Durkheim is to see that people are linked together symbolically and emotionally. Consider the following:

- In our society, we see family as a rather small unit linked together by blood ties. But other societies have seen family as something much vaster. I knew a man who came from a village of 3000 people, each of whom was related through family ties. Obviously, those ties weren't blood ties, at least not in the way we use the term. Yet those people saw themselves as related by family.
- On September 11, 2001, hijacked jetliners hit the World Trade Center in New York and the Pentagon outside Washington, D.C. News headlines around the world proclaimed "America Attacked," and people in San Diego, CA, Detroit, MI, and Cornville, AZ, wept openly.
- In 1997, Tiger Woods not only won the Golf Masters, he broke several tournament records in the process—youngest to win (21 years old), lowest score for 72 holes, and widest margin of victory. He was also the first African American to win the Masters. Or was he? Is Tiger Woods African American, Asian, Native American, or white?
- In 2001, Dale Earnhardt was killed after he crashed his car into a wall during the last lap of the Daytona 500. Many fans experienced personal loss and anguish. Many memorials have been erected, including on Internet sites. One of the Internet sites says that "it will stand on the Internet until the end of time." The person who wrote it said that as a result of Earnhardt's death, "my whole world was going to change."

Each of the above scenarios prompts a Durkheimian question. Family really isn't about blood. My wife is my closest family member and I'm dedicated to my brother-in-law unendingly but we aren't related by blood. The symbolic construction of family is especially true in societies other than ours where kinship is the primary social organization. Many people died on September 11, but most of the people who mourned and felt outrage didn't personally know those people—they weren't connected to them in any real way. The only connection was symbolic. Tiger Woods presents us with the question of race: Where are the boundaries? More specifically, *what* are the boundaries? They obviously aren't biological Whatever biology may or may not be there is of little importance when compared to the symbolic boundaries we create around the idea of race. For example, previous to 1997, the U.S. Census Bureau counted four racial categories: American Indian or Alaskan Native, Asian or Pacific Islander, black, and white. After 1997, the Census counted five: American Indian or Alaska Native; Asian; black or African American; Native Hawaiian or Other Pacific Islander; and white. Notice that prior to 1997, it was impossible to be statistically counted as an "African American." But the significant change is that people are now able to mark more than one racial category, which means that there are now 63 possible categories of race. Tiger Woods, who calls himself "cablinasian," could check black, American Indian, Asian, and white. If he did, he would fall into a different category than either of his parents. And with the Earnhardt example, how can someone's whole life change as the result of the death of another man whom he has never met? How are these connections made between people? Most are made symbolically. How that happens is a distinctly Durkheimian question.

So, to think like Durkheim is to be concerned about the symbolic and emotional connections among groups of people. And to think like Durkheim is to wonder about the level of cultural diversity a society can maintain and still be a single society. More than that, to think like Durkheim is to search for ways in which the apparent diversity is overridden by a larger, more general cultural system. The United States has more cultural diversity than ever, yet it still functions as a society. What are the ties that bind?

To think like Durkheim is to take seriously the Law of Durkheim: every time humans interact, they create culture. *Every* time. So, how are increases in communication and transportation technologies influencing the culture people have? The ability of people to interact with one another has increased exponentially over the past 20 years. People walk around with cell phones attached to their ears; they maintain multiple profiles on numerous email accounts; we have access to over 300 television stations; and the list goes on and multiplies every day. What difference does all this make? That is a question prompted by Durkheim's perspective.

To think like Durkheim is to also be attuned to the independent effects that culture can have. Culture has logic. It will create effects that we never intended. Let's think about gender. Does it matter that there is a culture of masculinity that says that men should be tough, goal oriented, independent, utilitarian, and so on? Does it matter how aggressive and macho men act when they are just with men, as when watching or participating in sports? It appears that it does: college campuses with strong fraternity and sports programs have high incidences of what's called "date rape" (being raped by someone the victim knows, such as the person she is dating). Also, the increasing incidence of eating disorders among women is a logical extension of a culture that emphasizes women as slender objects of gaze. Obviously, there is more going on in each of these examples, but to think like Durkheim is to look at culture and its logical consequences.

Our consideration of Durkheim's thought is divided into three main sections: *Religious Roots of Society; Social Diversity and Morality;* and *Individualism.* As we've already seen, Durkheim argues that society is produced and held together by the collective consciousness, both the cognitive and emotional aspects. In the *Religious Roots of Society* we will see that the wellspring for our primary categories and moralities is religion. In this cognitive sense of creating categories, religion even gave birth to science. Science has merely taken over one of religion's functions: to explain the universe. As a result of this and other social factors, Durkheim feels that the direct influence of religion on society is receding, and with it the strong moral basis for the collective consciousness. In *Social Diversity and Morality,* we will see that in place of strong morality, modern society is held together through more organic kinds of processes. There is, however, one particular moral value that all people share in a modern society, and it constitutes the third area of Durkheim we will consider: **Individualism.** In extremely diverse societies, the one value that all groups share is the right to be an individual. However, as we will see, this cult of the individual has its own set of problems, or "pathologies," as Durkheim calls them (based on the organismic analogy).

Religious Roots of Society

> Therefore, when I approach the study of primitive religions, it is with the certainty that they are grounded in and express the real. . . . The reasons the faithful settle for in justifying those rites and myths may be mistaken, and most often are; but the true reasons exist nonetheless. . . . Fundamentally, then, there are no religions that are false. (Durkheim, 1912/1995, p. 2)

I want to begin our consideration of Durkheim's theory with his study of religion. In some ways, it's an odd place to begin. His book, *The Elementary Forms of Religious Life,* was by no means his first, and other theory texts usually place this topic later on in their chapters on Durkheim. But, for me, his study of religion provides us with the clearest expression of his basic theoretical concerns, and based on his theory of religion, we can better understand his concerns with the problems of modernity and the collective consciousness.

Durkheim had two major purposes in writing his thesis on religion. Durkheim's primary aim was to understand the empirical elements present in all religions. He wanted to go behind the symbolic and spiritual to grasp what he calls "the real." Durkheim intentionally puts aside the issue of God and spirituality. He argues that no matter what religion is involved, whether Christianity or Islam, and no matter what god is proclaimed, there are certain social elements that are common to all religions. To put this issue another way, anytime a god does anything here on earth, there appear to be certain social elements always present. Durkheim is interested in discovering those empirical, social elements.

Religion and science: Durkheim's second reason for his writing his book is a bit trickier to understand. For Durkheim, religion is the most fundamental social institution.

He argues that religion is the source of everything social. That's not to say that everything social is religious, especially today. But Durkheim is convinced that social bonds were first created through religion. We'll see that in ancient clan societies, the symbol that bound the group together and created a sense of kinship (family) was principally a religious symbol.

Further, Durkheim argues that our basic categories of understanding are of religious origin. Humans divide the world up using categories. We understand things in terms of animal, mineral, vegetable, edible, inedible, private property, public property, male, female, and on and on. Each of these is a category that includes some things and excludes others. Some of these categories are obviously of cultural origin, like edible/inedible, for instance. Human beings can eat some pretty strange things, and what any one person may think of as falling under the category of edible is generally a function of his or her culture (I don't eat brains and eggs, for example).

How these categories exist and where they come from have been the subject of much philosophical debate. Idealists, for example, argue that the categories are made from ideas, whereas materialists argue that the elements that make categories are to be found in the objects themselves. Durkheim has a different point of view: he argues that certain elements of culture have an objective based in society.

> At the root of our judgments, there are certain fundamental notions that dominate our entire intellectual life . . . They are like solid frames that confine thought. Thought does not seem to be able to break out of them without destroying itself. (Durkheim, 1912/1995, pp. 8–9)

Durkheim says that many of the categories we use are of what one might call "fashionable" origin, that is, culture that is subject to change. Durkheim argues that fashion is a recent phenomenon and that its basic social function is to distinguish the upper classes from the lower. There is a tendency for fashion to circulate. The lower classes want to be like the upper classes and thus want to use their symbols (we want to drive their cars, wear the same kinds of clothes they do, and so on). This implies, in the end, that fashionable culture is rather meaningless: "Once a fashion has been adopted by everyone, it loses all its value; it is thus doomed by its own nature to renew itself endlessly" (Durkheim, 1887/1993, p. 87).

Durkheim has little if any concern for such culture (though quite a bit of contemporary cultural theory and analysis is taken up with it). He is interested in primary "categories of understanding." He argues that these categories—time, space, number, cause, substance, and personality—are of social origin, but not the same kind of social origin that fashion has. Fashion comes about as different groups demarcate themselves through decoration, and in that sense it isn't tied to anything real. It is purely the work of imagination. The primary categories of understanding, on the other hand, are tied to objective reality. Durkheim argues that the primary categories originate *empirically and objectively in society,* in what he calls social morphology.

Merriam-Webster defines *morphology* as "a branch of biology that deals with the form and structure of animals and plants." So, when Durkheim talks about **social morphology,** he is using the organismic analogy to refer to the form and structure of society, in particular the way in which populations are distributed in time and space. Let's take time, for example. In order to conceive of time, we must first conceive of differentiation. Time, apart from humans, appears like a cyclical stream. There is daytime and nighttime and seasons that endlessly repeat themselves. Yet that isn't how we experience time. For us, time is chopped up. Today, for example, is June 1, 2004. However, that date and the divisions underlying it are not a function of the way time appears naturally. So, from where do the divisions come? "The division into days, weeks, months, years, etc., corresponds to the recurrence of rites, festivals, and public ceremonies at regular intervals. A calendar expresses the rhythm of collective activity while ensuring that regularity" (Durkheim, 1912/1995, p. 10). The same is true with space. There is no up, down, right, left, and so on apart from the orientation of human beings that is itself social.

The important thing to see here is that Durkheim makes the claim that the way in which we divide up time and space, and the way we conceive of causation and number, is not a function of the things themselves, nor is it a function of mental divisions. Rather, the way we conceive of these primary categories is a function of the objective form of society—the ways in which we distribute and organize populations and social structures. Our primary categories of understanding come into existence through the way we distribute ourselves in time and space; they reflect our gatherings and rituals. Thus, all of our thinking is founded upon social facts, and these social facts, according to Durkheim, originated in religion.

> Of the two functions originally performed by religion, there is one, only one, that tends more and more to escape it, and that is the [cosmological] speculative function. What science disputes in religion is not its right to exist but its right to dogmatize about the nature of things, its pretensions to special expertise for explaining man and the world. (Durkheim, 1912/1995, p. 432)

Durkheim wants to make a point beyond social epistemology. He argues that if our basic categories of understanding have their roots in religion, then all systems of thought, such as science and philosophy, have their basis in religion. Durkheim (1912/1995) extends this theme, stating that "there is no religion that is not both a cosmology and a speculation about the divine" (p. 8). Notice that there are two functions of religion in this quote. One has to do with speculations about the divine—in other words, religion provides faith and ideas about God. Also notice the other function: to provide a cosmology. A cosmology is a systematic understanding of the origin, structure, and space–time relationships of the universe. Durkheim is right: every religion tells us what the universe is about—how it was created, how it works, what its purpose is, and so on. But so does science, and that's Durkheim's point. Speculations about the universe began in religion; ideas about causation began in religion. Therefore, the social world, even in its most logical of pursuits, was set in motion by religion.

In the end, Durkheim argues, "religion gave birth to all that is essential in society." What he means is that "society is not constituted simply by the mass of individuals who comprise it, the ground they occupy, the things they use, or the movements they make, but above all by the idea it has of itself" (Durkheim, 1912/1995, p. 425). And this idea is born out of the religious. Religion gives us our strongest sense of the collective consciousness; we are most fully aware of our collective soul in religion. That is, religion gives us our strongest sense of identity, values, norms, symbols, solidarity, and so forth. But more than that, even without religion *per se,* as in modern society, the interactions that are necessary to produce and preserve the collective consciousness are religious in nature.

Defining religion: But how did religion begin? Here we turn back again to Durkheim's principal purpose: to explain the origins of religion. In order to get at his argument, we will consider the data Durkheim uses, his definition of religion, and, most importantly, how the sacred is produced. The data that Durkheim employs is important because of his argument and intent. He wants to get at the most general social features underlying all religions—in other words, apart from doctrine, he wants to discern what is common to all religions. To discover those commonalities, Durkheim contends that one has to look at the most primitive forms of religion.

Using contemporary religion to understand the basic forms of religion has some problems, most notably the natural effects of history and storytelling. You've probably either played or heard of the game where a group of people sits in a circle and whispers a story to one another. What happens, as you know, is that the story changes in the telling. The same is true with religion, at least in terms of its origins. Basically what Durkheim is saying is that the further we get away from the origins of religion, the greater will be the confusion around why and how religion began in the first place. Also, because ancient religion was simpler, using the historical approach allows us to break the social phenomenon down into its constituent parts and identify the circumstances under which it was born. For Durkheim, the most ancient form of religion is totemism. He uses data on totemic religions from Australian Aborigines and Native American tribes for his research.

Conceptually, prior to deciding what data to use, Durkheim had to create a definition of religion. In any research, it is always of utmost importance to clearly delineate what will count and what will not count as your subject. For example, if you were going to study the institution of education, one of the things you would have to contend with is whether home schooling or Internet courses would count, or do only accredited teachers in state-supported organizations constitute the institution of education? The same kinds of issues exist with religion. Durkheim had to decide on his definition of religion before choosing his data sources—he had to know ahead of time what counts as religion and what doesn't, especially since he wanted to look at its most primitive form.

> A religion is a unified system of beliefs and practices relative to sacred things, that is to say, things set apart and forbidden—beliefs and practices which unite into one single moral community called a Church, all those who adhere to them. (Durkheim, 1912/1995, p. 44)

Note carefully what is missing from Durkheim's definition: there is no mention of the supernatural or God. Durkheim argues that before humans could think about the supernatural, they first had to have a clear idea about what was natural. The term *supernatural* assumes the division of the universe into two categories: things that can be rationally explained and those that can't. Now, think about this—when was it that people began to think that things could be rationally explained? It took quite of bit of human history for us to stop believing that there are spirits in back of everything. We had gods of thunder, forests, harvest, water, fire, fertility, and so on. Early humans saw spiritual forces behind almost everything, which means that the idea of *nature* didn't occur until much later in our history. The concept of nature—those elements of life that occur apart from spiritual influence—didn't truly begin until science. So, the idea of *supernatural* is a recent invention of humanity and therefore can't be included in a definition of religion, since there has never been society without religion—early societies would have had religion but no concept of supernatural.

Durkheim also argues that we cannot include the notion of God in the definition. His logic here concerns the fact that there are many belief systems that are generally considered religion that do not require a god. Though he includes other religions such as Jainism, his principal example is Buddhism. The focus of Buddhist faith is the Four Noble Truths, and "salvation" occurs apart from any divine intervention. There are deities acknowledged by Buddhism, such as Indra, Agni, and Varuna, but the entire Buddhist faith can be practiced apart from them. The practicing Buddhist needs no god to thank or worship, yet we would be hard pressed to not call Buddhism a religion.

> Since, in themselves, neither man nor nature is inherently sacred, both acquire sacredness elsewhere. Beyond the human individual and the natural world, then, there must be some other reality. (Durkheim, 1912/1995, pp. 84–85)

Thus, three things constitute religion in its most basic form: the sacred, beliefs and practices, and a moral community. The important thing to notice about Durkheim's definition is the centrality of the notion of the sacred. Every element of the definition revolves around it. The beliefs and practices are relative to sacred things and the moral community exists because of the beliefs and practices, which of course brings us back to sacred things. So at the heart of religion is this idea of the sacred.

Creating the sacred: But what are sacred things? By that I mean, what makes something sacred? For most of us in the United States, we think of the Cross or the Bible as sacred objects. It's easy for us to think they are sacred because of some intrinsic quality they possess. The Cross is sacred because Jesus died on it. However, the idea of the intrinsic worth of the object falls apart when we consider all that humanity has thought of as sacred. Humans have used crosses, stones, kangaroos, snakes, birds, water, swastikas, flags, yellow ribbons, almost anything to represent the sacred. Durkheim's point is that sacredness is not a function of the object; sacredness is something that is placed *upon* the object.

We could argue that sacredness comes through association, and in part at least that's true. We think the Cross is sacred because of its association with Jesus. But what makes the image of an owl sacred? It can't just be its association with the owl, so there must be something else in back of it. What, then, can make both the owl and the Cross sacred? (Remember, with a positivist like Durkheim, we are looking for *general* explanations, ones that will fit all instances.) Durkheim wouldn't accept the answer that there is some general spiritual entity in back of all sacred things. First, the sacred things and their beliefs are too varied. But more importantly, Durkheim is interested in the *objective* reality behind religion. So, how can we explain the power of the sacred using objective, general terms? Durkheim begins with his consideration of totemic religion.

Totems had some interesting functions. For instance, they created a bond of kinship among people unrelated by blood. Each clan was composed of various hunting and gathering groups. These groups lived most of their lives separately, but they periodically came together for celebrations. The groups weren't related by blood, nor were they connected geographically. What held them together was the name of the totem—all the members of the clan carried the same name, and the members of these groups acted toward one another as if they were family. They had obligations to help each other, to seek vengeance, to not marry one another, and so forth based on family relations.

In addition to creating a kinship name for the clans, the totem acted as an emblem that represented the clan. It acted as a symbol both to those within the clan and outside it. The symbol was inscribed on banners and tents and tattooed on bodies. When the clan eventually settled in one place, the symbol was carved into doors and walls. The totem thus formed bonds, and it represented the clan.

In addition, the totem was used during religious ceremonies. In fact, Durkheim (1912/1995) tells us, "Things are classified as sacred and profane by reference to the totem. It is the very archetype of sacred things" (p. 118). Different items became sacred because of the presence of the totem. For example, the clans both in daily and sacred life would use various musical instruments; the only difference between the sacred and mundane instruments was the presence of the totemic symbol. The totem imparted the quality of being sacred to the object.

This is an immensely important point for Durkheim: The totem represents the clan and it creates bonds of kinship. It also represents and imparts the quality of being sacred. Durkheim then used a bit of algebraic logic: If A = B and B = C, then A and C are equal. So, "if the totem is the symbol of both the god and the society, is this not because the god and the society are the same?" (Durkheim, 1912/1995, p. 208). Here Durkheim begins to discover the reality behind the sacred and thus religion. The empirical reality behind the sacred has something to do with society, but what exactly?

> The very act of congregating is an exceptionally powerful stimulant. Once the individuals are gathered together, a sort of electricity is generated from their closeness and quickly launches them to an extraordinary height of exaltation. Every emotion expressed resonates without interference in consciousnesses that are wide open to external impressions, each one echoing the others. The initial impulse is thereby amplified each time it is echoed, like an avalanche that grows as it goes along. (Durkheim, 1912/1995, pp. 217–218)

One of the primary features of the sacred is that it stands diametrically opposed to the profane. In fact, one cannot exist in the presence of the other. Remember the story of Moses and the burning bush? Moses had to take off his shoes because he was standing on sacred ground. These kinds of stories are repeated over and over again in every religion. The sacred either destroys the profane or the sacred becomes contaminated by the presence of the profane. So, one of Durkheim's questions is, how did humans come to conceptualize these two distinct realms? The answer to this will help us discover the reality behind sacredness.

Durkheim found that the aborigines had two cycles to their lives, one in which they carried on their daily life in small groups and the other in which they gathered in large collectives. In the small groups, they would take care of daily needs through hunting and gathering. This was the place of home and hearth. Yet each of these small groups saw themselves as part of a larger group: the clan. Periodically, the small groups would gather together for large collective celebrations.

During these celebrations, the clan members were caught up in collective **effervescence**, or high levels of emotional energy. They found that their behaviors changed; they felt "possessed by a moral force greater than" the individual. "The effervescence often becomes so intense that it leads to outlandish behavior. . . . [Behaviors] in normal times judged loathsome and harshly condemned, are contracted in the open and with impunity" (Durkheim, 1912/1995, p. 218). These clan members began to conceive of two worlds: the mundane world of daily existence where they were in control, and the world of the clan where they were controlled by an external force greater than themselves. "The first is the profane world and the second, the world of sacred things. It is in these effervescent social milieux, and indeed from the very effervescence, that the religious idea seems to have been born" (Durkheim, 1912/1995, p. 220).

In some important ways, Durkheim is describing the genesis of society. Let's assume, as many people do and Durkheim does, that human beings are by nature self-serving and individualistic. How, then, is society possible? One answer is found here: In Durkheimian thought, humans are linked emotionally. Undoubtedly these emotions, once established, mediate human connections unconsciously. That is, once humans are connected emotionally, it isn't necessary for them to rationally see or understand the connections, though we will always come up with legitimations. This emotional soup that Durkheim is describing is the stuff out of which human society is built. Initially, emotions run wild and so do behaviors in these kinds of primitive societies. But with repeated interactions, the emotions become focused and specified behaviors, symbols, and morals emerge.

> Feeling possessed and led on by some sort of external power that makes him think and act differently than he normally does, he naturally feels he is no longer himself. It seems to him that he has become a new being. . . . And because his companions feel transformed in the same way at the same moment . . . it is as if he was in reality transported into a special world entirely different from the one in which he ordinarily lives, a special world inhabited by exceptionally intense forces that invade and transform him. (Durkheim, 1912/1995, p. 220)

In general, Durkheim is arguing that human beings are able to create high levels of emotional energy whenever they gather together. We've all felt something like what Durkheim is talking about at concerts or political rallies or sporting events. We get swept up in the excitement. At those times, we feel "the thrill of victory and the agony of defeat" more poignantly than at others. It is always more fun to watch a game with other people such as at the stadium. Part of the reason is the increase in emotional energy. In this case, the whole is greater than the sum of the parts and something emerges that is felt outside the individual. This dynamic is what is in back of mob behavior and what we call "emergent norms" (and what is called "hooliganism" in Britain). People get caught up in the overwhelming emotion of the moment and do things they normally wouldn't do.

Randall Collins (1988), a contemporary theorist, has captured Durkheim's theory in abstract terms. Generally speaking, there are three principal elements to the kind of interactions that Durkheim is describing (see Figure 4.1): **co-presence,**

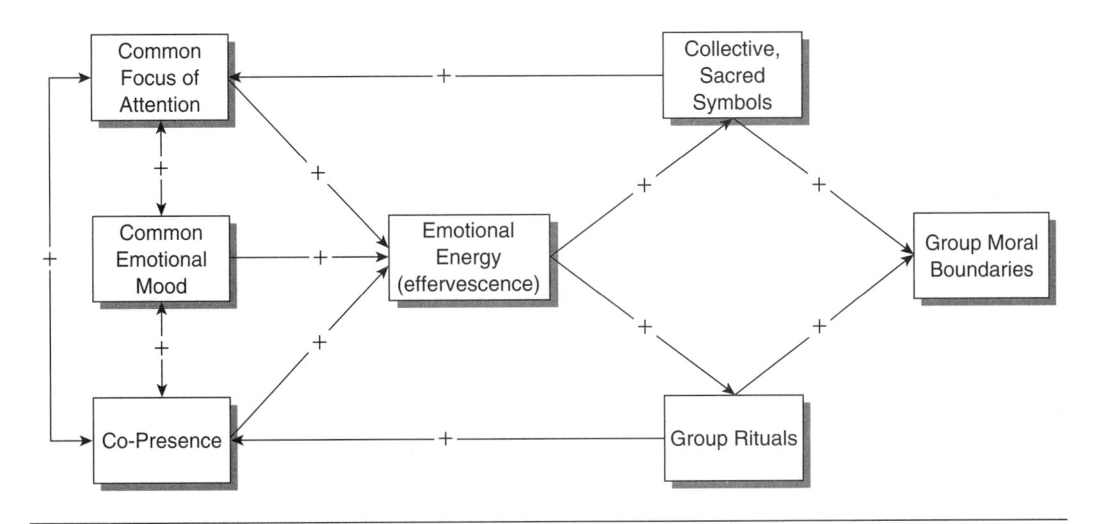

Figure 4.1 Durkheim's Ritual Theory

which describes the degree of physical closeness in space (we can be closer or further away from one another); **common emotional mood,** the degree to which we share the same feeling about the event; and **common focus of attention,** the degree to which participants are attending to the same object, symbol, or idea at the same time (a difficult task to achieve, as any teacher knows).

When humans gather together in intense interactions—with high levels of co-presence, common emotional mood, and common focus of attention—they produce high levels of emotional energy. People then have a tendency to symbolize the emotional energy, which produces a sacred symbol, and to create **rituals** (patterned behaviors designed to replicate the three interaction elements). The **symbols** not only allow people to focus their attention and recall the emotion, they give the collective emotion stability. These kinds of rituals and sacred symbols lead a group to become morally bounded; that is, many of the behaviors, speech patterns, styles of dress, and so on associated with the group become issues of right and wrong.

Groups with high **moral boundaries** are difficult to get in and out of. Street gangs and the Nazis are good examples of groups with high moral boundaries. One of the first things to notice about our examples is the use of the word *moral.* Most of us probably don't agree with the ethics of street gangs. In fact, we probably think their ethics are morally wrong and reprehensible. But when sociologists use the term *moral,* we are not referring to something that we think of as being good. A group is moral if its behaviors, beliefs, feelings, speech, styles, and so forth are controlled by strong group norms and are viewed in terms of right and wrong. In fact, both the Nazis and street gangs are probably more moral, in that sense, than you are, unless you are a member of a radical fringe group.

This theory of Durkheim's is extremely important. First of all, it gives us an empirical, sociological explanation for religion and sacredness. One of the problems that we are faced with when we look across the face of humanity is the diversity of belief systems. How can people believe in diverse realities? The Azande of Africa seek spiritual guidance by giving a chicken a magic potion brewed from tree bark and seeing if the chicken lives or dies. Christians drink wine and eat bread believing they are drinking the blood and eating the body of Christ. How can we begin to explain how people come to see such diverse things as real? Durkheim gives us a part of the puzzle. The issues of reality and sacredness and morality aren't necessarily based on ultimate truth for humans. Our experience of reality, sacredness, and morality are based on Durkheimian rituals and collective emotion. Let me put this another way. Let's say that the Christians are right and the Azande are wrong. How is it, then, that both the Christians and the Azande can have the same experience of faith and reality? Part of the reason is that human beings create sacredness in the same way, regardless of the correctness of any ultimate truth. One of the common basic elements of all religions, particularly during their formative times, is the performance of Durkheimian rituals. These kinds of rituals create high levels of emotional energy that come to be invested in symbols; such symbols are then seen as sacred, regardless of the meaning or truth-value of the beliefs associated with the symbol.

Another reason that this theory is so important is that it provides us with a sociological explanation for the experience that people have of something outside of and greater than themselves. All of us have had these kinds of experiences, some more than others. Some have experienced it at a Grateful Dead concert, others at the Million Man March, others watching a parade, and still others as we conform to the expectations of society. We feel these expectations not as a cognitive dialog, but as something that impress itself upon us physically and emotionally. Sometimes we may even cognitively disagree, but the pressure is there nonetheless.

Durkheim thus provides us with part of the answer for the moral basis of society. Have you ever wondered about extremist groups? How do they produce such conformity? Or, why is it that people who live in small, rural communities tend to be politically conservative ? How is bigotry created and preserved? Or, perhaps on a more positive note, how is national pride and identity created? How can Americans, Communists, Nazis, the Ba'ath, the Taliban, and so on all believe they are right?

I generally won't go into such detail about how a theorist goes about creating the theory. However, in this case, I think it's important for a couple of reasons. First, religion is something that most people have strong feelings about, either pro or con, and because Durkheim's theory is powerful, it, too, can elicit strong reactions. So I thought it important for you to see just how and why Durkheim came up with his theory.

Second, Durkheim's theory of religion contains the strongest statements of his general theory and concerns. We see his interest in categories and his assurance that the basis of cognitive categories is the physical form of society (in the next section, we'll explore the idea of social morphology further). We also have his strongest statement about how the collective consciousness takes on objective characteristics and a life of its own: the collective consciousness is born out of and maintained by emotion-producing rituals. In addition, we see the importance of symbols for collective life.

Social Diversity and Morality

One of the big questions that drove Durkheim is, what holds modern industrial societies together? Up until modern times, societies stayed together because most of the people in the society believed the same, acted the same, felt the same, and saw the world in the same way. However, in modern societies. people are different from each other and they are becoming more so. What makes people different? How can all these different people come together and form a single society? In order to begin to answer these questions, Durkheim created a typology of societies.

A theoretical typology is a scheme that classifies a phenomenon into different categories. We aren't able to explain things directly by using a typology, but it does make things more apparent and more easily explained. Herbert Spencer—the one to whom Durkheim compares his own theory throughout *The Division of Labor in Society*—categorized societies as either industrial or militaristic. In order to construct his typology, he focused on the state. So, for example, when a society is in a militaristic phase, the state is geared toward defense and war and social control is centralized: information, behavior, and production are tightly controlled by government. But when the same society is in an industrial phase, the state is less centralized and the social structures are oriented toward economic productivity: freedom in information exchange, behaviors, and entrepreneurship are encouraged. But Durkheim focuses on something different. His typology reflects his primary theoretical concern: social solidarity.

Mechanical and organic solidarity: **Social solidarity** can be defined as the degree to which social units are integrated. According to Durkheim, the question of solidarity turns on three issues: the subjective sense of individuals that they are part of the whole; the actual constraint of individual desires for the good of the collective; and the coordination of individuals and social units. It is important for us to notice that Durkheim acknowledges three different levels of analysis here: psychological, behavioral, and structural. Each of these issues becomes itself a question for empirical analysis: How much do individuals feel part of the collective? To what degree are individual desires constrained? And, how are activities coordinated and adjusted to one another? As each of these vary, a society will experience varying levels of social solidarity.

Durkheim is not only interested in the *degree* of social solidarity, he is also interested in the way social solidarity comes about. He uses two analogies to talk about these issues. The first is a mechanistic analogy. Think about machines or motors. How are the different parts related to each other? The relationship is purely physical and involuntary. Machines are thus relatively simple. Most of the parts are very similar and are related to or communicate with each other mechanistically. If we think about the degree of solidarity in such a unit, it is extremely high. The sense of an absolute relationship to the whole is unquestionably there, as every piece is connected to every other piece. Each individual unit's actions is absolutely constrained by and coordinated with the whole.

The other analogy is the organismic one. Higher organisms are quite complex systems, when compared to machines. The parts are usually different from one another, fulfill distinct functions, and are related through a variety of diverse subsystems. Organismic structures provide information to one another using assorted nutrients, chemicals, electrical impulses, and so on. These structures make adjustments because of the information that is received. In addition, most organisms are open systems in that they respond to information from the environment (most machines are closed systems). The solidarity of an organism when compared to a machine is a bit more imprecise and problematic.

By their very nature, analogies can be pushed too far, so we need to be careful. Nonetheless, we get a clear picture of what Durkheim is talking about. Durkheim says that there are two "great currents" in society: similarity and difference. Society begins with the first being dominant. In these societies, which Durkheim terms *segmented,* there are very few personal differences, little competition, and high egalitarianism. These societies experience **mechanical solidarity.** Individuals are mechanically and automatically bound together. Gradually, the other current, difference, becomes stronger and similarity "becomes channeled and becomes less apparent." These social units are held together through mutual need and abstract ideas and sentiments. Durkheim refers to this as **organic solidarity.** While organic solidarity and difference tend to dominate modern society, similarity and mechanical solidarity never completely disappear.

In Table 4.1, I've listed several distinctions between mechanical and organic solidarity. In the first row, the principal defining feature is listed. In mechanical solidarity, individuals are directly related to a group and its collective consciousness. If the individual is related to more than one group, there are very few and the groups tend to overlap with one another: "Thus it is entirely mechanical causes which ensure that the individual personality is absorbed in the collective personality" (Durkheim, 1893/1984, p. 242). Remember all that we have talked about concerning culture and morality (Durkheim's Law and ritual performance). People are immediately related to the collective consciousness by being part of the group that creates the culture in highly ritualistic settings. In these groups, the members experience the collective self as immediately present. They feel its presence push against any individual thoughts or feelings. They are caught up in the collective effervescence and experience it as ultimately real. The clans that Durkheim studies in *The Elementary Forms of the Religious Life* are a good illustration.

When an individual is mechanically related to the collective, all the rest of the characteristics we see under mechanical solidarity fall into place. In Table 4.1, the common beliefs and sentiments and the collective ideas and behavioral tendencies represent the collective consciousness. The collective consciousness varies by at least four features: *The degree to which culture is shared*—how many people in the group hold the same values, believe the same things, feel the same way about things, behave the same, and see the world in the same way; *the amount of power the culture has to guide individual's thoughts, feelings, and actions*—a culture can be shared but not very powerful (A group where the members feel they have options doesn't have a very powerful culture.); *the degree of clarity*—how clear the prescriptions and prohibitions are in the culture (For example, when a man and a woman approach a door at the same time, is it clear what behavior is expected?); and the *collective*

Table 4.1 Mechanical and Organic Solidarity

Mechanical Solidarity	*Organic Solidarity*
Individuals directly related to collective consciousness with no intermediary	Individuals related to collective consciousness through intermediaries
Joined by common beliefs and sentiments (moralistic)	Joined by relationships among special and different functions (utilitarian)
Collective ideas and behavioral tendencies are stronger than individual	Individual ideas and tendencies are strong and each individual has own sphere of action
Social horizon limited	Social horizon unlimited
Strong attachment to family and tradition	Weak attachment to family and tradition
Repressive law: crime and deviance disturb moral sentiments; punishment meted out by group; purpose is to ritually uphold moral values through righteous indignation	Restitutive law: crime and deviance disturb social order; rehabilitative, restorative action by officials; purpose is to restore status quo

consciousness varies by its content. Durkheim is referring here to the ratio of religious to secular and individualistic symbolism. Religiously inspired culture tends to increase the power and clarity of the collective consciousness.

As we see from Table 4.1, in mechanical solidarity, the social horizon of individuals tends to be limited. Durkheim is referring to the level of possibilities an individual has in terms of social worlds and relationships. The close relationship the individual has to the collective consciousness in mechanical solidarity limits the number of possible worlds or realities the individual may consider. In modern societies, under organic solidarity, we have almost limitless possibilities from which to choose. Media and travel expose us to uncountable religions and their permutations. Today you can be a Buddhist, Baptist, or Bahai, and you can choose any of the varied universes they present. This proliferation of possibilities, including social relationships, is severely limited under mechanical solidarity. One of the results of this limiting is that tradition appears concrete and definite. People in segmented societies don't doubt their knowledge or reality. They hold strongly to the traditions of their ancestors. And, at the same time, people express their social relationships using family or territorial terms.

But even under mechanical solidarity, not everyone conforms. Durkheim acknowledges this and tells us that there are different kinds of laws for the different types of solidarity. The function of these laws is different as well, corresponding to the type of solidarity that is being created. Under mechanical solidarity, punitive law is more important. The function of **punitive law** is not to correct, as we usually think of law today; rather, the purpose is expiation (making atonement). Satisfaction must be given to a higher power, in this case the collective consciousness. Punitive law is exercised when the act "offends the strong, well-defined states of the collective consciousness" (Durkheim, 1893/1984, p. 39). Because this is linked to morality, the punishment given is generally greater than the danger represented to society, such as cutting off an individual's hand for an act of thievery.

Punitive law satisfies moral outrage and clarifies moral boundaries. When we respond to deviance with some form of "righteous indignation," and we punish the offender, we are experiencing and creating our group moral boundaries. In punishing offenses, we are drawing a clear line that demarks those who are in the group and those who are outside. Punitive law also provides an opportunity for ritual performance. A good example of this principle is the past practice of public executions. Watching an execution of a murderer or traitor was a public ritual that allowed the participants to focus their attention on a single group moral norm and to feel the same emotion about the offense. In other words, they were able to perform a Durkheimian ritual that recreated their sacred boundaries. As a result, the group was able to feel their moral boundaries and experience a profound sense of "we-ness," which increases mechanical solidarity.

Organic solidarity, on the other hand, has a greater proportion of *restitutive law,* which is designed to restore the offender and broken social relations. Because organic social solidarity is based on something other than strong morality, the function of law is different. Here there is no sense of moral outrage and no felt need to ritualize the sacred boundaries. Organic solidarity occurs under conditions of complex social structures and relations. Modern societies are defined by high structural differentiation, with large numbers of diverse structures necessitating complex interconnections of communication, movement, and obligations. Because of the diversity of these interconnections, they tend to be more rational than moral or familial—which is why we tend to speak of "paying one's debt to society." The idea of "debt" comes from rationalized accounting practices; there is no emotional component, as there would be with moral or family connections. And the interests guarded by the laws tend to be more specialized, such as corporate or inheritance laws, rather than generally held to by all, such as "thou shalt not kill." As an important side note, we can see that restitutive and punitive laws are material social facts that help us see the nonmaterial organic and mechanical solidarity within a society.

Organic solidarity thus tends to be characterized by weak collective consciousness: fewer beliefs and sentiments, and ideas and behavioral expectations tend to be shared. There is greater individuality and people and other social units (like organizations) are connected to the whole through utilitarian necessity. In other words, we need each other to survive, just like in an organism (my heart would die without its connection to my lungs and the rest of my biological system).

The division of labor: Earlier I mentioned that Durkheim says that similarity and mechanical solidarity gradually become channeled and less apparent. That statement gives the impression that the change from mechanical to organic solidarity occurred without any provocation, and that's not the case. The movement from mechanical to organic solidarity, from similarity to difference, from traditional to modern was principally due to increases in the division of labor. The concept of the **division of labor** refers to a stable organization of tasks and roles that coordinate the behavior of individuals or groups that carry out different but related tasks. Obviously, the division of labor may vary along a continuum from simple to complex. We can build a car in our garage all by ourselves from the ground up (as the first automobiles were built), or we can farm out different manufacturing and assembly tasks to hundreds of subcontractors worldwide and simply complete the construction in our plant (as it is done today). Our illustration illustrates the poles of the continuum, but there are multiple steps in between.

What kinds of processes tend to increase the division of labor in a society? Bear with me for a moment; I'm going to put together a string of rather dry-sounding concepts and relationships. The answer to what increases the overall division of labor is competition. Durkheim sees competition not as the result of individual desires (remember they are curtailed in mechanical solidarity) or free markets, but rather as the result of what Durkheim variously calls **dynamic, moral, or physical density.** These terms capture the number and intensity of interactions in a collective taken as a whole. The *level* of dynamic density is a result of increasing population density, which is a function of population growth (birth rate and migration) and ecological barriers (physical restraints on the ability of a population to spread out geographically).

Durkheim's theory is based on an ecological, evolutionary kind of perspective. The environment changes and thus the organism must change in order to survive. In this case, the environment is social interaction. The environment changes due to identifiable pressures: population growth and density. As populations concentrate, people tend to interact more frequently and with greater intensity. The rate of interaction is also affected by increases in communication and transportation technologies. As the level of interaction increases, so does the level of competition. More people require more goods and services, and dense populations can create surplus workers in any given job category. The most fit survive in their present occupation and assume a higher status; the less fit create new specialties and job categories, thus creating a higher degree of division of labor.

The problem of modernity: Now, let me ask you a question. What kind of problem do you think that increasing the division of labor might cause for the collective consciousness? The answer to this question is Durkheim's problem of modernity. Recall Durkheim's Law, his notion of ritual performance, and how people are connected to the collective consciousness. If people are interacting in different situations, with different people, to achieve different goals (as would be the case with higher levels of the division of labor), then they will produce more particularized than collective cultures. Therefore, because they contain different ideas and sentiments, the presence of **particularized cultures** threatens the power of the collective consciousness.

There's an old saying, "Birds of a feather flock together." Well, what Durkheim is telling us just the opposite: "Birds become of a feather because they flock together." In other words, the most prominent characteristics of people come about because of the groups they interact with. As we internalize the culture of our groups, we learn how to think and feel and behave, and we become socially distinct from one another. We call this process **social differentiation.** As people become socially differentiated, they, by definition, share fewer and fewer elements of the collective consciousness. This process brings with it the problem of integration. It's a problem that we here in the United States are very familiar with: How can we combine diverse populations into a whole nation with a single identity? But there's more to this problem of modernity: as the division of labor increases, so does the level

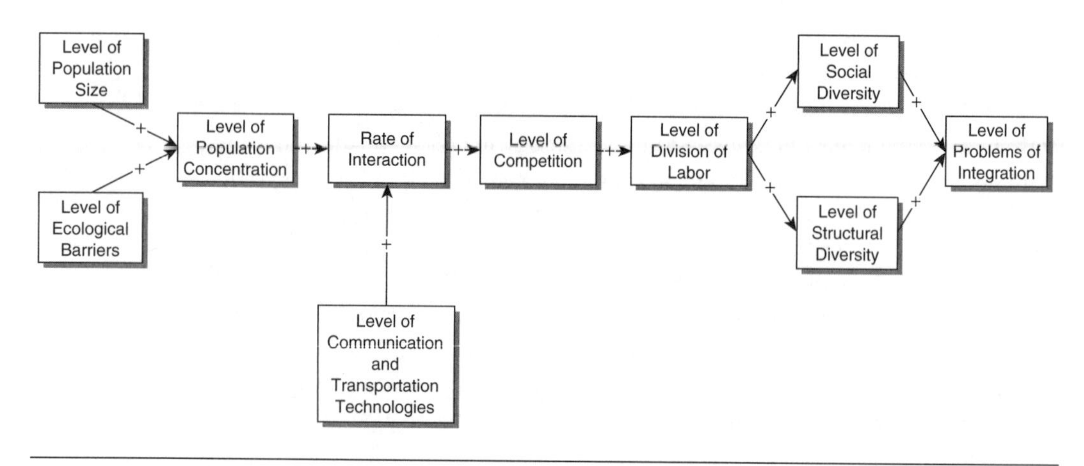

Figure 4.2 The Division of Labor and Problems of Integration

of structural differentiation—the process through which the needs of society are met through increasingly different sets of status positions, roles, and norms. Spencer made this the central issue for his social theorizing. Here we see the same concern. Durkheim argues much like Spencer—structural interdependency creates pressures for integration—but he adds a cultural component: value generalization.

Thus we are confronted with the problem of modernity. Because groups are more closely gathered together, the division of labor has increased. In response to population pressures and the division of labor, social structures have differentiated to better meet societal needs. As structures differentiate, they are confronted with the problem of integration. In addition, as the division of labor increases, people tend to socially differentiate according to distinct cultures. As people create particularized cultures around their jobs, they are less in tune with the collective consciousness and face the problem of social integration. I've illustrated these relationships in Figure 4.2.

Organic solidarity and social pathology: As social and structural differentiation create problems of integration, they simultaneously produce social factors that counterbalance these problems: intermediary group formation, culture generalization, **restitutive law** and centralization of power, and structural interdependency. Together, these factors form organic solidarity. First, structures become more dependent upon one another as they differentiate (**structural interdependency**). For example, your heart can't digest food, so it needs the stomach to survive. It is the same for society as it differentiates. The different structures become dependent upon one another for survival. Further, in order to provide for the needs of other structures, they must be able to interact with one another. Thus, structural and social differentiation also create pressures for a more generalized culture and value system (**culture generalization**). Let's think about this in terms of communication among computers. I have a PC and my friend Jamie has an Apple. Each has a completely different platform and operating system. Yet almost every day my computer communicates with his. How can it do this? The two different systems can communicate with one another because there is a more general system that contains values broad enough to encompass both computers (i.e., the Internet).

Societies thus produce more general culture and values in response to the need for different subsystems and groups to communicate. In contemporary structural analysis, this is an extremely important issue. Talcott Parsons termed this focus the "generalized media of exchange"—symbolic goods that are used to facilitate interactions across institutional domains. So, for example, the institutional structure of family values love, acceptance, encouragement, fidelity, and so on. The economic structure, on the other hand, values profit, greed, one-upmanship, and the like. How do these two institutions communicate? What do they exchange? How do they cooperate in order to fulfill the needs of society? These kinds of questions, and their answers, are what make for structural analysis and the most sociological of all research. They are the empirical side of Durkheim's theoretical concern.

Of course, in the United States we have successively created more generalized cultures and values. The idea of citizen, for example, has grown from white-male-Protestant-property-owner to include people of color and women. Yet generalizing culture is a continuing issue. The more diverse our society becomes, the more generalized the culture must become, according to Durkheim. For example, while we in the United States may say "in God we trust," it is now a valid question to ask "which god?" We have numerous gods and goddesses that are worshiped and respected in our society. To maintain this diversity, Durkheim would argue there has to be in the culture a concept general enough to embrace them all. According to Durkheim, if the culture doesn't generalize, we run the risk of disintegration.

The fact is that generalized culture is often too broad to invest much moral emotion, so more and more of our relations, both structural and personal, are mitigated by law. This law has to be rational and focused on relationships, not morals. For example, I don't know my neighbors. The reasons for this have a lot to do with what Durkheim is talking about: increases in transportation technologies, increasing divisions of labor, and so on. But if I don't know my neighbors, how can our relationship be managed? Obviously, if I have a problem with their dog barking or their tree limb falling on my house, there are laws and legal proceedings that manage the relationship. Increases in social and structural diversity thus create higher levels of restitutive law (in comparison to restrictive/moral law) and more centralized government to administer law and relations. Of course, one of the things we come to value is this kind of law and we come to believe in the right of a centralized government to enforce the law.

Both social and structural diversity also push for the formation of intermediary groups. Remember, Durkheim always comes back to real groups in real interaction: culture can have independent effects but it requires interaction to be produced. Thus, the problem becomes, how do individual occupational groups create a more general value system if they don't interact with one another? Durkheim theorizes that societies will create intermediary groups—groups between the individual occupational groups and the collective consciousness. These groups are able to simultaneously carry the concerns of the smaller groups as well as the collective consciousness.

For example, I'm a sociology faculty member at the University of North Carolina. Because of the demands of work, I rarely interact with faculty from other disciplines (like psychology), and I only interact with medical doctors as a patient. Yet I am a member of the American Sociological Association (ASA), and the ASA interacts with the American Psychological Association and the American Medical Association, as well as many, many others. And all of them interact with the United States government as well as other institutional concerns. This kind of interaction amongst intermediary groups creates a higher level of value generalization, which, in turn, is passed down to the individual members. As Durkheim (1893/1984) says, "A nation cannot be maintained unless, between the state and individuals, a whole range of secondary groups are interposed. These must be close enough to the individual to attract him strongly to their activities and, in so doing, to absorb him into the mainstream of social life" (p. liv).

As we've seen, structural interdependency, culture generalization, intermediary group formation, and restitutive law and centralization of power together create organic solidarity (see Figure 4.3). Increasing division of labor systemically pushes for these changes: "Indeed, when its functions are sufficiently linked together they tend of their own accord to achieve an equilibrium, becoming self-regulatory" (Durkheim, 1893/1984, p. xxxiv). As a side note, part of what we mean by functional analysis is this notion of system pressures creating equilibrium. Notice the dynamic mechanism: it is the system's need that brings about the change. It's kind of like pulling your car into the gas station because it needs gas. Society is a smart system, regulating its own requirements and bringing about changes to keep itself in equilibrium. However, if populations grow and/or become differentiated too quickly, the system can't keep up and these functions won't be "sufficiently linked together." Society can then become pathological or sick.

Durkheim elaborates two possible pathologies. (Actually, Durkheim mentions three—anomie, forced division of labor, and "lack of coordination"—but he clearly elaborates only the first two.) The first is **anomie**—social instability and personal unrest resulting from insufficient normative regulation of individual activities. Durkheim argues that social life is impossible apart from normative regulation.

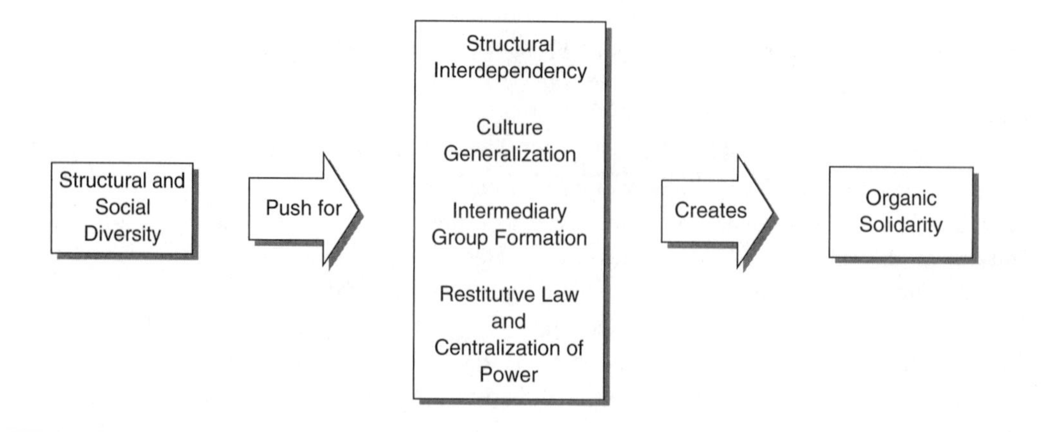

Figure 4.3 Organic Solidarity

People are naturally driven by individual appetites, and without norms to regulate interactions, cooperation is impossible. If it were necessary to "grope de novo for an appropriate response to every stimulus from the environing situation, threats to its integrity from many sources would promptly effect its disorganization . . . to this end, it is altogether necessary that the person be free from an incessant search for appropriate conduct" (Durkheim, 1903/1961, p. 37). The production of norms requires interaction. However, overly rapid population growth with excessive division of labor and social diversity hinders groups from interacting. Links among and between the groups cannot be formed when growth and differentiation happen too quickly. The result is anomie.

The other pathology that Durkheim considers at some length involves class inequality and the forced division of labor. Durkheim argues that the division of labor must occur "spontaneously," that is, apart from external constraint. Labor should divide because of organic reasons: population growth and density. The division of labor should not occur due to a powerful elite driven by profit motivations. According to Durkheim, it is dysfunctional to force people to work in jobs for which they are ill-suited. "We are certainly not predestined from birth to any particular form of employment, but we nevertheless possess tastes and aptitudes that limit our choice. If no account is taken of them, if they are constantly frustrated in our daily occupation, we suffer, and seek the means of bringing that suffering to an end" (Durkheim, 1893/1984, pp. 310–311), which of course would represent a threat to social solidarity.

This condition is similar to what Marx talks about, but it is different as well. For Marx, alienation is a state of existence that may or may not be subjectively experienced by the individual. Alienation for Marx is defined by separation from species-being, and we only become aware of it through critical and/or class consciousness. Durkheim, on the other hand, assumes that humans are egotistical actors without an essentially good nature from which to be alienated. Durkheim's alienation comes about only as a result of a pathological form of the division of labor—forced by capitalist greed rather than organic evolution. It exists as a subjective state—we are always aware of alienation when it occurs.

Durkheim's solution for this pathological state is "justice" enforced by the state, specifically, price controls by useful labor (equal pay for equal work) and elimination of inheritance. Two of the most powerful tools in producing a structure of inequality are ascription and inheritance. Ascription assigns different status positions to us at birth, such as gender and race, and apart from legislation guaranteeing equal pay for equal work, they strongly influence inequality. For example, studies done in the United States consistently show that women earn about 70% of what a man makes for the same job with the same qualifications. Inheritance is an obvious way to maintain structural inequality; estimates are that by 2055, at least $41 trillion will be inherited in the United States (Havens & Schervish, 2003). Both ascription and inheritance make the perpetuation of class differences appear natural: our position is ascribed to us and we are wealthy or poor by birth. America may be the land of opportunity, but it is not the land of *equal* opportunity: it matters if you are male or female, black or white, or go to school in Harlem or Hollywood. On the other hand, if inheritance is done away with and laws are implemented that bring equal pay for equal work, structured inequality would have a difficult time surviving.

Individualism

> Our social organization, then, must have changed profoundly in the course of this century, to have been able to cause such a growth in the suicide-rate. So grave and rapid an alteration as this must be morbid; for a society cannot change its structure so suddenly. (Durkheim, 1897/1951, p. 369)

Suicide: In early societies, the human self was an utterly social self. People were caught up in and saw themselves only in terms of the group. The self was an extension of the group just as certainly as your arm is an extension of your body. Yet as societies differentiated both structurally and socially, the self became more and more isolated and took on the characteristics of an individual. Thus in modern societies, the individual takes on increasing importance. We can think of many benefits from this shift. We have increased freedom of choice and individual expression under conditions of organic solidarity. Yet there are some dysfunctional consequences as well. Though Durkheim didn't phrase it in this way, in addition to the two pathologies of modern society listed above, we can include suicide.

> The fact is that we have a need to believe that our actions do not exhaust all their consequences in an instant . . . but that their consequences are extended for some distance in duration and scope. Otherwise they amount to very little . . . and they would hold no interest at all for us. (Durkheim, 1887/1993, p. 119)

There are two critical issues for the individual in modern society: the levels of group attachment and behavioral regulation. People need a certain level of group attachment. We are social creatures and much of our sense of meaning, reality, and purpose come from having interpersonal ties (both in terms of number and density) and a sense of "we-ness" or collective identity. I don't want to belabor the point, but imagine having something be meaningful to you apart from language and feeling—what Durkheim would call collective representations and sentiments. You might object and say, "My feelings are my own." That's true, but what do you feel? Do you feel "anger"? Do you feel "love"? Or, do you simply, purely *feel*? We rarely, if ever, simply and purely feel. What are we doing when we say we feel anger? We are labeling certain physiological responses and giving them meaning. The label is linguistic and the meaning is social.

Like I said, I don't want to dwell on this issue (we will do so when we get to Schutz), but I want us to see that it is extremely difficult for us to untangle personal meanings and realities from social ones. Certainly, because we are human, we can create utterly individualistic realities and meanings, although we usually see those realities and people as either strange or crazy. Most of us are aware, and even unconsciously convinced, that our meanings and realities have to be linked in some way to the social group around us—which is why, when group attachment is too low, Durkheim argues that **egoistic suicide** is likely: low group attachment leads to extreme individualism and the loss of a sense of reality and purpose.

However, extremely high group attachment isn't a good thing for the individual either. High attachment leads to complete fusion with the group and loss of individual identity, which can be a problem in modernity. Under conditions of high group attachment, people are more likely to commit **altruistic suicide.** The Kamikaze pilots during WWII are a good example. Some contemporary examples include religious cults such as The People's Temple and Heaven's Gate, and the group solidarity the United States government fosters in military boot camps. Under conditions of high group attachment, individual life becomes meaningless and the group is the only reality.

It's important to note here that modern societies are characterized by the presence of both mechanical and organic solidarity. It is certainly true that in general the society is held together by organic means—general values, restitutive law, and dependent opposites—yet it is also true that pockets of very intense, particularized culture and mechanical solidarity exist as well. These kinds of group interactions may in fact be necessary for us. This need may explain such intense interaction groups as rock music and sports fans. Both of these groups engage in the kind of periodic ritual gatherings that Durkheim explains in *The Elementary Forms of the Religious Life.* During these gatherings (shows or events), rock and sports fans experience high levels of emotional energy and create clear group symbols. Yet extremes of either organic or mechanical solidarity can be dangerous, as Durkheim notes.

The other critical issue for individuals in modern society is the regulation of their behaviors. Because in the advanced, industrialized nations we believe in individualism, the idea of someone or something regulating our behavior may be objectionable. But keep in mind Durkheim's view of human nature. Apart from regulation, our appetites would be boundless and ultimately meaningless. The individual by him- or herself "suffers from the everlasting wranglings and endless friction that occur when relations between an individual and his fellows are not subject to any regulative influence" (Durkheim, 1887/1993, p. 24). Further, our behaviors must have meaning to us with regard to time. Time, as we think of it, is a function of symbols (the past and future only exist symbolically), and, of course, symbols are a function of group membership.

Thus, there are a variety of reasons why the regulation of behaviors is necessary. Behaviors need to be organized according to the needs and goals of the collective, but the degree of regulation is important. Under conditions of rapid population growth and diversity, anomie may result if the culture is unable to keep pace with the social changes. Under these conditions, it is likely there will be an increase in the level of **anomic suicide.** The lack of regulation of behaviors leads to a complete lack of regulation of the individual's desires and thus an increase in feelings of meaninglessness. On the other hand, overregulation of behaviors leads to the loss of individual effectiveness (and thus increases hopelessness), resulting in more **fatalistic suicide.**

I've listed the suicide types in Table 4.2. It's important to keep in mind that the motivation for suicide is different in each case, corresponding to group attachment and behavior regulation. Also note that the kinds of social pathologies we are talking about here are different than the ones in the previous section. Here we are seeing how modernity can be pathological for the *individual,* in the extremes of attachment and regulation. In the previous section, we looked at how modernity can be pathological for society as a whole and its solidarity.

Table 4.2 Suicide Types

	Group Attachment	*Behavior Regulation*
High	Altruistic Suicide	Fatalistic Suicide
Low	Egoistic Suicide	Anomic Suicide

The cult of the individual: Both egoistic and anomic suicide can be seen as a function of high levels of individuality, but individuality itself is not the problem. Individualism is in fact necessary in modernity. Before we go on, I need to make a distinction between this idea of individuality and what might be called egoistic hedonism or materialism (what most of us think of when we hear the term *individualism*: "I gotta be me."). From a Durkheimian view, individuals who are purely and exclusively out to fulfill their own desires can never form the basis of a group. Group life demands that there be some shared link that motivates people to work for the collective rather than individual welfare. As we've seen in our discussion of Durkheim, this kind of group life and awareness is dependent upon a certain degree of collective consciousness. The idea of the individual that we have in mind here is therefore not pure ego. Rather, what is at stake might better be understood as the idea of "individual rights" and how it has progressed historically.

To do the concept justice, we should really go back to early Greek and then Roman times. But in actuality, we need only go back to the founding of the United States. What does the following sentence mean? "We hold these truths to be self-evident, that all men are created equal, that they are endowed by their Creator with certain unalienable Rights, that among these are Life, Liberty, and the pursuit of Happiness." The history of the United States is the tale of working out the meaning of that line. Obviously, the central struggle concerns the term "all men." Initially, "all men" referred only to white, property-owning, heterosexual, Protestant males. Somewhere along the line, the United States government decided (often in response to fierce civil struggle, such as the fight for women's rights) that you didn't have to own property, and that you didn't have to be male, or white, or Protestant, though you still have to be heterosexual to have full access to civil rights. The reason there has been struggle over this sentence is the basis that is given for these rights. According to the quoted line, the basis for civil rights is simply being human. These rights can't be earned nor can they be taken away. They are yours not because of anything you've done or any group membership, but simply because of your birth into the human race. Thus, what matters isn't your group membership; it's you, as an *individual* human being. It is this moral idea of individualism that Durkheim has in mind and that has the potential for creating social solidarity.

Durkheim calls this new moral basis for society the "cult of the individual." The individual, as he or she is historically separated from the group, becomes the locus of social concern and solidarity. The *individual* becomes the recipient of social rights and responsibilities, rather than castes or lineages. Today we see the individual as perhaps the single most important social actor. Even our legal system here in the United States is occupied with preserving the civil rights of the perpetrator of a crime because it is the individual that is valued. The individual becomes the focus of our idea of "justice," which Durkheim sees as the "medicine" for some of the problems that come with pathological forms of the division of labor. For example, in a society such as the United States, the problems associated with labor issues, poverty, deviance, and depression (all of which can be linked to Durkheim's pathologies) are generally handled individually through the court system or counseling. Remember that Durkheim sees culture as the unifying force of society, so the importance of these kinds of cases for Durkheimian sociology has more to do with the culture that the practices create and reproduce than the actual legal or psychological effects. In this way, the individual becomes a ritual focus of attention, the symbol around which people can seek a kind of redress for the forced division of labor, inequality, and anomie that we noted above. Today the individual has taken on a moral life. But it is not the particular person *per se*, with all of her or his idiosyncrasies, that has value. Rather, it is the ethical and sacred *idea* of the individual that is important.

This notion of the individual has become perhaps *the* collective symbol of modernity and postmodernity. What moral factor can link together extremely diverse groups? What can street gangs and medical doctors hold in common? Perhaps it's the belief that individuality is supremely important. This cultural idea of the individual becomes the focus of both public and private rituals and part of the moral fiber that holds modern society together. As Durkheim (1957) says,

> From being limited and of small regard, the scope of the individual life expands and becomes the exalted object of moral respect. . . . If he be the moral reality, then it is he who must serve as the pole-star for public as well as private conduct. (p. 56)

> This cult, moreover, has all that is required to take the place of the religious cultures of former times. It serves as well as they to bring about the communion of minds and wills which is a first condition of any social life. (p. 69)

Thinking About Modernity and Postmodernity

Grand narratives, doubt, and civil religion: Durkheim's perspective is and always has been central to sociology. The issues of culture, group cohesion, and religion have never left the sociological landscape, though the importance we grant them has varied. As a result of the social unrest in the 1960s and early 1970s, the ideas of culture and functionalism took a backseat to Marx and conflict theory. However, since the mid 1980s, culture has become more and more important for sociological thinking, due in large part to postmodernism. Culture is at the heart of the postmodern critique, and sociologists, whether they agree with the critique or not, have had to consider the issue of culture. As Hall and Neitz (1993) state, "At the end of the twentieth century, culture once again has become a central focus of sociological theorizing" (p. 1).

Durkheim's theory is motivated by his concerns about culture in modernity. Durkheim sees modernity as driven by increasing levels of population density and division of labor. These two dynamics create distinctly modern societal effects: increasing importance of the state and restitutive law, decreasing importance of religion and moral solidarity, increasing structural complexity and interdependency, increasing generalization of culture and media of exchange, and increasing levels of individualization. We have also seen that Durkheim is concerned about possible pathological states coming from these effects, particularly anomie and the forced division of labor. I believe that Durkheim had an additional lingering doubt about social cohesion in modernity that we haven't talked about yet. It has to do with symbolic unification.

The kinds of symbols that Durkheim is interested in are religious. He isn't arguing that religion itself is needed, only the kinds of symbolic meanings that religion produces. "Thus there is something eternal in religion that is destined to outlive the succession of particular symbols in which religious thought has clothed itself" (Durkheim, 1912/1995, p. 429). Pay particular attention here, because this is the point at which Durkheim expresses some doubt about purely organic solidarity and where postmodernism comes into play. Durkheim says that no replacement for these religious-type symbols has yet been created, which he explains in this way: "If today we have some difficulty imagining what the feasts and ceremonies of the future will be, it is because we are going through a period of transition and moral mediocrity" (1912/1995, p. 429). Nevertheless, he is convinced that the kinds of symbols and symbolic activities that society needs will be created: "A day will come when our societies once again will know hours of creative effervescence during which new ideals will again spring forth and new formulas emerge to guide humanity for a time" (p. 429).

There is therefore a gap in Durkheim's theory and confidence. Even though he argues for organic solidarity, he nonetheless senses that society needs something more. It is at this very point that postmodern theory finds its niche. Many postmodernists argue that the need for, or legitimacy of, cultural unification has past, in particular the need for what is termed grand narratives. A narrative is a story, and a **grand narrative** is an abstract, symbolic story that transcends and links together diverse cultures.

One of the earmarks of modernity is the presence of or belief in grand narratives—overarching themes that make things look monolithic. The social world of modernity is defined primarily through capitalism, science and technology, and the nation-state. A specific culture arose that served to reinforce these social factors: a secular and liberal grand narrative of equality and freedom, given hope by the advances of science and embodied in Western colonialism. The social world of postmodernity, on the other hand, is characterized by increasing doubt concerning the validity of these grand narratives. There is a way in which we could talk about this facet of postmodernity as the *institutionalization of doubt*. Part of the doubt originates in the failure of Western colonialism and disappointment in the inability of technically advanced nations to create true equality. The ubiquitous doubt of postmodernity is also facilitated by universal education in science and the commodification of religion, as explained further below.

Western colonialism was based on belief in one best system, and the foundation stone of equality was faith in the cultural melting pot. We no longer believe in forcing a dominant culture on others in our society and we no longer think that a melting pot that puts forth one national identity (which turned out to be white-Protestant-heterosexual-male) is the way to go. But notice what these changes imply: there are few if any unifying themes on which we can agree. But perhaps we shouldn't agree on any culturally unifying themes, because to do so would be to deny or silence the voices of others. This is postmodernism's point exactly: postmodernism advocates the legitimation of *polyvocality* (many voices).

This doubt about any kind of ultimate story is also, interestingly enough, facilitated by science and religion. Science teaches us to doubt. Prior to science, knowledge was held to be absolute. Science, however, is based on knowledge testing and accumulation: all information is held in doubt, tested, and replaced. Since the scientific model informs most of our schooling, we come away with a culture of doubt. Further, religion has also contributed to our doubt about grand narratives, as it has come under the influence of mass media and marketing, which we saw when we considered Luckmann's theory in Chapter 2. Religions use billboards, TV spots, newspaper ads, and so on to convey their message. They create sayings and logos that capture their mission, differentiate them from other churches, and make them easy to relate to. Thus, as with everything that is subjected to the trivializing effects of commodification and advertising, the religious individual is faced with a marketplace of fragmented ultimate meanings, competing churches and sects, and substitute religious communities (such as TV evangelists). According to some postmodernists, this kind of meaning market results in an individual consciousness that is unstable, doubtful about general legitimating myths, and is only concerned with immediate sensations and emotions.

On the other hand, there are those who argue from a Durkheimian perspective that we are not postmodern and that the solidarity functions of religion have moved to the sphere of the state. This point of view is in keeping with Durkheim's hope and is championed by Robert Bellah (1970, 1980; Bellah et al., 1985, 1991). Bellah acknowledges the contemporary debate over the cohesiveness of national culture, but it's his position that moral debates over the dysfunctions of institutions are an intrinsic element of the culture of any society. Societies have always argued over the economy, education, religion, and so forth. He makes the case that these debates actually constitute culture. He sees culture as a dynamic conversation or argument, something like a Durkheimian ritual. Thus, for Bellah, the debate over the condition of culture is itself a vehicle through which a kind of collective consciousness is produced.

However, Bellah also argues that religion has evolved in modernity to the point where the dualism between the sacred and the profane that was crucial to past religions and morals has collapsed. The collapse of the sacred is due to the loss of an institutional sphere where collective morals are self-consciously formed—institutional differentiation and size have grown to the point where individuals have lost the ability to collectively inform the general moral culture and have become passive recipients of institutional control.

Bellah decries the results of capitalistic exploitation and the colonization of everyday life and the public sphere by institutions that have failed to live up to their moral obligation. He makes the same point as many postmodernists: capitalist greed and exploding advertising have eroded the collective moral base. According to Bellah et al. (1991), what we need is to "move away from a concern for maximizing private interests" and to move toward "justice in the broadest sense—that is, giving what is due to both persons and the natural environment" (p. 143). This can be accomplished through the cultivation of face-to-face communities where people are "involved in the creation of regional cultures in some degree of harmony with the natural environment, where individuals, families, and local communities could grow in moral and cultural complexity" (p. 265). Bellah's argument clearly echoes Durkheim's concern with the creation of intermediate moral groups and the need to interact in face-to-face situations to produce moral culture.

These interactions will revolve around civil rather than religious moral discourses. Central to these civil moral discourses are "representational characters." Representational characters are a kind of symbol. They are a way through which people define what is good and legitimate in a society. In the United States, Abraham Lincoln, Helen Keller, Eleanor Roosevelt, John Kennedy, Jacqueline Kennedy Onassis, Rosa Parks, and Martin Luther King Jr. are examples of such characters. They have assumed mythic significance and have the ability to generate a degree of social solidarity. Today we might also add such media and sports figures as Oprah Winfrey and Lance Armstrong. What's important about these characters is that they provide us with a focus for intense interactions around central themes in our society. As such, they are the touch point for Durkheimian rituals that produce social solidarity.

Central themes (such as, "America as the chosen people, the champions of freedom") along with representative characters can form a **civil religion.** In civil religion, sacred qualities that in previous times were the custody of religion are attached to certain civil institutional arrangements, historical events, and people. The theme of Bellah's (1975) work is a call to create such a civil religion because "only a new imaginative, religious, moral, and social context for science and technology will make it possible to weather the storms that seem to be closing in on us in the late 20th century" (p. xiv).

It is certainly the case that the United States has a great deal of cultural diversity and institutional differentiation. But are we left without a unifying story or grand narrative that can give us a collective identity? Or are we instead a nation with a civil religion that can bring social solidarity? Or are there layers of each, like Durkheim's continuum of mechanical and organic solidarity? If there are pockets of each, how do they relate to each other? How do they form a collective consciousness? If we are indeed as diverse and differentiated as we seem, what is holding us together? What values do we hold in common? How can large-scale collective projects, such as war, be carried out? Or is the diversity only skin deep? Are we modern or postmodern? I don't know the answers, but these are good Durkheimian (and important) questions.

> For if society lacks the unity that derives from the fact that the relationships between its parts are exactly regulated, that unity resulting from the harmonious articulation of its various functions assured by effective discipline, and if, in addition, society lacks the unity based upon the commitment of men's wills to a common objective, then it is no more than a pile of sand that the least jolt or the slightest puff will suffice to scatter. (Durkheim, 1903/1961, p. 102)

Summary

- Durkheim is extremely interested in what holds society together in modern times. In order to understand this problem, he constructs a perspective that focuses on three issues: social facts, collective consciousness, and the production of culture in interaction. Durkheim argues that society is a social fact, an entity that exists in and of itself, which can have independent effects. The facticity of society is produced through the collective consciousness, which contains collective ideas and sentiments. The collective consciousness is seen as the moral basis of society. Though it may have independent effects, the collective consciousness is produced through social interaction.

- Durkheim argues that the basis of society and the collective consciousness is religion. Religion first emerged in society as small bands of hunter–gatherer groups assembled periodically. During these gatherings, high levels of emotional energy were created through intense interactions. This emotional energy, or effervescence, acted as a contagion and influenced the participants to behave in ways they normally wouldn't. So strong was the effervescent effect that participants felt as if they were in the presence of something larger than themselves as individuals, and the collective consciousness was born. The emotional energy was symbolized and the interactions ritualized so that the experience could be duplicated. The symbols and behaviors became sacred to the group and provided strong moral boundaries and group identity.

- Because of high levels of division of labor, modern society tends to work against the effects of the collective consciousness. People in work-related groups and differentiated structures create particularized cultures. As a result, society has to find a different kind of solidarity than one based on religious or traditional collective consciousness. Organic solidarity integrates a structurally and socially diverse society through interdependency, generalized ideas and sentiments, restitutive law and centralized power, and through intermediary groups. These factors take time to develop, and if a society tries to move too quickly from mechanical to organic solidarity, it will be subject to pathological states, such as anomie and the forced division of labor.

- At the center of modern society is the cult of the individual. The ideal of individuality, not the idiosyncrasies of individual people, becomes one of the most generalized values a society can have. However, the individual can also be subject to pathological states, depending on the person's level of group attachment and behavioral regulation. If a society produces the extremes of either of these, then the suicide rate will tend to go up. Suicide due to extremes in group attachment is characterized as either egoistic or altruistic. Suicide due to extremes in behavioral regulation is characterized as either anomic or fatalistic.

- Durkheim's argument for social solidarity is a moral one. He holds the hope that one day a religious-like grand narrative will once again provide the firm center for society. Some thinkers, like Robert Bellah, agree with Durkheim and argue that civil religion can provide just such a base. However, postmodernists are unsure that such a grand narrative is possible or even desirable. Postmodernists point to the pervasive doubt that has come about through the de-centering of religion, the failure of Western colonialism, the effects of mass media and markets, and the doubt intrinsic within science itself as reasons why a Durkheimian grand narrative is unlikely. Further, grand narratives themselves are based on privileged knowledge: in order to produce a grand narrative, a particular kind of knowledge has to be seen as more important than all others. Grand narratives are thus morally objectionable, under postmodern values.

Building Your Theory Toolbox

Conversations With Émile Durkheim—Web Research

In each of these chapters, I offer a short review of the theorist's background, both socially and intellectually. However, I purposely don't dwell on these issues. These sections on *Conversations With . . .* are your opportunity to get to know our theorists a little better. As a good start, use an Internet search engine, or read through one of the biographies on the recommended Web sites below, to find the answers to the following questions:

- Quite a bit of information can be gleaned from a person's name and childhood. What was Durkheim's full name? Why do you think he changed it? What was Durkheim's early education like? What were his family's expectations about Durkheim's choice of career?
- Education, of course, plays an important role in the life of an academic. In addition to the person's actual intellectual education, we can also pick up insightful information from someone's life experiences at school. How many times did Durkheim have to take his entrance exams? Why did his classmates at *École* call him "the metaphysician"? What can we learn about Durkheim's career plans from his nickname?
- Durkheim was politically involved, and not just in his career as an educator. What was the Dreyfus Affair and how was Durkheim related to it?
- As is the case with most of the people in this book, Durkheim had a profound and broad effect on academic and social thinking. Who was Marcel Mauss? How was he related to Durkheim? What is Mauss considered "the father of"?

Passionate Curiosity

Seeing the World (using the perspective)

- Often in theory class I will take the students on a walk. We walk through campus, through a retail business section, past a church, and through a residential area.

Either think about such a walk or go on an actual walk yourself. Based on Durkheim's perspective (not necessarily his theory), what would he see? How would it be different than what Marx or Spencer would see?

- Given Durkheim's perspective, what do you think ought to be social theory's central questions?

Engaging the World (using the theory)

- I'd like for you to go to a sporting event—football, basketball, or hockey would be best. Analyze that experience using Durkheim's theory of rituals. What kind of symbols did you notice? What kinds of rituals? Did the rituals work as Durkheim said they would? What do you think this says about religious rituals in contemporary society? Can you think of other events that have the same characteristics?
- There are a lot of differences between gangs and medical doctors. But there might also be some similarities. Using Durkheim's theory and perspective, how are gangs and medical doctors alike?
- In the beginning of this chapter, I posed a few Durkheimian questions. Now is the time to try and answer a couple of them. Explain this event and the reactions from a Durkheimian perspective: On September 11, 2001, hijacked jetliners hit the World Trade Center in New York and the Pentagon outside Washington, D.C. News headlines around the world proclaimed "America Attacked" and people in San Diego, CA, Detroit, MI, and Cornville, AZ, wept openly. What Durkheimian processes must have been in place for such an event to happen? And, how would Durkheim explain the fact that Americans had such a strong emotional reaction to the loss of people unknown to them? Also, explain the subsequent use of flags and slogans, and the "war on terrorism," using Durkheim's theory.
- In 1997, Tiger Woods not only won the Golf Masters, he broke several tournament records in the process—youngest to win (21 years old), lowest score for 72 holes, and widest margin of victory. And he was also the first African American to win the Masters. Or was he? Why is this a Durkheimian question? What can Durkheim's theory tell us about the ways through which we construct race?

Weaving the Threads (synthesizing theory)

Durkheim adds to a number of our growing themes. We are also getting to a point where we can begin to compare and contrast what our theorists are saying. In particular, consider the following:

- Spencer gave us a list of requisite needs for society. What does Durkheim give us as the requisite need for society? Does this need add anything new to Spencer's list? Having both Spencer's and Durkheim's lists, do you think anything else is needed? In other words, have Spencer and Durkheim listed everything necessary for society's survival? If not, what needs to be added? Do they list anything you think is *not* necessary for society's survival?

- Both Spencer and Durkheim consider the problem of social and structural integration. What does Durkheim add to our understanding of how societies differentiate and integrate? Based on this combined theory, do you think our society is experiencing more integration or disintegration?
- Think about what Marx and Durkheim say about religion. Marx talks about reification and ideology and Durkheim talks about the facticity (making something fact-like, real, and objective) and the moral core that religion provides for the collective consciousness. It's easy to see the differences, but can you think of any ways these two approaches are similar? Or do you think that they are simply irreconcilable?
- Compare and contrast Spencer and Durkheim on the issue of power and the state. How do they complement one another? How are they different?
- Both Durkheim and Marx are interested in the division of labor. What do they say that is similar? What do they say that it different? Do you think these two theories can be joined to give us a fuller understanding of the division of labor? Do you think that the division of labor is higher, lower, or the same as when Durkheim wrote? Why? What effects would you predict?

Further Explorations—Books

Alexander, J. C. (Ed.). (1988). *Durkheimian sociology: Cultural studies.* Cambridge: Cambridge University Press. (Excellent collection of contemporary readings concerning Durkheim's contribution to cultural sociology)

Giddens, A. (1978). *Émile Durkheim.* New York: Viking Press. (Brief introduction to Durkheim's work by one of contemporary sociology's leading theorists)

Lukes, S. (1972). Émile Durkheim, his life and work: A historical and critical study. New York: Harper & Row. (The definitive book on Durkheim's life and work)

Jones, R. A. (1986). *Emile Durkheim: An introduction to four major works.* Newbury Park, CA: Sage. (Part of the *Masters of Social Theory* series; short, book-length introduction to Durkheim's life and work)

Meštrovic, S. G. (1988). *Émile Durkheim and the reformation of sociology.* Totawa, NJ: Rowman & Littlefield. (Unique treatment of Durkheim's work; emphasizes Durkheim's vision of sociology as a science of morality that could replace religious morals)

Further Explorations—Web Links: Durkheim

http://durkheim.itgo.com (The Durkheim Archives; a site run especially for undergraduates; contains background information, glossary, and quotes)

http://www.relst.uiuc.edu/durkheim/ (The Durkheim Pages; contains reviews, summaries of Durkheim's work, timeline of Durkheim's life, glossary, and discussion threads that everyone can post to)

Further Explorations—Web Links: Intellectual Influences on Durkheim

Charles-Louis de Secondat, Baron de Montesquieu: http://plato.stanford.edu/entries/montesquieu/ (Site maintained by Stanford University; excellent source for Montesquieu's life, philosophy, influences, and published works)

Jean-Jacques Rousseau: http://www.wabash.edu/rousseau/ (Site maintained by the Rousseau Association; contains Rousseau's works, music, pictures from his time, as well as links to other sites)

Auguste Comte: http://en.wikipedia.org/wiki/Auguste_Comte (Site maintained by Wikipedia Internet Encyclopedia; short and concise introduction to Comte with hypertext)

CHAPTER 5

Authority and Rationality—Max Weber (German, 1864–1920)

> *Weber's writings are somewhat schizophrenic. . . . [I]n his voluminous works, one can find almost anything one looks for. There is plenty of material for Parsons' functionalism . . . and also for Schluchter or Habermas's rationalist evolutionism. Weber is a legitimate ally of the symbolic interactionists, as well as an influence upon Alfred Schutz, who in turn influenced social phenomenology and ethnomethodology. On the other hand, modern organization theory and stratification theory could reasonably emerge from Weber's work, and he could influence conflict sociologists . . . all these elements are in Weber.* (Collins, 1986, p. 11)

Max Weber is one of sociology's most intricate thinkers. Part of this complexity is undoubtedly due to the breadth of his knowledge. Weber was a voracious reader with an encyclopedic knowledge and a dedicated workaholic. In addition, Weber was in contact with a vast array of prominent thinkers from diverse disciplines. As Lewis Coser (2003) comments, "In leafing through Weber's pages and notes, one is impressed with the range of men with whom he engaged in intellectual exchanges and realizes the widespread net of relationships Weber established within the academy and across its various disciplinary boundaries" (p. 257). This social network of intellectuals in diverse disciplines helped create a flexible mind with the ability and tendency to take assorted points of view (for a theoretical explanation, see Collins, 1998, pp. 19–79).

Another reason why I think Weber's writings are complex is due to the way he views the world. Weber sees that human beings are animals oriented toward meaning, and meaning, as we've seen, is subjective and not objective. Weber also understands that all humans are oriented toward the world and each other through values. Further, Weber sees the primary level of analysis to be the social action of individuals; for Weber, individual action is social action only insofar as it is meaningfully oriented toward other individuals. Weber sees these meaningful orientations as produced within a unique historical context. Weber's (1949) theoretical questions, then, are oriented toward understanding "on the one hand the relationships and the cultural significance of individual events in their contemporary manifestations and on the other hand the causes of their being historically *so* and not *otherwise*" (p. 72). What this means is that Weber contextualizes individual social action within the historically specific moment. He then asks the question, why does this cultural context exist and not another one? How is it that out of all the possible cultural worlds, this one exists right here, right now, and not a different one?

Weber's perspective, then, is a cultural one that privileges individual social action within a historically specific cultural milieu. This orientation clearly sets him apart from Spencer, Durkheim, and Marx, who were much more structural in their approaches. It also means that Weber's (1949) explanations are far more complex and tentative: "There is no absolutely 'objective' scientific analysis of culture . . . [because] . . . all knowledge of cultural reality . . . is always knowledge from particular points of view" (pp. 72, 81). Yet at the same time, as we will see, Weber believes that we can create objective knowledge. This knowledge is created mainly through ideal types and historical comparisons. Yet, while it is possible to create objective knowledge about how things came to be historically so and not otherwise, Weber was extremely doubtful about prediction. But I'm getting ahead of myself.

My point here is that Weber's thinking is quite complex. And, as a result, Weber probably inspired our thinking in more areas than any other person. If we are going to study religion, bureaucracy, culture, politics, conflict, war, revolution, the subjective experience of the individual, historic trends, knowledge, or the economy, then we have to incorporate Weber. He is also a founding thinker in many distinct schools of sociological thought, such as ethnomethodology, interpretive sociology, geo-political theory, the sociology of organizations, and social constructivism.

But beyond these, there is still much to discover about Weber. All classics exist as such because of their ability to excite thought. For example, I read Durkheim's *Elementary Forms of the Religious Life* about once a year, and every time I do I come away with some new insights. Yet these are really only new twists about things I pretty much already knew. However, Weber's writings are so dense and expansive that when I reread him I come away with entirely new ideas, and some of these ideas are ones that haven't ever been written about. Weber thus remains a wellspring for original sociological thought.

For example, one of the questions that Weber's work has prompted is the problem of *voluntaristic action*: what are the conditions under which people make free choices about behavior in such a way that there is little uncertainty experienced in social life? Talcott Parsons, one of the most prominent social theorists of the twentieth century, took his cue from Weber and argued that human action involves cultural elements such as norms, values, and beliefs; situational factors such as peer pressure; known goals that are informed by both cultural and situational factors; and choices about means and ends that are likewise influenced.

Contemporary theorists, such as Anthony Giddens, see this complexity even more than Parsons. Giddens wants to get rid of the concepts of structure and agency entirely. He feels that they are false dichotomies, created to explain away the complexity of human practice. Giddens (1986) says that, "Human social activities, like some self-reproducing items in nature, are recursive" (p. 2). It's like the chicken and egg question. In some ways, it's a silly question: When you have the egg, you have the chicken. They are one and the same, just in different phases. So, it's a bit like this: social actors produce social reality, but the mere fact that you have "social actors" presumes an already existent social world. For example, you reproduce gender in all your actions, but in doing so, you are also producing the very conditions under which gender can exist at all. If we stop producing gender, the conditions under which gender can exist go away, too. In other words, structure and agency exist in the same moment, in the same actions, and they mutually constitute one another. Neither Giddens nor Parsons represents extensions of Weber's theory *per se*, but they do capture the complexity that troubled Weber. For Weber, there aren't any simple explanations.

Weber in Review

- Max Weber was born on April 21, 1865, in Erfurt, Germany (Prussia). His father, Max, was a typical bourgeois politician of the time; he was a man-about-town, drinking and making deals. His mother, Helene Fallenstein, was a devoutly religious (Calvinist) woman who had been sexually abused as a teen by an older friend of her parents. The elder Max and Helene were ill-suited for one another, which created constant strain in the Weber household. The young Weber was exposed to many of the prominent thinkers and power brokers of the day, as well as his mother's strict Calvinist upbringing.
- Young Weber was a voracious reader, having extensive knowledge of the Greek classics as a young boy and being fluent in such philosophers as Kant, Goethe, and Spinoza before entering college. In 1882, Weber entered the University of Heidelberg, where he studied law. By all accounts, Weber was the typical fraternity member, spending a good deal of time drinking beer and fencing.
- After a year's military service in 1883, Weber returned to school at the University of Berlin. While at Berlin, he completed his dissertation ("The History of Trading Companies in the Middle Ages"), wrote a postdoctoral thesis ("Roman Agrarian History"), and several other works on the plight of East Elbian farm workers, the longest of which was 900 pages. During this time, he also worked as a junior lawyer and lectured at the university.
- In 1893, Weber married Marianne Schnitger, his cousin. No children were born of this union. Weber continued to publish and in 1896 he returned to Heidelberg to become Professor of Economics. While at Heidelberg, Max and Marianne's home became a meeting place for the city's intellectual community. Marianne was active in these meetings, which at times became significant discussions of gender and women's rights. Georg Simmel frequently attended. During this time, Weber worked as professor, lawyer, and public servant.

- In 1897, Weber suffered a complete emotional and mental breakdown. He was unable to work and would sit staring out the window for hours on end. Weber was unable to write again until 1903; he left the university and didn't teach again for almost 20 years.
- During his convalescence, Weber read the works of Wilhelm Dilthey and Heinrich Rickert. Weber became convinced that values and research should be kept separate; and his writings, which had been more historical and legal, took on a sociological point of view. After his breakdown, Weber wrote the majority of the works that he is best known for: his methodological writings date from this period (now found in *The Methodology of the Social Sciences*), as well as *The Protestant Ethic and the Spirit of Capitalism, The Sociology of Religion, The Religion of China, The Religion of India,* and *Ancient Judaism.* (Several of these weren't finished until later, but were begun during this time.)
- During WWI, Weber administered a hospital in Heidelberg. As a German nationalist, Weber initially supported the war effort but eventually became critical of it.
- In 1918, Weber accepted a position at the University of Vienna, where he once again began to teach. He started working on *Economy and Society,* which was to be the definitive outline of interpretive sociology. His *General Economic History* came from lectures given during this time.
- On June 14, 1920, Max Weber died of pneumonia.

The Perspective: Complex Sociology

> An "objective" analysis of cultural events, which proceeds according to the thesis that the ideal of science is the reduction of empirical reality of "laws," is meaningless. . . . It is meaningless . . . because the knowledge of social laws is not knowledge of social reality but is rather one of the various aids used by our minds for attaining this end. (Weber, 1949, p. 80)

> Only a small portion of existing concrete reality is colored by our value-conditioned interest and it alone is significant to us. It is significant because it reveals relationships which are important to us due to their connection with our values. (Weber, 1949, p. 76)

Problems with social science: Much of Weber's concern about social science finds its roots in the idea of **culture.** Weber takes seriously the notion of culture. He doesn't see it as an epiphenomenon as Marx did, nor does he see it as the most important requisite function like Durkheim, nor is he captivated by the tension between objective and subjective culture as was Simmel. Rather, Weber sees culture as a historical process that at times leads social change and at others simply reinforces it. Culture for Weber (1949) is a value concept: "Empirical reality becomes 'culture' to us because and insofar as we relate it to value ideas" (p. 76). He sees culture as creating intrinsic difficulties for a scientific sociology, as introducing complexity around the issues of stratification and oppression, and as always influencing the subjective value orientation of social actors.

Weber's view of culture is not determinative—he doesn't see culture as determining human action. According to Weber, people are very much motivated by economic and cultural interests. But culture can act like a switch on railroad tracks and actually change the course of the train: "Not ideas, but material and ideal interests directly govern men's conduct. Yet very frequently the 'world-images' that have been created by 'ideas' have, like switchmen, determined the tracks along which action has been pushed by the dynamic of interest" (Weber, 1948, p. 280).

In all his writings, Weber is interested in explaining the relationships among cultural values and beliefs (generally expressed in religion), social structure (overwhelmingly informed by the economy), and the psychological orientations of the actors. None of these elements is particularly determinative for Weber. One of the tasks, then, of Weberian sociology is to *historically explain* which factor is more critical at any given time and why, rather than predicting an outcome. So, for example, we could never have truly predicted what *African American* would mean in 2004 based on what *Negro* meant in 1950, but we can explain the historical, social, and cultural processes through which it came about. That being the case, the kind of knowledge that Weberians construct about the world is decidedly different than that proposed by the general scientific method.

Weber also sees another culture-based problem for researchers. He recognizes that to ask a question about society or humans is itself a cultural act: it requires us to place value on something. In other words, for us to even see a problem to study, we must have a value that helps us to see it. For example, it would have been almost impossible for us to study spousal abuse 300 years ago (it would have been difficult even 50 years ago). It isn't that the behaviors weren't present; it is simply that the culture would not have allowed us to define them as abusive, at least not very easily. And the same is true about everything social scientists study (and laboratory scientists, too, for that matter). Humans can ask questions only insofar as they have a culture for it, and culture is a value orientation toward the world.

So, if scientific knowledge is defined as being empirical and non-evaluative, then you can see why creating a social science might be a problem. Human reality is meaningful, not empirical; it is historical and thus concerned with unique configurations of values, and all the questions we ask are strongly informed by our culture and thus are value-laden. However, Weber is also convinced that a social science is possible, but there are certain caveats. Because human existence is a subjective one, creating an objective science about people is difficult. Knowledge about people must be based upon an interpretation of their subjective experience. And because people are self-aware free agents, the law-like principles that science wants to discover are provisional and probabilistic at best (people can always decide to act otherwise). So the kind of knowledge that we produce about people will be different than that produced in the laboratory, though Weber feels that objective knowledge is still possible. One key in creating this objective-like knowledge is that social scientists have to be reflexively and critically aware of their values in forming and researching their questions.

Creating objective knowledge: Weber argues that knowledge about humans in society can be made object-like through the use of **ideal types.** Ideal types are analytical constructs that don't exist anywhere in the real world. They simply provide a logical touchstone to which we can compare empirical data. Ideal types act like a yardstick against which we can measure differences in the social world. These types provide objective measurement because they exist outside the historical contingency of the data we are looking at. According to Weber, without the use of some objective measure, all we can know about humans would be subjective.

There are two main kinds of ideal types: historical and classificatory. Historical ideal types are built up from past events into a rational form. In other words, the researcher examines past examples of whatever phenomenon he or she is interested in and then deduces some logical characteristics. Weber uses this form in *The Protestant Ethic and the Spirit of Capitalism.* In that work, Weber constructs an ideal type of capitalism in order to show that the capitalism in the West is historically unique. Classificatory ideal types, on the other hand, are built up from logical speculation. Here the researcher asks him- or herself, what are the *logically possible* kinds of ______________ (fill in the research interest)? Weber uses this approach in forming his ideal typology of action.

Remember that his intent is to create an objective measure that exists outside of the content. Weber asked himself, what are the possible different kinds of human action? He came up with four. In Figure 5.1, these ideal types are pictured as a block to which the variety of complex behaviors can be compared. When we use this ideal typology, we are able to offer an objective explanation of human behavior.

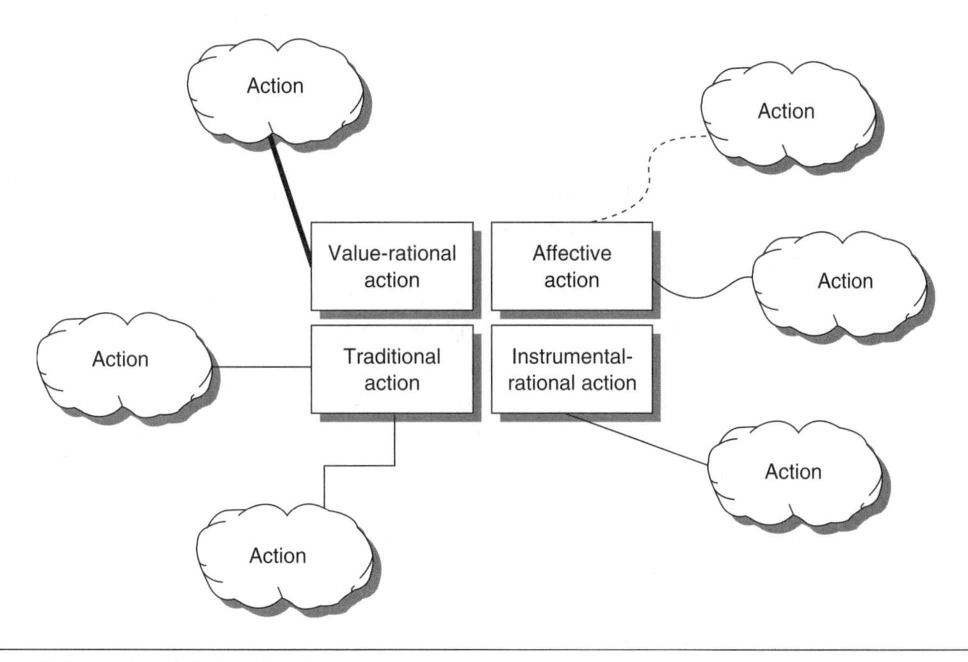

Figure 5.1 Social Action Typology

Instrumental-rational action is behavior in which the means and ends of action are rationally related to each other. So, your action in coming to the university is instrumental-rational in that you see it as a logical means to achieve an "end," that is, a good job or career. **Value-rational** action behavior is that is based upon one's values or morals. If there is no way you could get caught paying someone to write your term paper for you, then it would be instrumentally rational for you to do so. It would be the easiest way to achieve a desired end. However, if you don't do that because you think that it is dishonest, then your behavior is being guided by values or morals and is value-rational. **Traditional action** is action that is determined or motivated by habit, and **affectual action** is determined by people's emotions in a given situation. All of these different types of behavior can become social action insofar as we take into account the behaviors and subjective orientations of other actors, whether present or absent.

We can tease out an important element in Weber's thinking from this ideal typology. Value-rational behavior is distinguished from affective action because Weber sees some semblance of self-aware decision-making processes in value-rationality. Values are emotional, but when we consciously decided to act because of the logic of our values, it is rational behavior. Emotion, on the other hand, is irrational and proceeds simply from the feelings that an actor may have in a given situation. According to Weber (1922/1968), both affectual and traditional behaviors lay "very close to the borderline of what can justifiably be called meaningfully oriented action" (p. 25), because they are not based on explicit decisions.

In his methodology, Weber also emphasizes understanding of the subjective meanings of the actions to the actors by contextualizing it in some way. Weber advocates the use of ***Verstehen***, the German word meaning "to understand." It is important to note that when Weber talks about meaning in this context, he has in mind the motivations of the actor. These motives may be intellectual in the sense that the actor has an observable and rational motive for his or her actions in terms of means and ends; or they may be emotional in the sense that the behavior may be understood in terms of being motivated by some underlying feeling like anger. So we can understand it when Sam hits John if we know that Sam is angry with John for cheating him in a business deal—the meaning of the action comes out of our knowledge of the motivation.

Weberian sociologists, then, will see the world in terms of ideal types (abstract categorical schemes), broad historical and cultural trends, or from the point of view of the situated subject (interpretive sociology). Weber's causal explanations have to do with understanding how and why a particular set of historical and cultural circumstances came together, and his general explanation is always subject to case-specific variations. There are a couple of things that this implies.

First, Weber rarely gives us highly specified relationships among his concepts. He is more concerned with providing the historical preconditions for any phenomenon. Thus, it isn't as easy to create a dynamic model of Weber's theory as it is for Durkheim's, nor is it as true to the theorist's thinking to do so. What we *can* do is create a picture of the general historical processes that tend to produce an environment conducive to whatever issue it is we are trying to explain. The second implication of this Weberian approach is that the general theory will hold if and only if there are no mitigating circumstances. In other words, Weber will give us a general theory, but it will only hold if nothing gets in the way. So, with religion for example, Weber may give us a general account of how ethnical monotheism evolved from magic, but we can see histories where it didn't work the way the theory implies. That is one of the frustrating things about reading Weber's *The Sociology of Religion*; yet, it is also what makes the account more accurate than many simpler approaches.

Rationalization: The historical and cultural trends that interest Weber the most, and continue to be a focus of Weberian sociology, concern the broad sweeping movement toward rationalization and rational-legal legitimation. Weber argues that one of the prime forces bringing about modernity is the process of rationalization. He uses the word **rationalization** in at least three different ways: He uses it to talk about means–ends calculation, in which rationality is individual and specific. Rational action is action based on the most efficient means to achieve a given end. Secondly, Weber uses the term to talk about bureaucracies. The bureaucratic form is a method of organizing human behavior across time and space. Initially we used kinship to organize our behaviors, using the ideas of extended family, lineages, clans, moieties, and so forth. But as the contours of society changed, so did our method of organizing. Bureaucracy is a more rational form of organization than the traditional and emotive kinship system.

> The fate of our times is characterized by rationalization and intellectualization and, above all, by the 'disenchantment of the world.' (Weber, 1948, p. 155)

Finally, Weber uses the term *rationalization* in a more general sense. One way to think about it is to see rationalization as the opposite of enchantment. Specifically, an enchanted world is one filled with mystery and magic. **Disenchantment,** then, refers to the process of emptying the world of magical or spiritual forces. Part of this, of course, is in the religious sense of secularization. Peter Berger (1967) provides us with a good definition of *secularization*: "By secularization we mean the process by which sectors of society and culture are removed from the domination of religious institutions and symbols" (p. 107). Thus, both secularization and disenchantment refer to the narrowing of the religious or spiritual elements of the world. If we think about the world of magic or primitive religion, one filled with multiple layers of energies, spirits, demons, and gods, then in a very real way the world has been subjected to secularization from the beginning of religion. The number of spiritual entities has steadily declined from many, many gods to one; and the presence of a god has been removed from immediately available within every force (think of the gods of thunder, harvest, and so on) to completely divorced from the physical world, existing apart from time (eternal) and space (infinite). In our more recent past, secularization, and demystification and rationalization, have of course been carried further by science and capitalism.

This general process of rationalization and demystification extends beyond the realm of religion. Because of the prominence of bureaucracy, means–ends calculation, science, secularization, and so forth, our world is emptier. Weber sees this move toward rationalization as historically unavoidable; it is above all else the defining feature of modernity. Yet it leads inexorably to an empty society. The organizational, intellectual, and cultural movements toward rationality have emptied the world of emotion, mystery, tradition, and affective human ties. We increasingly relate to our world through economic calculation, impersonal relations, and expert knowledge. Weber (1948) tells us that as a result of rationalization the "most sublime values have retreated from public life" and that the spirit "which in former times swept through the great communities like a firebrand, welding them together" is gone (p. 155). Weber sees this not only as a condition of the religious or political institutions in society, he also sees the creative arts, like music and painting, as having lost their creative spirit as well. Even our food is subject to rationalization, whether it is the McDonaldized experience (Ritzer, 2004a) or the steak dinner that is subjected to the "fact" that it contains in excess of 2000 calories and 100 gram of fat. Thus, for Weber, the process of modernization brings with it a stark and barren world culture.

Legitimation: Another issue arises from seeing the social world culturally: legitimation. When you see things through a cultural lens, as does Weber, you realize that for society or a social structure to work, people have to *believe in* it. **Legitimation** refers to the process by which power is not only institutionalized but more importantly is given moral grounding. Legitimations contain discourses or stories that we tell ourselves that make a social structure appear valid and acceptable. The strongest legitimations will make social structure appear inevitable and beyond human control (for example, the essentialist arguments surrounding gender—women cannot think logically and men cannot nurture children because of the nature of their sex). Weber argues that all oppressive structures, and, in fact, all uses of power, must exist within a legitimated order.

A legitimated order creates a unified worldview and is based on a complex mixture of two kinds of legitimations: subjective (internalized ethical and religious norms) and objective (having the possibility of enforced sanctions from the social group [conventions] or an organizational staff [law]). Weber indicates that subjective legitimacy is assumed in the presence of the objective. Underlying both subjective and objective legitimacy are three different kinds of belief systems or authority (charismatic, traditional, and rational-legal). Legitimacy works only because people believe in the rightness of the system. So, for example, your professor tells you that you will be taking a test in two weeks. And in two weeks you show up to take the test. No one has to force you; you simply do it because you believe in the right of the professor to give tests. And that's Weber's point: social structures can function because of belief in a cultural system.

One particularly interesting type of legitimation that Weber mentions is theodicy. A *theodicy* is a vindication or legitimation of the justice and love of God in the face of suffering. The meaning of suffering or evil becomes a problem as people move from magic to rational religion. Magic is the direct manipulation of physical objects to bring about a desired end. Under a magical system, suffering exists simply because the proper ritual or incantation hasn't yet been found to counteract it. Suffering doesn't require an explanation *per se;* it is a natural part of the world as it exists. The only issue is its possible removal by effective practice. However, once human beings posit a loving, all-powerful God, the meaning of suffering becomes problematic, particularly when the good suffer and the evil prosper. Under a purely magical system, good and evil don't have the same meanings as under an ethical god. Under ethical monotheism, the good are obviously on God's side, so why do they suffer? The explanations of this suffering are the legitimations of theodicy (for example, God allows suffering in order to purify our soul so that we can gain eternal rewards).

To think like Weber is to take seriously the complexities created by people living at the intersection of culture and structure. To think like Weber is to understand social action as culturally defined through different subjective value orientations of the actors. But it also means seeing those subjective orientations as the result of historical processes. Every person acts within a cultural context that is historically specific. This emphasis on history and the individual is what C. Wright Mills would later term the *sociological imagination.* Weberian methodology, then, places emphasis on comparative historical analysis and *verstehen* through the use of ideal types to reduce complexity. To think like Weber also means seeing the modern world becoming more rationalized and disenchanted. And, finally, to think like Weber implies that we understand that belief in legitimating systems is in back of every social structure and use of power. In general, we might say that Weber provides us with a cultural sociology. In the way I'm using it here, cultural sociology is as much an approach to sociology as it is a particular subject matter (culture) within sociology. That is, it's as much a way of thinking about social life as it is a thing to study. This approach is something that we might term "the cultural imagination." While Weber spent little of his time studying culture as such, he did see everything social and historical through a cultural lens.

I have divided the material on Weber into four main sections. Throughout all four divisions, Weber's concern for cultural sociology is a binding thread. I begin our discussion of Weber's theory with religion because he, like many other theorists, sees religion as core to society. As we've noted before, humans are linked primarily through symbols, and the strongest symbols we have are religious. "[T]here is no communal activity . . . without its special god. Indeed, if an association is to be permanently guaranteed, it must have such a god" (Weber, 1922/1993, p. 14).

Weber is particularly interested in the interplay between the culture that religion produces and social structures. In the first section on Weber's theory, *The Evolution of Religion,* he shows us that changes in social structure brought about changes in religion. Yet in the section, *The Rise of Capitalism: Religion and States,* he reverses that influence. We see that Weber argues that religion (in addition to the state) actually paved the way for the rise of capitalism. In *Class, Authority, and Social Change,* we move from a focus on religion to cultural processes in general. Weber argues that structural inequality is a complex issue, and that cultural legitimation and authority are key issues in bringing about social change. Finally, *Rationality in Action* addresses the issue of social organization through bureaucratic structures. We will see that using bureaucracies as our chief organizing structure yields some very interesting social effects, some of them cultural.

The Evolution of Religion

Evolution versus revelation: Historically, society in general has moved from magic to polytheism, to pantheism, to monotheism, to ethical monotheism. There are two basic approaches to explaining this movement: progressive revelation and social evolution. Progressive revelation argues that there has always been one god who is concerned with the way we behave. But because of our inability to receive the full revelation of God, he (and these gods are always male) had to progressively reveal himself to us. This perspective also posits a god who is bound to human history. In traditional Christianity, for example, humankind had to reach bottom through the revelation of the law before it was possible for God to reveal grace. Social evolution, on the other hand, posits a link between religion and society. Religion changes under this model because society changes. Religion and society are locked in a kind of reciprocal relationship. As one changes, it brings about changes in the other, because each is necessary for the other's survival.

Weber gives us a social evolution model of religious change. However, these two models are not necessarily exclusionary. It is possible that God works through history. And it is possible that part of humankind's preparedness to receive God's revelation is linked to society, since humans are social beings. The reason I say this is to let you know that you aren't faced with an either/or decision. If you are religious, you don't have to reject your religion in order to see value in Weber's theory. Weber is only concerned with the empirical elements of religious change. We can either believe that there are spiritual forces in back of those changes or not. In either case, our beliefs are based on assumptions we make about how the universe works. And, as is always the case, assumptions are never tested or proven.

From magic to religion: Weber is interested in explaining how ethical monotheism came about. Monotheism, of course, is the belief in one god. *Ethical monotheism* is the belief in one god that cares about human behavior. In terms of history, such a god is a recent occurrence. For much of our history, we believed in many gods and goddesses, and we didn't feel that they much cared about how we behaved. What's more, most of the gods and goddesses were quite risqué in their own actions. Weber's approach is to look through history and see what kinds of social factors were associated with changes in the way people thought about God.

Weber first recognizes that the movement from magic to religion is the same as the change from naturism to symbolism. **Magic** is the direct manipulation of forces. These forces are seen as being almost synonymous with nature. So to assure a good harvest, a magician might perform a fertility ritual, because seeds and fertility are all wrapped up together. *Religion,* on the other hand, is more symbolic. Let's take Christian communion, for example. In Protestant Churches, the bread and grape juice *symbolize* the body and blood of Christ. They *represent* not only the atonement, but also the solidarity believers have in sharing the same body. The Catholic example is a bit more interesting. While there is quite a bit of symbolism in the Eucharist, there is also what is known as *transubstantiation.* In transubstantiation, the wine and bread *literally become* the blood and body of Christ. Taking the sacrament *brings about* union with Christ. Through it, venial sin and punishment are remitted.

The reason I compare the Protestant and Catholic versions is to bring out this point of Weber's. In Protestant communion, there are only symbolic elements present. In the Catholic Eucharist, there are actual elements present that are effectual in their own right. Weber would argue, then, that there are magical elements present in the Eucharist. To say something like this isn't meant as a slight against Catholicism (or a slight against Protestantism, depending on your perspective). It is simply meant to point out that these things work in different ways. With the Eucharist, there is a direct manipulation of forces to accomplish some purpose, which is Weber's definition of magic. We can also see the complexity of the social world in this example. Weber argues that *in general,* there is a movement from magic to religion, from physical manipulation to symbols. However, real life is more complex. Catholicism shows us that due to a religion's specific history, its may contain both symbolic and magical elements.

Generally speaking, humanity moved from magic to religion due to economic stability and professionalization. To understand these effects, we can picture a small hunter–gatherer society. In such a group, everyone lives communally, the division of labor is low, and life is uncertain. The tribe is subject to the whims of nature: game may or may not be there and the weather may or may not have been good enough for the berries to grow. In an attempt to make their world more stable, they perform rituals before the hunt and at season changes. They perform rituals at childbirth or when they need to find water. These rituals constitute magical manipulations, and the people performing the rituals would be the same who are involved in the actual work. For example, we can imagine the men gathering before a hunt, perhaps painting themselves with animal blood or putting on masks and acting out the hunt.

Soon our tribe learns about horticulture and they plant food. They become sedentary and tied to the land, and their life becomes a bit more predictable. After some time, they learn how to use metal and how to plow the land, and they learn how to irrigate. From these small technological advances come surplus, population growth, power, and a different kind of division of labor. Rather than living communally, certain people are able to work at specialized jobs. Thus, the uncertainty of life is further diminished.

What we want to glean from this little story is how the type of economy can influence the development of religion. Early in our story, many people were involved in the practice of magic. Because people were tied to nature, their view of how things work was naturalistic and not symbolic. They were concerned with day-to-day existence: what mattered was getting food, water, and shelter. It was necessary for them to have a method of control that each person could use.

But as life became more predictable, people did not have to be as concerned with immediate sustenance, and tasks became more specialized. Some people tilled the soil; others planted the seeds. Some were in charge of the irrigation; others transported the harvested grain to the grinders. And those who had a knack for it became what Weber characterizes as the "oldest of all vocations," professional necromancers (magicians).

Three things happened as the result of the *professionalization* of magic: individuals could devote all their time to experiencing it, they increased their knowledge surrounding the experience, and they acquired vested interests (their livelihood became dependent upon an esoteric knowledge). The experience factor is important in this case. Weber argues that ecstatic experiences are the prototypical religious experience. Religion always involves transcendence, the act of going above or outside of normal life, and ecstasy is a primal form of transcendence. Like Durkheim, Weber (1922/1993) argues that "ecstasy occurs in a social form, the orgy, which is the primordial form of communal religious association" (p. 3). For laymen, the experience is only available on occasion, but because the professional wizard is freed from daily concerns, she or he can develop practices that induce ecstatic states almost continually.

These experiences, along with increasingly available time, prompted the professional to develop complex belief systems. As a result, what had once been seen as something everyone could practice (magic) became part of secret lore and available to only a select few. Because the professional magician could afford to engage in continual transcendent experiences, he saw more behind the world than did others. The world thus became less empirical and more filled with transcendent beings, spirits that lived outside of and controlled daily life. The magician developed belief and ritual systems that reflected this growing complexity. In addition, professionals could consider their own thoughts and beliefs, and the very act of reflexive thought will tend to make ideas more abstract and complex. Reflexive thought by its nature is abstract because it isn't thinking about anything concrete. And because it isn't thinking about anything concrete, it can go anywhere it wants. Further, the connections among ideas that are "discovered" through this kind of thinking tend to be systematized around abstract ideas.

Let's think about sociology as an example of professionalization. In a very real way, everybody is a sociologist. We all have an understanding about how society works and what is involved in getting around it. That's one of the funny things about teaching an Introduction to Sociology course—most of the students have a "oh, I knew that" response to the stuff we talk about. But what do *professional* sociologists do? Well, because we get paid to do nothing but sit around and think about society, we have made things very abstract and complex. We use big words (like *dramaturgy* and *impression management*) to talk about pretty mundane things (picking out clothes to wear). And we've developed these ideas into some pretty complex theories that take years of education to understand.

Further, think about the book you are reading right now. It is the result of my thinking about other people's thinking (Weber's, for example). But it's really more than that. This chapter could not have been written right after Weber. No, what we have here is not only my reading of Weber, but my reading of other people's reading of Weber as well. And because I get paid to do this kind of thing, I can weave all this complexity into a systemic whole. More than that—and this is an important point for Weber—we have made it so that you can't understand sociology without the help of a professional. You had to pay to get this book, and you had to pay to get another professional to stand up in front of the class and explain this explanation of Weber. Thus, professionalization pushes for symbolic complexity because of practice, reflexive thinking, and vested interests (I have to make sociology complex to protect my job). This kind of process is exactly what created symbolic religion.

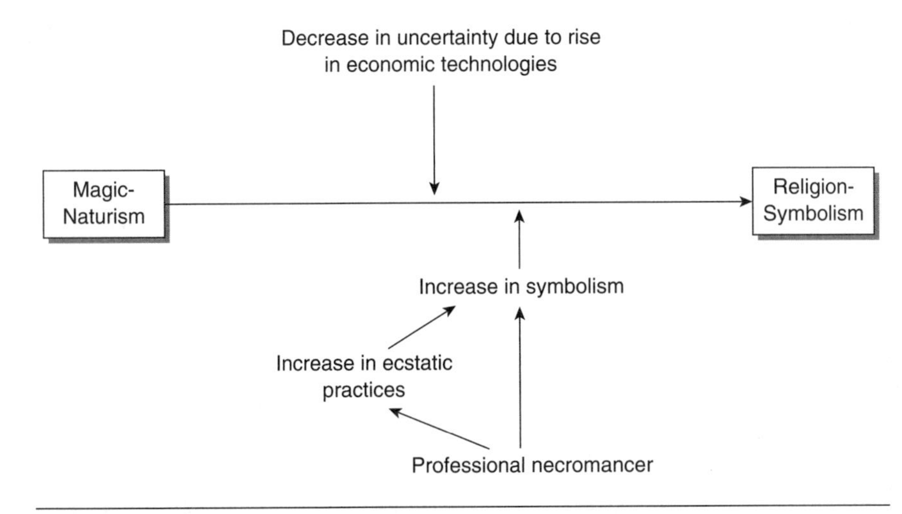

Figure 5.2 Evolution to Symbolism

I bring the ideas of professionalization and economic changes together in Figure 5.2. As technology allows people to be less tied by direct relations to the environment, they are less dependent upon magic for manipulating nature. As the economy produces surplus and creates more complex divisions of labor, necromancers can be relieved of other duties and paid simply to practice magic—they become professionals. These two factors work together to increase the level of abstract symbolism, which in turn moves magic toward religion.

From polytheism to ethical monotheism: In the beginning, religious practice was oriented around local gods and goddesses. These gods lived in specific locations and were connected to specific collectives. The next step in religious evolution was to place these local gods into organized pantheons. Organized pantheons are not simply clusters of various gods and goddess. Rather, in a pantheon, each deity is given a specific sphere of influence, and the activities and deities are related to one another.

Professionalization and symbolism play important roles in the religious evolution to pantheons. Professionals affected religion primarily through increasing the level of abstraction (analogous thinking, religious stereotyping, and the use of symbols in ritual rather than actual things—Weber points out that the oldest use of paper money was to pay off the dead, not the living) and creating coherent systems of knowledge out of localized beliefs. In short, as religion became more professionalized, it tended to become more rationally and abstractly organized.

At this point in our evolution, politics has come to play a very important role as well. One of the things the idea of God does is unite different groups into a single community. A kin- or tribal-based god provides the symbolic ties that small groups need, but at some point in our history we began to bring these different collectives together through conquest or voluntary association. Different groups, each with its own god, began to form larger collectives. In order to link these groups, a more abstract and powerful god was needed; one that could be seen above all other gods,

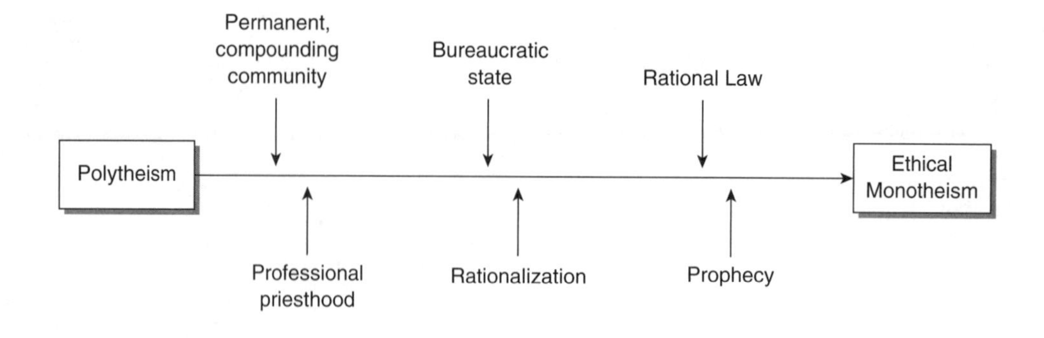

Figure 5.3 Polytheism to Ethical Monotheism

thus linking all the people. The gods that were highest on the polytheistic hierarchy began to be seen as less and less attached to the earth and more as part of the heavens. As populations with even greater diversity came together, more abstract symbols were needed to link them. These universal gods became seen as more powerful than all other gods. They became the god of gods and lord of lords.

This development toward monotheism was aided by the need of monarchs to break the hold that the priesthood had on the populace through the "multiplicity of sacerdotal gods" (*sacerdotal* meaning "pertaining to priests"). As long as there was a multiplicity of gods still linked to the daily needs of the people, there would be need for specific rites to approach those gods. In order to consolidate power, it was necessary for the king to eliminate those various paths, because each one represented a power that he didn't control. Monarchs thus began to monetarily and politically privilege the universal god and his priests. And the idea of the one god was thus born.

Yet one more aspect needs to be added here: morality. The gods had typically never been concerned with the behaviors of their cults. Weber argues that ethical monotheism came about in response to the increasing rationalization of the state and the social control of human behavior. Because of the changes in the way government was organized, from traditional to rational-legal authority, a culture developed that proclaimed that individual behaviors should be controlled. Of course, this was in response to the needs of a large population under a centralized state: the larger and more diverse a population, the greater the need for rational, centralized control. And, according to Weber (1922/1993), this need was reflected in religion: as the state became more interested in the actions of the populace, so did the ruling god. "[T]he personal, transcendental and ethical god is a Near-Eastern concept. It corresponds so closely to that of an all-powerful mundane king with his rational bureaucratic regime that a causal connection can scarcely be overlooked" (p. 56).

In Figure 5.3, we still see the influence of professionalization. This remains an important factor in all areas of modern life. But rather than symbolization, rationalization now plays an important role in the movement toward an ethical god. As state governments become more and more bureaucratized and rational law becomes the way in which relationships are managed, people begin to see their lives as subject to rational control. Of course, today we feel this extended to almost every area of our lives. We not only sense that our bodies, emotions, and minds *can* be rationally controlled, we believe they *should* be controlled.

Religion contributed to this belief through the role of the prophet. Weber categorizes prophecy as two kinds: exemplary and ethical. The exemplary type is found in India and other Eastern civilizations. This kind of prophet *shows* the way by being an example. The emphasis in this case is on a kind of self-actualization. There is no real sense of right and wrong, but only of a better way. And neither is there a god to whom the individual or collective owes allegiance. The Buddha is a good example of this. The ethical prophet, on the other hand, is vitally concerned with good and evil and with bringing a wayward people back to the right relationship with the dominant god. Just as a centralized state needs and creates a single national identity and story, so the ethical prophet presents the cosmos as a "meaningful, ordered totality" (Weber, 1922/1993, p. 59). Thus, God and the state come together to rationally control human behavior.

The Rise of Capitalism: Religion and States

For Weber, there are three main factors that influenced the rise of capitalism as an economic form: religion, nation-states, and transportation and communication technologies. But, as with most of Weber's work, it is difficult to disengage their effects. There is not only quite a bit of overlap when he talks about these issues, he is also interested in explicating the pre-conditions for capitalism rather than determining a causal sequence. So, what we have are a number of social factors that overlap to create the bedrock out of which capitalism could spring, but did not necessarily cause capitalism. Picking up from our last section, we'll begin with religion.

The religious culture of capitalism: The Protestant Ethic and the Spirit of Capitalism is probably Weber's best-known work. It is a clear example of his methodology. In it, he describes an ideal type of spirit of capitalism, he performs an historical-comparative analysis to determine how and when that kind of capitalism came to exist, and he uses the concept of *verstehen* to understand the subjective orientation and motivation of the actors. Weber had three interrelated reasons for writing the book. First, he wanted to counter Marx's argument concerning the rise of capitalism—Weber characterizes Marx's historical materialism as "naïve." The second reason is very closely linked to the first: Weber wanted to argue *against* brute structural force and argue *for* the effect that cultural values could have on social action.

> [T]he capitalist economic order of today is a vast cosmos into which a person is born. It simply exists, to each person, as a factually unalterable casing (*unabänderliches Gehäuse*) in which he or she must live. (Weber, 1904–1905/2002, p. 18)

The third reason that Weber wrote *The Protestant Ethic* was to explain why rational capitalism had risen in the West and nowhere else. Capitalism had been practiced previously. But it was traditional, not rational capitalism. In **traditional capitalism**, traditional values and status positions still held; the elite would invest but would spend as little time and effort doing so in order to live as they were "accustomed to live." In other words, the elite invested in capitalistic ventures in order to maintain their lifestyle. It was, in fact, the existence of traditional values and status positions that prevented the rise of rational capitalism in some places. **Rational capitalism,** on the other hand, is practiced to increase wealth for its own sake and is based on utilitarian social relations.

Weber's argument is that there are certain features of Western culture that set it apart from any other system, thus allowing capitalism to emerge. As we talk about this culture, it is important to keep in mind that Weber is describing the culture of capitalism in its beginning stages. In many ways, the United States is now experiencing a form of late capitalism. And some of the spirit that Weber is describing has been lost to one degree or another. Also keep in mind that what he describes is an ideal type.

Weber's first task was to define the ***spirit of capitalism.*** The first thing I want you to notice is the word *spirit.* Weber is concerned with showing that a particular cultural milieu or mindset is required for rational capitalism to develop. This culture or mindset is morally infused: the spirit of capitalism exists as "an *ethically*-oriented maxim for the organization of life" (Weber, 1904–1905/2002, p. 16). This culture, then, has a sense of duty about it, and its individual components are seen as virtues.

> People do not wish "by nature" to earn more and more money. Instead, they wish simply to live, and to live as they have been accustomed and to earn as much as is required to do so. (Weber 1904–1905/2002, p. 23)

As the above quote indicates, modern capitalism contains principles for the way in which people organize or live out their lives. For example, you woke up this morning and you will go to sleep tonight. In between waking and sleeping you will live your life, but how will you use that time? From an even broader position, you were born and you will die. What will you do with your life? Does it matter? According to Weber, under capitalism, it does matter. How you live your life is not simply a matter of individual concern; the culture of modern capitalism provides us with certain principles, values, maxims, and morals that act as guideposts telling us how to live. In my reading of Weber, I see three such prescriptions in the spirit of capitalism.

Weber begins his consideration of the spirit of capitalism with a lengthy quote from Benjamin Franklin:

> Remember, that time is money . . . that credit is money . . . that money is of the prolific, generating nature. . . . After industry and frugality, nothing contributes more to the raising of a young man in the world than punctuality and justice in all his dealings. . . . The sound of your hammer at five in the morning, or eight at night, heard by a creditor, makes him easy six months longer. . . . [Keep] an exact account for some time, both of your expenses and your income. (Franklin, as quoted in Weber, 1904–1905/2002, pp. 14–15)

From these sayings, Weber gleans the first maxim of rational capitalism: life is to be lived with a specific goal in mind. That is, it is good and moral to be honest, trustworthy, frugal, organized, and rational because it is *useful* for a specific end: making money, which has its own end, "the acquisition of money, and more and more money." The culture of modern capitalism says that money is to be made but not to be enjoyed. Immediate gratification and spontaneous enjoyment are to be put off so that money can be "rationally used." That is, it is invested to earn more money. The making of money then becomes an end in itself and the purpose of life.

The second prescription is that each of us should have a vocational calling. Of course, another word for vocation is job, but Weber isn't simply saying that we should each have a job or career. The emphasis is on our attitude toward our vocation, or the way in which we carry out our work. There are two important demands to this attitude. First is that we are obligated to pursue work: we have a *duty* to work. In the spirit of capitalism, work is valued in and of itself. People have always worked. But generally speaking, we work to achieve an end. The spirit of capitalism, however, exalts work as a moral attribute. We talk about this in terms of a work ethic, and we characterize people as having a strong or weak work ethic. We can further see the moral underpinning when we consider that the opposite of a strong work ethic is being lazy. We still see laziness as a character flaw today. In the culture of capitalism, work becomes an end in itself, rather than a means to an end. More than that, it becomes the central feature of one's life, overshadowing other areas such as family, community, and leisure.

> The development of the concept of the calling quickly gave to the modern entrepreneur a fabulously clear conscience—and also industrious workers; he gave to his employees as the wages of their ascetic devotion to the calling and of co-operation in his ruthless exploitation of them through capitalism the prospect of eternal salvation. (Weber, 1923/1961, p. 269)

We not only have duty to work, we also have duty *within* work. Weber (1904–1905/2002) says that "*competence and proficiency* is the actual alpha and omega" of the spirit of capitalism (p. 18). Notice the religious reference: in the New Testament, Jesus is referred to as the alpha and omega. Weber is again emphasizing that this way of thinking about work is a moral issue. Our duty within work is to organize our lives "according to *scientific* vantage points." That is, under the culture of capitalism, we are morally obligated to live our lives rationally.

The rationally organized life is one that is not lived spontaneously. Rather, all actions are seen as stepping stones that bring us closer to explicit and valued goals. Let me give you a contrary example to bring this home. A number of years ago, a (non-Hawaiian) friend of mine managed a condominium complex on the island of Maui. In the interest of political and cultural sensitivity, he hired an all-Hawaiian crew to do some construction. He tells of the frustrations of having a crew come to work whenever they got up, rather than at the prescribed 8 AM. What's more, periodically during the day the crew would leave at the shout of "surf's up!" The work got done and it was quality work, but the Hawaiians that he supervised did not organize their lives rationally. They lived and valued a more spontaneous and playful life. In contrast, most of us have Day Runners and Palm Pilots that guide our life and tell us when every task is to be performed in order to reach our lifetime goals.

The third prescription or value of the spirit of capitalism, according to Weber, is that life and actions are legitimized "on the basis of strictly *quantitative* calculations." Weber makes an interesting point with regard to legitimation or rationalization—humans can rationalize their behaviors from a variety of ultimate vantage points. And we always *do* legitimize our behaviors. We can all tell stories about why our behaviors or feelings or prejudices are right. And those stories can be told from various religious, political, or personal perspectives. Weber's point here is that the culture of capitalism values quantitative legitimations. That is, capitalist behaviors are legitimized in terms of bottom-line or efficiency calculations. So, for example, *Roger and Me,* a film by Michael Moore, depicts the closing of the General Motors plants in Flint, Michigan, resulting in the loss of over 30,000 jobs and the destruction of Flint's economy. The film asks about GM's social responsibility; but from GM's position, the closing was legitimated through bottom-line, financial portfolio management.

These cultural directives find their roots in Protestant doctrine and practice. But Weber doesn't mean to imply that these tenets of capitalist culture are themselves religious—far from it. What we can see here is how culture, once born, can have unintended and independent effects from its creating group. Protestantism did not directly produce capitalism; but it did create a culture that when cut loose from its social group, influenced the rise of capitalism.

The most important religious doctrine behind the spirit of capitalism is Luther's notion of a *calling*. Prior to Luther and the advent of Protestantism, a calling was seen as something peculiar to the priesthood. Men could be called out of daily life to be priests and women to be nuns, but the laity was not called. Luther, however, taught that every individual can have a personal relationship with God; people didn't need to go through a priest. Luther also taught that individuals were saved based on personal faith. Prior to this, the church taught that salvation was a property of the church—people went to heaven because they were part of the Bride of Christ, the Church. With this shift to the individual also came the notion of a calling. If each person stood before God individually, and if priests aren't called to intercede, then the ministry belongs to the laity. Each person will stand before God on Judgment Day to give an account of what he or she did in this life. That means that God cares about what each individual does and has a plan for each life. God's plan involves a calling. Luther argued that every person in the church is called to do God's will. So it is not simply the case that ministers are called to work for God—carpenters are called to work for God as well, as carpenters. One calling is not greater than another; each is a religious service. This doctrine, of course, leads to a *moral* organization of life.

This idea of a calling was elaborated upon and expanded by John Calvin. Calvin took the idea of God's omniscience seriously: If God knows something, then God has always known it. Calvin also took the "sin nature" of humanity seriously. According to the sin nature doctrine, every human being is born in sin, reckoned sinful under Adam. That being the case, there is nothing anyone can do to save him- or herself from hell. We are doomed because of our very nature. Salvation, then, is from start to finish a work of God. Taking these ideas together, we come up with the Calvinistic idea of predestination. People are born in sin, there is nothing they can do to save themselves (neither faith nor good works), salvation is utterly a work of grace, and God has always known who would be saved. We are thus predestined to heaven or hell.

This doctrine had some interesting effects. Let's pretend you're a believer living in the sixteenth century under Calvin's teaching. Heaven and hell are very real to you, and it is thus important for you know where you are headed. But there isn't anything you can do to assuage your fears. Because salvation is utterly of God, joining the church isn't going to help; neither is being baptized or evangelizing. Confessing faith isn't going to help either, because you are either predestined for heaven or you are not. And you can't go by whether or not you *feel* saved—feelings were seen as promoting "sentimental illusions and idolatrous superstition."

"*Restless work in a vocational calling* was recommended as the best possible means to *acquire* the self-confidence that one belonged among the elect" (Weber, 1904–1905/2002, p. 66). Good works weren't seen as a path to salvation, but they were viewed as the natural fruit: If God has saved you, then your life will be lived in the relentless pursuit of His glory. In fact, only the saved would be able to dedicate their entire lives in such a way. The emphasis was thus not on singular works, but on an entire life organized for God's glory.

Thus diligent labor became the way of life for the Calvinist. Every individual was called to a job and was to work hard at that job for the Lord. Even the rich worked hard, for time belonged to the Lord and glorifying Him was all that mattered. Everything was guarded and watched and recorded. Individuals kept journals of daily life in order to be certain that they were keeping good works. Their entire lives became rational and systematically ordered.

Further, if one was truly chosen for eternal salvation, then God would bless the individual and the fruits of one's labor would multiply. In other words, in response to your labor, you could expect God to bless you economically. Yet, aestheticism became the rule, for the world was sinful and the lusts of the flesh could only lead to damnation. The pleasures of this world were to be avoided. So the blessings of God were reinvested in the work that God had called you to.

Taken together, then, the Protestant ethic commits each individual to a worldly calling, places upon her or him the responsibility of stewardship, and simultaneously promises worldly blessings and demands abstinence. This religious doctrine proved to be fertile ground for **rational capitalism:** a money-generating system that values work, rational management of life, and the delay of immediate gratification for future monetary gain.

Structural influences on capitalism: Thus, Weber argues that rational capitalism in the West found a seedbed in a culture strongly influenced by Protestantism. Yet there are other preconditions for the emergence of capitalism. I've illustrated these preconditions in Figure 5.4. They are divided into institutional, structural, and cultural influences, but these demarcations are not clear-cut. I've pictured them as rather shapeless, overlapping preconditions, as that's how Weber talks about them. All of these processes mutually reinforce one another. With capitalism in particular, there is movement back and forth between culture and structure in terms of causal influence.

We've seen the influence of Protestantism, but in terms of institutions there also had to be a centralized, bureaucratized state, as well as significant changes in modes of transportation and communication. I will be spending far less time explaining these structural influences than I did the effects of religious culture. The structural issues are more clear-cut than the cultural ones (which is generally the case). But don't let brevity fool you: these effects are equally important and you must understand how each of them works to comprehend the base for rational capitalism.

Nation-states are relatively recent inventions. Up until the nineteenth century, the world was not organized in terms of nation-states. People were generally organized ethnically, with fairly fluid territorial boundaries. They didn't have nations as we think of them today. A *nation* is a collective that occupies a specific territory, has a common history and identity, and sees itself as sharing a common fate. The

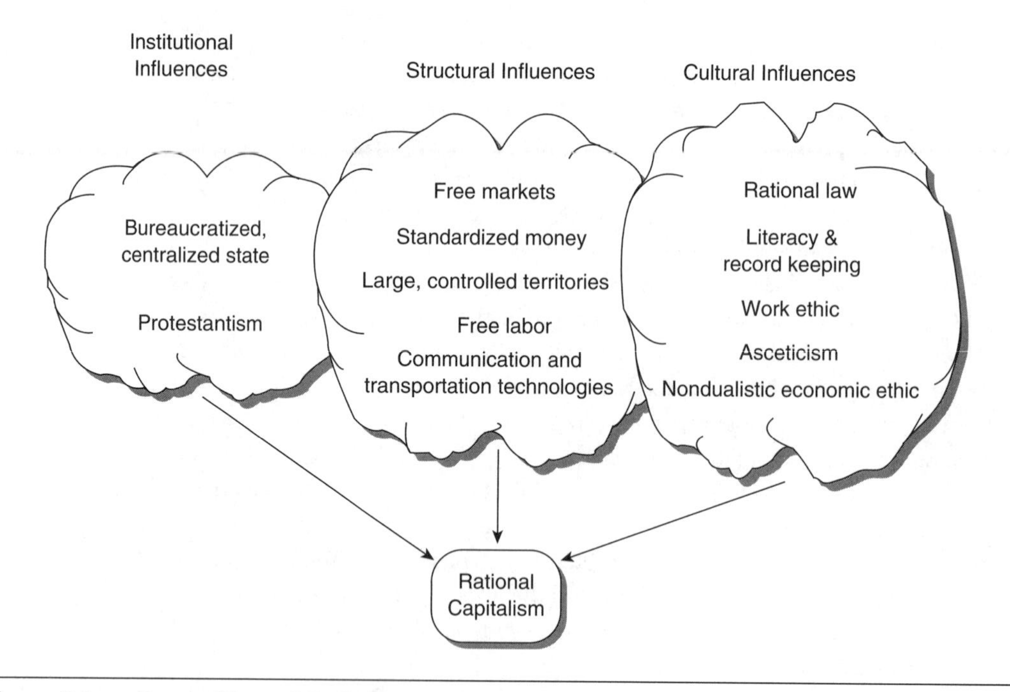

Figure 5.4 Preconditions of Capitalism

widespread use of the idea of a nation for organizing people was necessary for capitalism. Nations were responsible for controlling large territories, standardizing money, organizing social control, and facilitating free and open markets. All of these factors allowed for easier exchanges of goods and services involving large populations of people. These kinds of exchanges are, of course, necessary for rational capitalism to exist. Labor also had to be freed from social and structural constraints. Capitalism depends on a labor force that is free to sell its labor on the open market. Workers can't be tied down to apprenticeships or guild obligations, nor can they be attached to land obligations as in feudalism.

Along with increases in communication and transportation technologies, nation-states and Protestantism helped form an objectified, rationalized culture, wherein written records were kept, people practiced a strong work ethic coupled with asceticism, and the traditional dualistic approach to economic relations would break down. This latter issue is particularly important for Weber. In traditional societies, groups usually had two different kinds of ethics when participating in exchanges. There were restrictions having to do with ritual and fairness when dealing with group members, but people not within the group could be exploited without measure. Both of these frameworks needed to be lifted in order for capitalism to flourish. All business and loans needed to be rationalized so that there could be continuity.

Class, Authority, and Social Change

As we can see with capitalism, social change for Weber is a complex issue, one involving a number of variables coming together in indeterminate ways. Social change always involves culture and structure working together, and it involves complex social relations. We turn our attention now to stratification and change. Again we will see the interplay of culture and structure as well as complex social categories.

Weber's understanding of stratification is more complex than Marx's. Marx hypothesized that in capitalist countries there will be only one social category of any consequence—class. And in that category, Marx saw only two types: owners and workers. For Weber, status and power are also issues around which stratification can be based. Most importantly, class, status, and power do not necessarily covary. That is, a person may be high on one of those dimensions and low on another (like a Christian minister, typically high in terms of status but low in terms of class). These crosscutting life circumstances or affiliations can prevent people from forming into conflict groups and bringing about social change. For example, in the United States, a black man and a white man may both be in the poor class, but their race (status) may prevent them from seeing their life circumstances as being determined by similar factors.

Weber also argues that all systems are socially constructed and require people to believe in them. Marx did say that capitalism has a cultural component that holds it together (if not for ideology, the proletariat would immediately overthrow the system); but true communism requires no such cultural reinforcement, because it corresponds to our species-being nature. For Weber, **legitimation** is the glue that holds not only society together but also its systems of stratification. Thus, for Weber, issues of domination and authority go hand in hand (in fact, Weber used just one German word to denote them both—*Herrschaft*). So, people must have some level of belief in the authority (culture) of those who are in charge and they must cooperate with the system to some degree in order for it to work. Though the concept may sound a bit strange, we could call this aspect of stratification cooperative oppression.

Class: But let's begin our consideration of these issues by taking a closer look at Weber's complex idea of stratification. Weber's definition of class is different than Marx's. Marx defines class around the ownership of the means of production. Weber (1922/1968), on the other hand, says that a "class situation" exists where there is a "typical probability of 1. procuring goods 2. gaining a position in life and 3. finding inner satisfaction, a probability which derives from the relative control over goods and skills and from their income-producing uses within a given economic order" (p. 302). In other words, Weber defines class based on your ability to buy or sell goods and/or services that will bring you inner satisfaction and increase your life chances (how long and healthfully you will live).

Weber also sees class as being divided along several dimensions as compared to Marx's two. Marx acknowledges that there are more than two class elements, but he also argues that the other classes (such as the lumpenproletariat or petite bourgeoisie) become less and less important due to the structural squeeze of capitalism. Weber also speaks of two main class distinctions, yet they are constructed around completely different issues, each with "positively privileged," "negatively privileged," and "middle class" positions. The **property class** is determined by property differences, either owning (positively privileged) or not owning (negatively privileged).

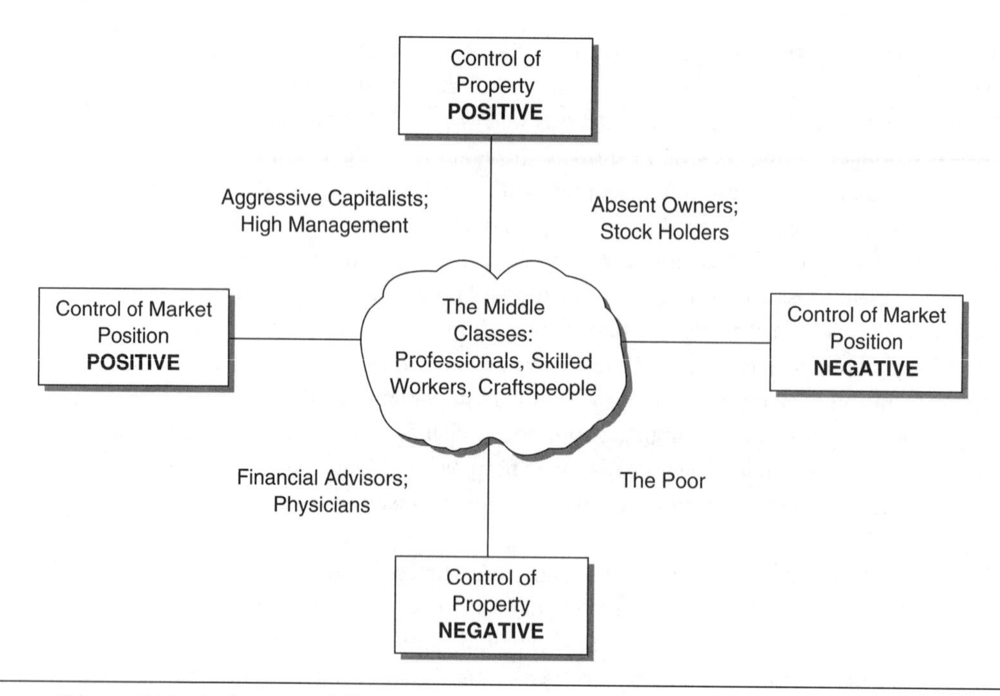

Figure 5.5 Weber's Concept of Class

Rentiers—people who live off property and investments—are clear examples of owners; whereas *debtors*—those who have more debt than assets—are good examples of negatively privileged. The middle property classes are those who do not acquire wealth or surplus from property, yet they are not deficient either.

The **commercial class** is determined by the ability to trade or manage a market position. Those positively related to commercial position are typically entrepreneurs who can monopolize and safeguard their market situation. Negatively privileged commercial classes are typically laborers. They are dependent on the whim of the labor market. The in-between or middle classes that are influenced by labor market positions are those such as self-employed farmers, craftspeople, or low-level professionals who have a viable market position yet are not able to monopolize or control it in any way.

> A *uniform* class situation prevails only when completely unskilled and propertyless persons are dependent on irregular employment. Mobility among, and stability of, class positions differs greatly; hence, the unity of a social class is highly variable. (Weber, 1922/1968, p. 302)

In Figure 5.5, I've given us a picture of Weber's ideas about class. We can see that there are two axes to class: property and market position. People can have a positive, negative, or middle position with respect to each of these issues. I've conceptualized this as a typology, because Weber spoke of people holding a position on both. I have also provided some contemporary examples in this typology. Today, those in upper management, such as CEOs, are not only paid large salaries, they are also given stock options that translate into ownership, which places them high in both the property and market dimensions (they hold a monopoly with the skills they have). For example, *Forbes* magazine lists Jeffrey C. Barbakow of Tenet

Healthcare as having been the highest paid American CEO in 2003. He received about $5.4 million in salary, plus he had stock gains of $111 million. Even the lowest paid major CEO in 2003 received $1 million in salary and owned $28.2 million in stock. Weber's typology of class also allows us to conceptualize certain kinds of knowledge as a resource or market position and thus a class indicator. Those who monopolize such skills and knowledge in our society include medical doctors and other such professionals. Stock owners as well as Weber's rentiers use the control of property to attain wealth, and we find them in the upper right corner of the picture. Those who have little or no control over property and market are the poor.

Weber sees that holding a specific class position does not necessarily translate into being a member of a group. Unless and until people develop a similar identity that focuses on their class differences, they remain a statistical aggregate. In other words, the census bureau may know you are in the middle class, but you may not have a group identity around your class position. So, in addition to the property and commercial classes, Weber also talks about social class. His emphasis here seems to be on the social aspect—the formation of unified group identity.

Weber identifies at least three variables for the formation of social class. He argues that class-conscious organization will most easily succeed if the following criteria are met:

It is organized against immediate economic groups. There must be some immediate market position or control issue, and the other group must be perceived as relatively close. Thus, in today's economy, workers are more likely to organize against management rather than ownership.

Large numbers of people are in the same class position. The size of a particular group can give it the appearance, feel, and unavoidability of an object.

The technical conditions of organization are met. Weber recognizes that organization doesn't simply happen; there's a technology needed to organize people. People must be able to communicate and meet with one another; there must be recognized and charismatic leadership; and there must be a clearly articulated ideology in order to organize.

Keep in mind that these are all variables. A group may have more or less of any of them. As each of these increases, there will be a greater likelihood of the formation of a social class.

Status and party: But Weber sees stratification as more complex than class. Money isn't the only thing in which people are interested. People also care about social esteem or honor. Weber termed this issue *status.* Status, for Weber (1922/1968), entails "an effective claim to social esteem in terms of positive or negative privileges" (p. 305). Thus, status groups are hierarchically ranked by structural or cultural criteria and imply differences in honor and privilege.

Status may be based or founded on one or more of three things: a distinct *lifestyle,* formal *education,* or differences in *hereditary or occupational prestige.* Being a music fan may entail a distinctive lifestyle—people who listen to jazz are culturally different than those who listen to rock. And there are positive and negative privileges involved as well: you can get a degree in jazz at my university, but not in rock. Homosexuals can also be seen in terms of lifestyle status groups. One's education level can provide the basis of status as well. Being a senior at the university is better, status wise, than being a freshman, and you have greater privileges as a senior as well. Professors are also examples of prestige by education. They are good examples because they usually rate high on occupational prestige tables but low in class and power. Race and gender are also examples of status founded on perceived heredity.

Status groups maintain their boundaries through particular practices and symbols. Boundary maintenance is particularly important for status groups because the borders are more symbolic than actual. The differences between jazz and rock, for example, are in the ear of the listener. Generally speaking, Weber argues, we maintain the boundaries around status through marriage and eating restrictions, monopolizing specific modes of acquisition, and different traditions—all of which are described in greater detail below.

The norm of marriage restriction is fairly intuitive; we can think of religious and ethnic groups who practice "endogamy" (marriage within a specific group as required by custom or law). But eating restrictions may seem counterintuitive. It may help us to realize that most religious groups have special dietary restrictions and practice ritual feasting. Feasts are always restricted to group members and are usually seen as actively uniting the group as one (the traditional Jewish Passover Seder is a good example). We can also think of special holidays that have important feast components, like Thanksgiving and Christmas here in the United States. And, if you think back just a few years, I'm sure you can recall times in Junior High when someone your group didn't like tried to sit at your lunch table. For humans, eating is rarely the simple ingestion of elements necessary for biological survival. It is a form of social interaction that binds people together and creates boundaries.

We also maintain the symbolic boundaries around status groups through monopolizing or abhorring certain kinds and modes of acquisition, and by having certain cultural practices or traditions. The kinds of things we buy obviously set our status group apart. That's what we mean when we say that a BMW is a status symbol. Status groups also try and guard the modes of acquisition. Guilds, trade unions, and professional groups can function in this capacity. And, of course, different status groups have different practices and traditions. Step concerts, pride marches, Fourth of July, Kwanzaa, Cinco de Mayo, and so on are all examples of status-specific traditions and cultural practices.

When sociologists talk about Weber's three issues of stratification, they typically refer to them as class, status, and power. However, power is not the word that Weber uses; instead he talks about *party*. What he has in mind, of course, is within the sphere of power, "the chance a man or a number of men to realize their own will in a social action even against the resistance of other who are participating in the action" (Weber 1922/1968, p. 926). Yet, it is important to note that power always involves social organization, or, as Weber calls it, party. Weber uses the word *party* to capture the social *practice* of power. The social groups that Weber would consider parties are those whose practices are oriented toward controlling an organization and its administrative staff. As Weber puts it, a party organizes "in order to attain ideal or material advantages for its active members" (1922/1968, p. 284). The Democratic and Republican parties in the United States are obvious examples of what Weber intends. Other examples include student unions or special interest groups like the tobacco lobby, if they are oriented toward controlling and exercising power.

Crosscutting stratification: One of Weber's enduring contributions to conflict theory is this tripartite distinction of stratification. Marx rightly argues that overt conflict is dependent upon bipolarization. The closer an issue of conflict gets to having only two defined sides, the more likely is the conflict to become overt and violent. But Weber gives us an understanding of stratification that shows the difficulty in achieving bipolarization.

Every individual sits at a unique confluence of class, status, and party. I have a class position, but I also have a variety of status positions and political issues that concern me. While these may influence one another, they may also be somewhat different. To the degree that these issues are different, it will be difficult to achieve a unified perspective. For example, let's say you're white, gay, male, the director of human resources at one of the nation's largest firms, and Catholic. Some of these social categories come together, like being white, male, and in upper management. But some don't. If this is true of you as an individual, then it is even more so in your association with other people. You'll find very few white, gay, male, Catholic upper managers to hang around with. And if you add in other important status identities, such as Southern Democrat, it becomes even more complex. Thus, Weber is arguing that the kind of conflict that produces social change is a very complex issue. Different factors have to come together in unique ways in order for us to begin to formulate groups and identities capable of bringing about social change.

> Domination in the most general sense is one of the most important elements of social action. Of course, not every form of social action reveals a structure of dominancy. But in most of the varieties of social action domination plays a considerable role, even where it is not obvious at first sight. (Weber, 1922/1968, p. 941)

Authority and social change: At the heart of the issue of stratification and social change is legitimacy, and here we see Weber's emphasis on culture. In order for a system of domination to work, people must believe in it. Part of the reason behind this need is the cost involved in the use of power. If people don't believe in authority to some degree, they will have to be forced to comply through coercive power. The use of coercive power requires high levels of external social control mechanisms, such as monitoring (you have to be able to watch and see if people are conforming) and force (because they won't do it willingly). To maintain a system of domination not based on legitimacy costs a great deal in terms of technology and manpower. In addition, people often ultimately respond to the use of coercion by either rebelling or giving up—the end result is thus contrary to the desired goal.

Authority, on the other hand, implies the ability to require performance that is based upon the performer's belief in the rightness of the system. Because authority is based on socialization, the internalization of cultural norms and values, authority requires low levels of external social control. We can thus say that any structure of domination can exist in the long run if and only if there is a corresponding culture of authority. Weber identifies three ideal types of authority. (Keep in mind that these are ideal types and may be found in various configurations in any society.)

- **Charismatic authority**—belief in the supernatural or intrinsic gifts of the individual. People respond to this kind of authority because they believe that the individual has a special calling. (Examples of this type of authority include Susan B. Anthony, Adolf Hitler, Martin Luther King Jr., John F. Kennedy, Golda Meir, and Jesus—notice that it is people's belief in the charisma that matters; thus, we can have Hitler and Jesus on the same list.)
- **Traditional authority**—belief in time and custom. People respond to this kind of authority because they honor the past and they believe that time-proven methods are the best. (Good examples of this type of authority are your parents and grandparents, the Pope, and monarchies.)
- **Rational-legal authority**—belief in procedure. People respond to this kind of authority because they believe that the requirements or laws have been enacted in the proper manner. People see leaders as having the right to act when they obtain positions in the procedurally correct way. (A good example of this type is your professor—it does not matter who the professor is, as long as he or she fulfills the requirements of the job.)

These diverse types of authority interact differently in the process of social change. According to Weber, the only kind of authority that can instigate social change is charismatic. Traditional and rational-legal authorities bring social stability—they are each designed to maintain the system. Charismatic individuals come to bring social change, yet charismatic authority is also inherently precarious. Because charisma is based on belief in the special abilities of the individual, every instance of charismatic authority will fail within that person's lifetime—the gifts die with the person. Thus, every charismatic authority will someday have to face the **problem of routinization** (making something routine and thus predicable). Every social movement based upon charismatic authority—and Weber argues that they all are—routinizes the changes by either using traditional authority or rational-legal authority.

> Charisma, on the other hand, may effect a subjective or internal reorientation born out of suffering, conflicts, or enthusiasm. It may then result in a radical alteration of the central attitudes and directions of action with a completely new orientation of all attitudes toward the different problems of the "world." (Weber, 1922/1968, p. 245)

The case of the Christian church might give us insight. Jesus was a charismatic leader. When he died, the church was faced with the problem of continuing his leadership. Though somewhat a gloss, it can be said that the Catholic Church is based upon the traditional authority of the Pope and that Protestant churches are based upon rational-legal authority. You cannot study to become a Pope, but you can study to become a Protestant minister. As I said, this is a gloss: people do study to become priests, and ministers are perceived as charismatically ordained by God. So, these authority systems are mixed (as is always the case with ideal types). But their systems are *stabilized* through the use of tradition and bureaucracy.

Combining Weber's ideas of class, status, and party with his argument concerning authority and social change, we can put together a theory of conflict and social

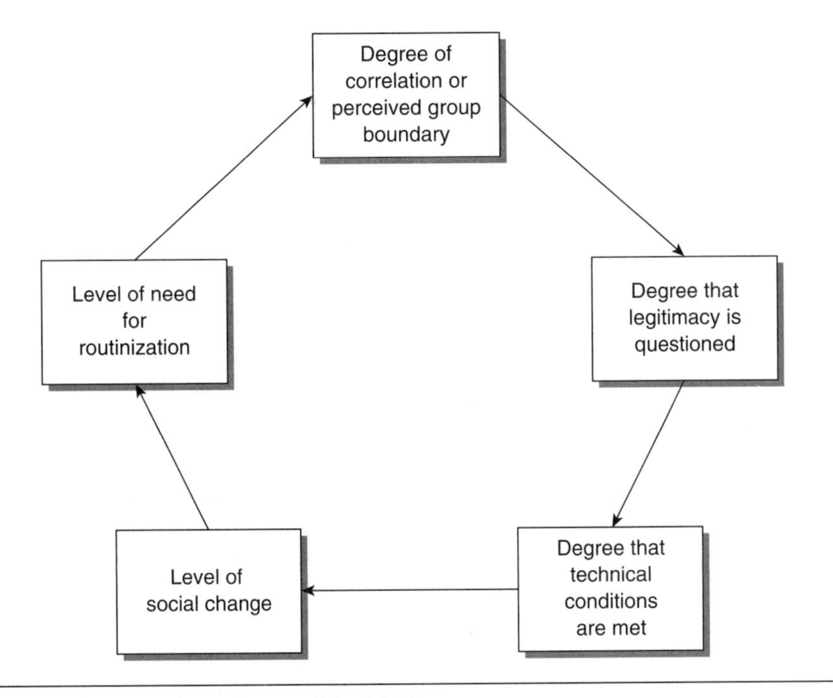

Figure 5.6 Weber's Theory of Social Change

change (see Figure 5.6). Social change will occur only if the legitimacy of the system of stratification is questioned. Conflict and change are likely to occur when there are clear breaks between or limited upward mobility in one of the systems—I've depicted this in the figure as "perceived group boundary." Groups will tend to perceive their boundaries when in close proximity to another competing group or when mobility is limited. It is possible for change to occur in only one of three areas, but that change will be limited due to the crosscutting influences of the other systems. In the United States, for example, many people questioned the legitimacy of the systems of race and gender during the 1960s and 70s. These systems are primarily built around and understood through status. So, while there have been some real structural changes, it appears that *most* of the change in race and gender systems is cultural—the status of each has been improved. (Please note the "appears" in my statement: this issue needs close **empirical** research.)

In a Weberian sense, significant social change will occur if and when there is perceived correlation among class, status, and power ("degree of correlation" in Figure 5.6). In other words, if people perceive that certain groups are high on all three stratification systems and certain groups are low, then they are more likely to question the legitimacy of the whole system rather than just one part. This *delegitimation* is particularly likely when the correlation is perceived as arbitrary. As legitimacy is questioned, the technical conditions of group conflict tend to be met. These include charismatic leadership, clearly articulated goals and ideology, and the ability to meet and communicate. Note that there is a reciprocal effect between questioning legitimacy and technical conditions: each will tend to reinforce the other. As technical conditions are met, the group will become more effectual in bringing about social change. The degree of social change in turn will impact the need for routinization.

Weber argues that routinization will in the long run lead to a kind of stratification that again sets up the conditions for conflict and change. In this respect, Weber's model is more dialectical than Marx's, which stops with the advent of communism, but Weber sees conflict and change as ubiquitous features of every social system. For Weber, social systems will move cyclically through routinization and charismatic change. At different times, for different reasons, social groups will question the legitimacy of domination.

Rationality in Action

Historical shift to bureaucratic organization: To be human is to be social, and to be social is to be organized. Thus, organization is primary to what it means to be human. Human beings have been organized over time by different social forms, and these different organizational forms have clear consequences for us. In the not too distant past, we organized much of our lives around affective (emotionally based) systems, such as kinship. Kinship linked people through blood and marriage. In such organizations, people saw each other in terms of familial obligations and rights, which they felt as emotional ties.

But today we are organized through mostly bureaucratic means. From the time you were put in school at the age of five or six, bureaucracies have been organizing your life. You know how to stand in line, how to use time and space, and how to relate to people (the grocery clerk, the teacher, the minister, the insurance agent) as a result of spending most of your life in a bureaucracy. Even our dining experience has been strongly influenced by bureaucracy, thanks to McDonald's (Ritzer, 2004a).

Bureaucracies tend to rationalize and **routinize** all tasks and interactions. Tasks and interactions are rationalized in terms of means–ends efficiency, and they are routinized in the sense that they may be carried out without thought or planning or dependency on individual talents. Thus, each person and task is met in the same efficient and equal manner; personal issues and emotions do not have a place in a bureaucracy.

While bureaucracies have been around for quite awhile, for much of their history they were not the primary way in which humans organized. Certain social factors came about that simultaneously pushed out affective systems of organization (traditional) and set the stage for systems based more on reason than emotion. For Weber, these social factors are preconditions for bureaucracy, not causal forces. As these different features lined up, they created an environment ripe for rational organization.

According to Weber (1922/1968, pp. 217–226; 956–958), there are at least six preconditions for bureaucracy, and they don't necessarily occur in any specific order. These include increases in the following:

- the size and space of the population being organized
- the complexity of the task being performed

- the use of markets and the money economy
- communication and transportation technologies
- the use of mass democracy
- the volume of complicated and rationalized culture

All these factors created needs for more objective and rational social relations and culture, thus driving a society to use rational-legal authority and bureaucracy almost exclusively. As society became larger and spread out over vast spans of geographic space, it became increasingly difficult to use personal relationships as a method of organization. However, increases in population size and geographic space aren't sufficient to make bureaucratic organization necessary. For example, China was able to primarily use an elaborate kinship system rather than bureaucracy to organize most of their behaviors for many years. Other factors are necessary to push a society toward bureaucratic organization, such as increasing complexity of the task being performed. In England and elsewhere, this occurred through the Industrial Revolution, urbanization, and high division of labor. The increasing use of money and open markets for exchange also increased the need for rationalized and speedy calculations, thus adding to rational culture.

Increasing levels of communication and transportation technologies created the demand for faster and more predictable reactions from the governing state. Insightfully, Weber (1922/1968) says that with "traditional authority it is impossible for law or administrative rule to be deliberately created by legislation" (p. 227). Remember that traditional authority is based on history and time, so the only way a new rule or law can be legitimized through tradition is to say that it is according to the wisdom of the ages. Rational-legal authority, on the other hand, can create new rules simply because it is expedient to do so. So as states came into more frequent contact with one another and dealt with more complex problems, it became necessary to respond quickly to new situations with new rules, thus pushing forward a rational-legal authority and bureaucracy.

> Bureaucratic administration means fundamentally domination through knowledge. (Weber, 1922/1968, p. 225)

Two additional forces created demands for rational and objective standards. The first was democracy for the masses, rather than simply the elite, which created demands for equal treatment before the law. Initially, the idea of democracy was limited to the educated and powerful. Including the masses created the demand for equal treatment regardless of power or prestige. This, of course, led to the idea that rules and laws should be blind to individual differences. The other additional social factor that pushed for bureaucratic organization was the increase in complex and rational culture. Due to the increases in knowledge that came with science and technology, a new social identity came into existence and rose to prominence—the expert, bureaucracy's manpower.

Ideal-type bureaucracies: Weber gives us an ideal type for bureaucracy. Remember that this ideal type is not intended to tell us what the perfect bureaucracy should look like; rather, Weber uses the ideal type as an objective yardstick against which we may measure different subjective and cultural states. Weber talks about bureaucracy in a number of different places. But if we combine his lists, we come up with six important features to the ideal type of **bureaucracy**: an explicit division of labor with delineated lines of authority, the presence of an office hierarchy, written rules and communication, accredited training and technical competence, management by rules that is emotionally neutral, and ownership of both the career ladder and position by the organization rather than the individual.

Each of these characteristics is a variable; organizations will thus be more or less bureaucratized. For example, the first job I ever had was with Pharmaseal Laboratories. I worked as a lead man, which meant I kept the production lines stocked with raw materials and moved the finished product to the warehouse. After about six months of working there, I was promoted to management. During my week-long orientation, I was told about AVOs (Avoid Verbal Orders) that documented in triplicate any communication or disciplinary action. I was introduced to the Human Resources department and shown the employee files and where each of those AVO copies went. I was told of my span of control, given an organizational chart of the company, and a complete job description for myself and every person with whom I would have business. I was also informed about the company's internal promotion policy (the career ladder)—to get past the level I was at would require additional schooling and credentialing.

On the other hand, my wife's experience at a local firm indicates a more patrimonial and less bureaucratized organization. When she first arrived at this company, she asked for the organizational chart and job descriptions. She also wanted to know exactly what her span of control was and to whom she reported. But what she found was a 70-person company that was run on personal relationships. Most of the communication was not documented, there was no organizational chart or policy defining the span of control or communication, and there were no job descriptions. Thus, there are differences in the level of bureaucratization, even in a society such as ours. However, that company has changed over the years and has become more bureaucratized. These changes are due to increasing pressures for rationalization and objectification, coming from much the same preconditions of which Weber initially spoke. (A side note: It's good to keep in mind that all the theoretical issues that Weber, and our other theorists, brings up continue to be important social variables. While we may talk about them historically, like the preconditions for bureaucracy, they continue to influence society and our lives. So, keep your eyes open.)

> As we have seen, bureaucratic rationalization, too, often has been a major revolutionary force. . . . But it revolutionizes with *technical means*. . . . It *first* changes the material and social orders, and through them the people.(Weber, 1922/1968, p. 116)

Effects of bureaucratic organization: One of the things that should become clear as we move through the different theorists is that social processes and factors are not innocuous. For example, we see from Marx and Simmel that the use of money to facilitate exchanges in the end changes people. A good analogy might be putting gasoline in a car. We put gas in the car to make it run, but the accidental and almost inevitable consequence is pollution. Bureaucracies have unintended and largely unavoidable consequences as well. They influence both the people in the bureaucracies and the social system as a whole.

Contemporary theorists point out that living in a society that organizes through bureaucracy can produce the **bureaucratic personality.** There are at least four characteristics of this kind of temperament. One, individuals tend to live more rationally due to the presence of bureaucracy, and not just at work. People generally become less and less spontaneous and less emotionally connected to others in their lives. They understand goals, the use of time and space, and even relationships through rational criteria. Two, people who work in bureaucracies also tend to identify with the goals of the organization. Workers at levels that are less bureaucratized tend to complain about the organization; on the other hand, management who exist at more bureaucratized levels tend to support and believe in the organization. Again, this isn't something that we just put on at work—we *become* bureaucratic ourselves.

> The decisive reason for the advance of bureaucratic organization has always been its purely technical superiority over any other form of organization . . . Precision, speed, unambiguity, knowledge of the files, continuity, discretion, unity, strict subordination, reduction of friction and of material and personal costs. (Weber, 1948, p. 217)

Three, because bureaucracies are based on technical knowledge, people in bureaucratic societies tend to depend on expert systems for knowledge and advice. In traditionally based societies, people would trust the advice of those they loved, or those who had extensive experience, or those who stood in a long lineage of oral discipleship. Conversely, in societies like the United States, we look to those who have credentials to help us. Honor in modern society is given to those with credentials; age and experience are of no consequence. And, lastly, bureaucracies lead to sequestration of experience. By that I mean that different life experiences are separated from one another, such as dying from living. In traditional society, life was experienced holistically. People would see birth, sickness (emotional, mental, and physical), and death as part of their normal lives. Children were conceived and grandparents died in the same home. Today, most of those experiences are put away from us and occur in bureaucratic settings where we don't see them as part of our normal and daily life; such settings include hospitals, rest homes, asylums, and so on. The world has thus become tidy, clean, and rational.

Society is affected as well by bureaucracy. There are two main effects. The first is the **iron cage of bureaucracy.** Once bureaucracies are in place, they are virtually inescapable and indestructible for several reasons: they are the most efficient form of organizing large-scale populations; they are value free; and they are based upon expert knowledge. We've just seen that individuals within a rationalized society become more and more dependent upon expert knowledge; the same is true with leaders. Whether the leaders are in charge of political, religious, or economic organizations, they become increasingly dependent upon rationally trained personnel and expert knowledge in the bureaucratic information age. As we noted under professionalization, the experts themselves engage in secrecy and mystification in order to avoid inspection and secure their position. Further, one of the definitions of a professional is self-administration, which means that bureaucratic experts become a self-recruiting and self-governing class, existing apart from any other

> When those subject to bureaucratic control seek to escape the influence of the existing bureaucratic apparatus, this is normally possible only by creating an organization of their own which is equally subject to bureaucratization. (Weber, 1922/1968, p. 224)

organizational control. Thus, neither the people ruled, nor the rulers, nor the experts themselves can escape the domination of the bureaucratic form.

There is one further factor to note: bureaucracies are value free, which means they can be used for any purpose, from spreading the gospel to the eradication of ethnic minorities. This implies that bureaucracies are quite good at co-optation. To *co-opt*, in this context, means to take something in and make it part of the group, which on the surface might sound like a good thing. But because bureaucracies are value and emotion free, there is a tendency to downplay differences and render them impotent. For example, one of the things that our society has done with race and gender movements is to give them official status in the university. One can now get a degree in race or gender relations. Inequality is something we now study, rather than the focus of social movements. In this sense, these movements have been co-opted.

Bureaucracies also accelerate the process of **credentialing.** Remember that position within a bureaucracy is achieved through diplomas and certification. It isn't supposed to be *who* you know but *what* you know that determines rank in the organization. That being the case, society needs a legitimated process through which credentials can be conferred. The United States uses the education system. But one of the effects of that decision is that education is no longer about education. The education system is used to credential technical expertise rather than cultivate an informed citizenry. Universities are thus becoming populated by professional schools—the school of business, nursing, social work, computer technology, criminal justice, and so on, and there is mounting tension between the traditional liberal arts and these professional schools. Many students express this tension (and preference for credentialing) when they ask, "How will this course help me get a job?"

The emphasis on credentials coupled with the American view of mass education and the use of education to give credentials has created **credential inflation.** Every year, more and more people are going to college to get a degree so that they can be competitive in the job market. Using data from the National Center for Education Statistics, we can get a sense of how this is working. In the United States, between the mid-eighties and mid-nineties, there was a 20% increase in the number of bachelor degrees, 37% increase in master's degrees, and 44% increase in doctorates. The result is that there are too many people with advanced degrees, which, in turn, decreases the value of those degrees.

Thinking About Modernity and Postmodernity

Bureaucracy, disorganized capitalism, and new social classes: The chief characteristic of modernity for Weber is the progression of rationalization. Due to such social processes as Protestantism, state formation, markets, money, democracy, population size, the complexity of tasks, and the level of abstract and rational knowledge, modern societies are organized more in terms of rational-legal than traditional authority. Thus, bureaucracy is the chief organizational form in modernity, increasing the levels of rationalization and routinization in society at large, as well as creating an iron cage for society and the bureaucratic personality. In many ways, then, Weber was disillusioned about modernity.

Weber would also argue that modern societies are complex entities. Class remains important, as Marx would agree, but modern societies are also noted for their differing and intricate status and power relations. Weber's emphasis on status comes, of course, from the place he gives culture in his analysis. Weber understands that humans are symbolic as well as material creatures. As such, we seek to be gratified through symbols and not just money. Weber also sees that social control is wrapped up in authority—the belief in the rightness of any system of domination. Thus, the exercise of power is based on cultural issues as well. Together, class, status, and power create a complex of social positions and societal control.

Weber was always interested in the relation between structure and culture and the interplay among class, status, and power. You remember that he saw culture leading change at one time and structure at another. Most often, history is the result of the dialog between these two forces. Within that dialog, the relative importance of and interaction among class, status, and power are played out. Social change can occur as the three are seen to correlate, or if the legitimacy of one or the other stratification system is questioned.

It is quite possible that as societies move toward what some call postmodernity, the issues and relative importance of class, status, and power change significantly. There are many factors that produce issues that influence these relations, but let's take a look at two of them: the increasing importance of symbolic gratification and life politics. Lash and Urry (1987), who conduct a Weberian analysis, argue that symbols and symbolic gratification become increasingly important for people 1) as the size and ability to organize of the working class decrease, 2) as the size of the professional and service classes increase, and 3) as ready-made (as opposed to socially emergent) cultural images and their meaning constitute a significant and growing social reality. In back of these three elements are what Lash and Urry term "disorganized capitalism" and the proliferation of media images.

The media we can understand and I'll come back to it in a moment, but disorganized capitalism needs some defining. Organized capitalism, the kind that Weber called rational capitalism, is characterized by increasing concentration of the means of production, distribution, and social reproduction. This type of capitalism brought large corporate and regional headquarters, massive factories, central distribution points, and workers living in concentrated areas within urban spheres. Disorganized capitalism, on the other hand, entails the de-concentration of the means and administration of production, commercial capital, collective consumption, and residential concentration of labor power. Capitalism has moved from large national corporations to larger international corporations to multinational conglomerates. Along with the movement of capital comes the movement of the work force—primary production moves from core nations to peripheral or third-world nations. Countries such as the United States are becoming less based on production and more based on consumption. For example, according to the Bureau of Labor statistics, the United States lost 2 million manufacturing jobs between 1991 and 2002. And during that same period of time, the number of jobs in retail trade rose by 4 million.

Not only did retail jobs increase while production-based jobs fell, but employment in the service (from 28 million to 41 million), government (from 18.5 million to 21 million), and finance/insurance (6.8 to 7.8) sectors all increased as well. These numbers give us a picture of the movement from a class- (working) based economy to a service-based one. In such an economy, identities around class are increasingly difficult to achieve. And, as capitalist relations have moved and have become decentralized, the significance of large collectives (such as corporations, workplaces, and cities) has a diminished role in the process of identity formation. Markets and media fill this void and provide a diversity of identities from which the individual may choose.

These shifts in the economy also create demands for greater education and extended periods of unstable identity. We can think of the period between childhood and adulthood as being indeterminate and thus unstable. Between those epochs in one's life there are few institutional obligations. You're no longer utterly dependent, yet at the same time, you're not responsible for the institutions of society. In traditional societies, there was little time between being a child and an adult. In fact, children were seen simply as little adults. However, as the need for cultural training and socialization increased, so did the time before one would take on the responsibilities of adulthood. During these periods of instability, people are prone to play with symbols and identities. For example, look around your college campus at the diverse array of identities and symbols, and then consider a large corporation like IBM. What you will see is that the kinds of identities and the diversity of cultural symbols that IBM tolerates is very, very low when compared to the university.

The extended period between adulthood and being a child is, of course, due to the increasing demand for education and credentials (effects of bureaucracy) in a postmodern society. The professional and service sectors are based on knowledge and credentials, not on work skills. Credentialing is seen by some postmodernists as increasing the power of culture—because differences between people and access to jobs, social standing, status groups, and other valued social niches are determined symbolically, not materially. Moreover, credentialing expands the educational system, which in turn produces new sets of symbols that elevate the overall level of cultural production in a society.

Along with these changes, there has been a shift from emancipatory politics to life politics (see Giddens, 1991). *Emancipatory politics* is concerned with liberating individuals and groups from the constraints that adversely affect their lives. In some ways, this was the theme of modernity—it was the hope that democratic nation-states could bring equality and justice for all. On the other hand, *life politics* is the politics of choice and lifestyle. It is not based on group membership and characteristics, as is emancipatory politics; it is based on personal lifestyle choices.

We have come to think of choice as a freedom we have in the United States. But it is more than that—it has become an obligation. Choice is a fundamental element in contemporary living, due in part to the media (all print and visual forms of communication). Think of it this way: The basic reality for people is found in social events—face-to-face interactions. If I have an experience with Duke Power Company—let's say they cut down the trees in my yard without my permission—and I tell you about it, the telling of it is once removed from the actual event. And if I were to tell the story, I would emphasize those elements that back up *my* concerns, rather than those of the power company. Now let's think about what the media does. If what is presented actually occurred in real life, by the time we see it in the media it has been removed many times over from the incident, especially if we are talking about global rather than local events. Each one of these levels is subject to its own politics and interpretations. That's why we can talk about the conservative or liberal media. But much of what we are presented with in the media is pure fiction or advertising, which further removes it from embedded, social reality.

In lifting out experience from its social embeddedness in face-to-face interaction., the media presents people with a vast array of diverse lifestyles and cultures. The plurality of lifestyles presented to an individual not only *allows* for choice, it *necessitates* choice. In other words, what becomes important is not the hard issue of group equality, but rather, the insistence on personal choice. In addition to mediated experience, the plurality of choice is also a function of a number of other influences, such as the de-institutionalization of expert oversight (think of the commercials telling us not only how to diagnose our illness but also what kind of medicine we need), the pluralization of life-worlds (there are many potential realities within our grasp), and the legitimation of doubt (science and education tell us nothing is certain).

What is at issue in this milieu is not so much political equality (as with emancipatory politics) as inner authenticity. In a world that is perceived as constantly changing and uprooted, it becomes important to be grounded in oneself. Life politics creates such grounding. It creates "a framework of basic trust by means of which the lifespan can be understood as a unity against the backdrop of shifting social events" (Giddens, 1991, p. 215). In some ways, we can see life politics as an extension of the feminist idea that "the personal is political."

A good example of life politics is veganism—the practice of not eating any meat or meat byproducts. Not only is eating flesh avoided, but also any products with dairy, eggs, fur, leather, feathers, or any goods involving animal testing. In the words of one Web site for vegans, veganism "is an integral component of a cruelty-free lifestyle." It is a political statement against the exploitation of animals, and for some it is clearly a condemnation of capitalism—capitalism is particularly responsible for the unnatural mass production of animal flesh as well as commercial animal testing. Yet for most vegans, it is simply a lifestyle, one that brings harmony between the outside world and inner beliefs, and not a collective movement to bring about social change.

What we can see through this discussion is that in postmodernity, material class becomes less important, while cultural issues become more important. Further, status and party are not single, focused social identities around which groups can form. Rather, status identities for more and more of the population have become like playful masks that we can put on and take off at will. Yet, the importance of status in identity formation has exceeded that of class, at least in terms of this way of viewing the world, and cultural signs, symbols, and images, as well as symbolic gratification, have become paramount. For many, politics has moved from group considerations of equality and political participation to individual practices designed to create a sense of personal authenticity in a shifting landscape. In terms of the bipolarization of conflict and the probability of social change, we can see that the effects of bureaucracy and Weber's notion of crosscutting influences become more and more prominent as we move deeper into postmodernity.

Summary

- To think like Weber is to take seriously the ramifications of culture. Weberians focus on the historical, cultural, and social contexts wherein the subjective orientation of the actor takes place. To think like Weber, then, means to use ideal types and *verstehen* to explain how these contexts came to exist rather than others. To think like Weber also means paying attention to the process of rationalization and the need for legitimation.

- Religion began with the movement from naturalistic to symbolic ways of seeing the world. The movement toward symbolism and religion was influenced by increases in economic technologies and professionalization. Religion was initially practiced in kinship-based groups with local deities. These local gods became hierarchically organized into pantheons due to the political organization of kinship groups into larger collectives, and the abstracting and rationalizing effects of the professional priesthood. Eventually, these same forces produced the idea of a monotheistic god under which all the other gods were subsumed and finally disappeared. Monotheism became ethical monotheism in response to the need of polity to control behavior on a large scale.

- The cultural foundations of rational capitalism were laid by Protestantism. This religious movement (through the doctrines of predestination and abstention, and the idea of a calling) indirectly created a rationalizing, individuating culture wherein money could be made for the purpose of making *more* money, rather than for immediate enjoyment. The establishment of nation-states structurally paved the way for rational capitalism by creating a free labor force, controlling large territories, standardizing money, and protecting free global markets.

- Social stratification is a complex of three scarce resources: class, status, and power. These three systems produce crosscutting interests that make social change difficult and multifaceted. Large-scale social changes become increasingly likely only as class, status, and power are seen to correlate; the legitimacy of the system is questioned; and the technical conditions of organization are met. Since social change is led by charismatic authority, each change will need to be routinized through traditional or rational-legal authority, which, in the long run, will once again set up conditions for conflict and social change.

- Bureaucratic forms of organization became prominent as societies became larger and more democratic, as tasks and knowledge became more complex, as communication and transportation technologies increased, and as markets became more widespread through the use of money. The extent of bureaucratic organization can be measured through an ideal type consisting of six variables: explicit division of labor, office hierarchy, written rules and communication, accreditation for position, affectless (without emotion or emotional connection) management by rule, and the ownership of career ladders and position by the organization. The use of bureaucracy as the chief organizing technology of a society results in the bureaucratic personality, the iron cage of bureaucracy, and social emphasis on credentials.

- A Weberian understanding of postmodern society focuses on the complex relationships among class, status, and party, with particular emphasis on the rise of the professional or credentialed class. Weber explicitly argues against a simple Marxian understanding of class and social change. For Marx, class in modern society would eventually devolve to only two classes who would face off over the means of production. Weber argues that social change is more complex, with cross-cutting influences from multiple class levels along with diverse status groupings and political parties. Postmodern theorists, such as Lash and Urry, argue that the class and status structures have become even more complex than Weber imagined, due to disorganized capitalism. Lash and Urry posit that postmodern culture takes on increasing importance due to disorganized capitalism, particularly for middle-class youth and for the expressive professions (such as flight attendants, newscasters, actors, models, and so forth) in the service class. This shift in class structure toward symbolic gratification is coupled with increasing differentiation of cultural images presented through the mass media to create an emphasis on life politics rather than emancipatory politics. Increasing differentiation of the class and status structures, along with the proliferation of media images, makes it increasingly difficult for political groups to effectively form, and personal choice has become more important than political change.

Building Your Theory Toolbox

Conversations With Max Weber—Web Research

All of our theorists are pretty interesting people, I think. But Weber was *really* an interesting person. Use your Internet search engine or review some of the recommended Weber Web sites to answer the following questions:

- Weber was truly an amazing scholar. Hints of his abilities showed early in his life. For example, what did Weber give his family as Christmas presents when he was about 12 years old? What kinds of books was he reading in his preteen years?
- Most scholars feel that Weber's family relationships played a role in his life and work. What kind of people were his mother and father? What was Weber's relationship to his parents? How is the death of Weber's father associated with his work?
- Some scholars think that Weber's relationship with his wife was indicative of some of his internal conflicts. How would you describe Weber's personal relationship with his wife? Speaking of his wife, what kind of person was Marianne Weber? (You might try putting her name in your Internet search engine.)

- Rather than being one of the founders of sociology, it appears that Weber wanted to do something else with his life. What kind of occupation or career did Weber want to have? How did he view academic work?
- At one point in his career, Weber visited the United States. What did he do during that visit and what effect did it have on his work?

Passionate Curiosity

Seeing the World (using the perspective)

- Weber presents us with our first critique of social science. He takes seriously the ideas of culture, value, and free will. However, Weber is still convinced that we can create object knowledge through the use of ideal types. His most famous one is bureaucracy. I'd like for you to create an ideal type with at least five points of comparison, and use it to think about the social world—use gender or another topic that interests you (perhaps one as important as "criminal behavior" or as mundane as "professor"). How did you form your ideal type (remember, Weber said there are two ways)? What were some of the difficulties you ran into? What do you think you can learn about the social world using your ideal type? For example, if you used gender, how do people vary from the ideal type? Under what external conditions do the variations take place? What do the ideal type and the empirical variations tell you about the cultural expectations concerning gender? What does this tell you about the way society is organized? (Notice I didn't ask what it told you about the person.) What are the drawbacks to using ideal types?
- One of the central themes in Weberian theory is rationalization (As a contemporary example, see George Ritzer's *The McDonaldization of Society*). Take a look at your life: In what ways has rationality influenced you? Do you think your life is more or less rationalized than your parents' was at your age? Let's take this one step further. Go to a place of business, like a fast-food restaurant or mall, and observe behaviors for at least two hours. How rationalized were the actions you observed? Overall, do you think that life is becoming increasingly rationalized? What are the benefits and drawbacks to rationalization? How do you think Weber felt about this process?

Engaging the World (using the theory)

- Get a sense of the kinds of jobs you can get today with a college education and the jobs available with the same education 50 years ago. You can do this by using your Internet search engine, going to http://nces.ed.gov/ and searching the data, or by asking your parents and grandparents. Explain your findings using Weber's theory. What do you think society can or should do in response to these changes? Using Weber's theory, do you think this trend will continue or abate? Do you think the purpose of education has changed in this country? To what level is education completely funded by government (that is, to what level is education free); why to that level, do you think?

- Think about Weber's ideal type of the spirit of capitalism. Are those traits more or less present in the United States today? Does this imply anything about capitalism in this country? Are we perhaps practicing a different kind of capitalism? If so, what would you call it?
- Weber gives us a robust theory concerning the rise of ethical monotheism. One of the interesting things that Weber's theory of religion tells us is that religion is clearly related to a political regime's interest in social control. Do you think there have been any changes in the structure or kind of religion practiced since Weber's time? If so, what kinds of changes? What kinds of social factors do you think are responsible for these changes (think about how Weber associated professionalism and economic changes with shifts in religion)? What, if anything, do these changes indicate about the state and social control?
- In addition to being spiritual centers, churches are social organizations. As such, Weber would argue that the type of authority and the concurrent organizational type that a church uses will influence the church and its parishioners. Using either your own experiences or by calling various churches in your area, what kind of authority and organization do you find to be most prevalent? Using Weber's theory, how do you think the church is being affected?

Weaving the Threads (synthesizing theory)

We are getting to the point in our studies where we are asking big questions and have the potential to create robust (strong and substantial) theories. It can be hard work, but it can also be exciting. When we move above the level of simply knowing what Marx said about capitalism, to the level where we are bringing together two or more theorists on the same issue, we are then beginning to create a theory that is robust enough to actually explain that part of society.

- Compare and contrast Durkheim, Marx, and Weber on the central features of modernity. What do they say makes a society modern? What problems do they see associated with modern society and modern lives?
- How does Weber expand Spencer's theory on the evolution of religion? Compare and contrast Spencer, Durkheim, and Weber on the origins and functions of religion.
- Compare and contrast Marx and Weber on the origins of capitalism. Can these theories be reconciled in any way? Do you think one is more correct than the other? Why or why not?
- Compare and contrast Marx and Weber on structures of inequality and social change. What does information does Weber give us about inequality that Marx doesn't? According to Weberian theory, what class did Marx miss entirely? Do you think that negates Marx's theory? How are Marx's and Weber's theories of social change the same and different? Can we bring them together to form a more powerful theoretical understanding of the conditions under which social change or conflict are likely to occur?
- Compare and contrast Spencer and Weber on the foundation of state power. How does Marx's theory critique both Spencer and Weber?

Further Explorations—Books

Bendix, R. (1977). *Max Weber: An intellectual portrait.* Berkeley: University of California Press. (The standard Weber reference)

Collins, R. (1986). *Weberian sociological theory.* New York: Cambridge University Press. (Systemization of Weber's theory by a prominent contemporary theorist)

Turner, B. S. (1992). *Max Weber: From history to modernity.* New York: Routledge. (Explanation of Weber's theories of modernity)

Turner, S. P. (2000). *The Cambridge companion to Max Weber.* New York: Cambridge University Press. (Excellent resource concerning Weber's works and influence)

Weber, M. (1988). *Max Weber: A biography.* (H. Zohn, Trans.) New Brunswick: Transaction Books. (Definitive biography, written by Weber's wife)

Further Explorations—Web Links

http://www.faculty.rsu.edu/~felwell/Theorists/Weber/Whome.htm (Site maintained by Frank W. Elwell at Rogers State University; site specifically oriented toward undergraduates and contains good explanations of some of Weber's concepts)

http://www2.fmg.uva.nl/sociosite/topics/weber.html (Internet resource for Weber's writings; some of the pages are in German)

Further Explorations—Web Links: Intellectual Influences on Max Weber

As a well-read individual, Weber was influenced by many thinkers. Wilhelm Dilthey and Heinrich Rickert both influenced Weber's work, but, unfortunately, at this time there are no helpful Web sites for either theorist. Two other important influences were Karl Marx, whose Web sites were noted previously, and Friedrich Nietzsche.

Friedrich Nietzsche: http://plato.stanford.edu/entries/nietzsche/ (Site maintained by Stanford University; contains extensive information on Nietzsche's life and work)

Karl Marx: http://en.wikipedia.org/wiki/Karl_Marx (Site maintained by Wikipedia Internet Encyclopedia; Marx's life, theory, and influences; additional links also provided)

CHAPTER 6

Society and the Individual—Georg Simmel (German, 1858–1918)

In some ways, Simmel is the most underappreciated theorist of the classical era, and given the time in which we live, his neglect is our loss. As we will see, Simmel focuses on the relationship between the individual and culture, which is the heart of late- or postmodern theory. He is particularly concerned with the potential effects of what he calls "objective culture," culture that stands outside of the individual and faces her or him as an alien force. As we have seen, the postmodern age is one in which the individual and culture have become the foci of social theory and our daily life. Simmel's theory lies right at the intersection of these two issues. As such, his insights and perspective are extremely important for us. Guy Oakes (1984) wrote concerning Simmel, "the discovery of objectivity—the independence of things from the conditions of their subjective or psychological genesis—was the greatest achievement in the cultural history of the West" (p. 3).

However, I don't mean to imply that Simmel has had no influence on contemporary sociology. Quiet the contrary. Simmel's thought forms the basis of what has become known as **exchange** or rational choice theory. Simmel and exchange theorists argue that exchange is the basis of society: people seeking to satisfy individual drives are motivated in encounters to exchange valued assets to meet those needs. Simmel also has had direct influence on contemporary conflict theory through Lewis Coser (1956). Before Simmel, conflict had been understood as a source of social change and disintegration. Simmel was the first to acknowledge that conflict is a natural and necessary part of society. Coser brought Simmel's idea to mainstream sociology, at least in America. From that point on, sociologists have had to acknowledge that "groups require disharmony as well as harmony" and that "a certain degree of conflict is an essential element in group formation and the persistence of group life" (Coser, 1956, p. 31). In addition to Coser, we can find Simmel's influence in the works of such people as Robert K. Merton, Kurt Wolff, and Donald Levine. Further, Simmel's work is continuing to gain an audience in arenas like the sociology of knowledge, gender studies, cultural sociology, and formal sociology. Yet, what Kurt Wolff said concerning Simmel almost 50 years ago is still true today—"we hardly know him yet."

Simmel in Review

- Simmel was born in the heart of Berlin on March 1, 1858, the youngest of seven children. His father was a Jewish businessman who had converted to Catholicism before Georg was born. He died when Georg was an infant. Georg was thereafter mentored by a neighboring businessman (in music publishing), Julius Friedlanger, who upon his death left Simmel a substantial inheritance.
- In 1876, Simmel began his studies (history, philosophy, psychology) at the University of Berlin, taking some of the same courses and professors as Max Weber would a few years later. In 1881, Simmel defended his dissertation on Kant's philosophy of nature.
- In 1885, Simmel became an unpaid lecturer at the University of Berlin; he was dependent upon student fees. He was unable to attain a full-time academic position until almost the end of his career. At Berlin, he taught philosophy and ethics, as well as some of the first courses ever offered in sociology. In all probability, George Herbert Mead was one of the foreign students in attendance. During his life, a great many important intellectuals attended Simmel's lectures or were influenced by his thought; among them are Robert Park, Georg Lukács, Max Scheler, and Karl Mannheim.
- Many of Simmel's writings and public lectures addressed what were then considered nonacademic issues, such as love, flirtation, scent, and women. His diverse subject matter was due not only to his position as an outsider, but also to his intellectual bent—he was extremely interested in cultural trends and new intellectual movements. He was little involved in politics, and quite involved in the avant-garde philosophies of the day.
- Though he wrote many sociological essays and articles, his most important work of sociology was published in 1900, *The Philosophy of Money.* All together, Simmel published 31 books and several hundred essays and articles. In 1910, he, along with Max Weber and Ferdinand Tönnies, founded the German Society for Sociology.
- In 1914, Simmel was offered a full-time academic position at the University of Strasbourg. However, as WWI broke out, the school buildings were given over to military uses and Simmel had little lecturing to do.
- On September 28, 1918, Simmel died of liver cancer. He was aware of his terminal illness and spent his remaining days working on his philosophy of experience (*Lebensanschauung*)—a thesis on the tension between the individual's drive for creativity and its creations.

The Perspective: The Individual and Objective Culture

> Society exists where a number of individuals enter into interaction. . . . Strictly speaking, neither hunger nor love, work nor religiosity, technology nor the functions and results of intelligence, are social. They are factors in sociation only when they transform the mere aggregation of isolated individuals into specific forms of being with and for one another, forms that are subsumed under the general concept of interaction. (Simmel, 1971, pp. 23–24)

Simmel's thinking is possibly more complex and more in tune with modern and postmodern culture than any other classical theorist. He is quite comfortable maintaining dualisms and multiple effects in his theorizing, an important perspective when considering the diversification and possible fragmentation of modern life. Like Weber, he is convinced that culture is important when studying social behavior, and, like Weber, he is concerned with the subjective experience of people. But, unlike Weber, Simmel is troubled by the complexity of effects that modern culture can have on the individual. For example, the cultural effects of urbanization can be both advantageous (it increases individuality) and detrimental (it decreases emotional connectedness).

His thinking also addresses issues that some may think exist on the boundary between the center and periphery of society. Rather than theorizing about the structure of the economy *per se,* like Marx and Weber, Simmel is fascinated by the influence of money or the "social form" of exchange on human experience. He also addresses what may be seen as mundane or non-sociological concerns: his interests include such things as strangers, love, dyads, art, sexuality, music, fashion, and so on. Simmel was the first to take common, everyday social experiences as a focus of study (he was later followed in this by Erving Goffman). Part of the reason for his interest in such things is that Simmel does not think that society exists as macro-level structures; rather, he argues that society is recreated at the micro-level through the production of social forms.

The individual and social forms: One of the major questions that Simmel addresses is, "How is society possible?" In his answer, Simmel follows one of his favorite philosophers, Immanuel Kant. Kant did not ask about the possibility of society. Instead, he wondered how nature could exist as "nature" to science. Basically, Kant argued that the universe could exist as "nature" to scientists only because of the *category* of nature. Objects in the universe can only exist *as objects* because the human mind orders sense perception in a particular way. But Kant didn't argue that it is all in our heads; rather, he argued for a kind of synthesis: the human mind organizes our perceptions of the world to form objects of experience. So, nature can only exist as the object "nature" because scientists are *observing* the world through the *a priori* (existing before) *category* of nature. In other words, a scientist can see weather as a natural phenomenon, produced

> All the various ways in which man lives by his actions, knowledge, feelings, and creativity might be regarded as types or categories that are imposed on existence. (Simmel, 1997, p. 139)

through processes that we can discover, only because she or he assumes beforehand (*a priori*) that weather does not exist as a result of the whim of a god.

Simmel wants to discover the *a priori* conditions for society. This was a new way of trying to understand society. At the time, there were two main theories about society: mechanical-atomistic and organic. The mechanical-atomistic approach argues that individuals are the only reality: individuals are self-sustaining and independent and society is simply the summation of their activities (somewhat like Mead and very similar to many exchange or rational choice theories today). The organismic approach sees society as an independent entity, distinct from and sometimes subjugating of the individual (Marx and Durkheim are two examples of theorists with this perspective). Simmel wants to maintain the integrity of the individual (in this, Simmel is rather rare amongst sociologists) but at the same time recognize society as a true force. What Simmel argues is that society is a process that comes about through human interaction. Society continues to exist and to exert influence over the individual through forms of interaction. These forms, Simmel argues, are the *a priori* conditions of society.

A **social form** is a regular shape or mold into which interaction is placed. Simmel argues that we interact for many reasons and the content of our interaction may be as varied as the individuals that are participating. Nevertheless, interactions also tend to be patterned and consistent across time because of social forms. For example, you and I may decide to trade baseball cards or buy and sell stocks on the New York Stock Exchange, or a school system in California may buy computers from IBM. Whatever the content of the interactions, they will be guided by the same values and rules because they are all of the same form: they are exchanges. The content of the exchange is less important for the production of society than the form. Thus, society is produced through patterned interactions, and social forms determine the pattern of those interactions.

Producing a record of these social forms is part of what Simmel accomplishes in his sociology. He delineates specific kinds of interactions and their functions. We will be looking at some of those forms. Simmel did not, however, produce an exhaustive catalog of forms, which means that there are many yet to discover. So, in thinking like Simmel, part of what you can do is look for these forms. For example, what is the social form of what we might call a "fan"? Is there such a form regardless of its content (such as sports, music, or film)? Is it distinct from what Simmel calls sociability?

Simmel argues that we become conscious of society—it becomes part of our subjective experience—as we become conscious of our participation in social forms. Through our participation in social forms, we come into specific relationships with other individual consciousnesses. Thus, Simmel does not argue for a separate entity called society with a kind of group mind, as does Durkheim, nor does he maintain that there are nothing but individuals and individual motives, as does John Stuart Mill (and George Herbert Mead to a degree). Rather, Simmel maintains the distinction of each while seeing them as mutually constitutive. According to Simmel, individual humans do exist apart from society and society does exist apart from individual humans. Individuals directly experience society, not through a collective consciousness, but through a collective experience with a social form. In this argument, Simmel goes beyond Kant's notion of nature and science. For Kant, nature only exists as a scientific entity because of a scientific observer. For Simmel, society does not need an outside observer to be present because it exists directly in the consciousness of the individual through the experience of social forms.

Subjective and objective cultures: At the core of Simmel's thought is the individual. He assumes that human nature is something with which we are born. As we will see, Simmel feels that we naturally have a religious impulse and that gender differences are intrinsic. He also assumes that in back of most of our social interactions are individual motivations. This position is somewhat unique in sociology, as most theorists assume that human nature is intrinsically social. An important ramification of this assumption of Simmel's is that he is vitally concerned with the influence objective culture has on the individual, or, as Simmel phrases it, the relationship between objective and subjective culture.

> Every day and from all sides, the wealth of objective culture increases, but the individual mind can enrich the forms and content of its own development only by distancing itself still further from that culture and developing its own at a much slower pace. (Simmel, 1907/1978, p. 449)

Simmel was the first social thinker to make the distinction between subjective and objective culture the focus of his research. Individual or **subjective culture** refers to the ability to embrace, use, and feel culture. Collectives can form group-specific cultures, such as those who sported the spiked Mohawk of early punk culture. To wear such an item of culture immediately links the individual to certain social forms and types, and a group member would subjectively feel those links. Individuals and dyads are able to produce such culture as well. An individual could have special incense that she or he blends just for extraordinary, ritual occasions; or a couple could create a picture that would symbolize their relationship. This culture is very close to the individual and her or his psychological experience of the world.

Objective culture is made up of elements that become separated from the individual or group's control and reified as separate objects. Think about tie-dye tee shirts, for example. You can now go to any department store and buy such a shirt. You do not have to be a hippie to wear it, nor are you necessarily identified as a hippie, nor do you necessarily feel the connection to the values and norms of the hippie culture. It exists as an object separate from the individuals who produced it in the first place. Once formed, objective culture can take on a life of its own and it can exert a coercive force over individuals. For example, many of us growing up in the United States believe in the ideology and morality of democracy, though in truth we are far removed from its crucial issues, ideals, and practices. Other systems of objective culture identified by Simmel include tools, means of transportation, technology, art, language, religion, law, and philosophical systems.

Simmel's unique focus of study is this tension point between subjective and objective cultures. According to Simmel, cultural forms are necessary to achieve goals in a social setting. However, if these forms become detached from the lived life of the individual, they present a potential problem for the subjective experience of that individual. The idea behind this is much like Marx's species-being. In an ideal world, there is an intimate connection between the personal experience of the individual and the culture that he or she uses. However, as the gap between the individual and culture increases, and as culture becomes more objective, culture begins to attain an autonomy that is set against the creative forces of the individual. An important difference between Simmel and Marx's idea is that for Simmel, this reflexive antagonism is not a problem to solve but a legitimate site of research. For Simmel, objective culture is the distinctive modern puzzle. The question becomes, how does money, urbanization, religion, the division of labor, and so on affect the relationship between subjective and objective culture?

It is important to note that Simmel's writings reveal elements of both the objective and subjective approach to understanding society. On one level, that might seem obvious, but it is important because it influences not only the way Simmel sees society but also the way he writes. Sometimes his writings sound scientific and other times they sound interpretist or philosophical. We will see some of that back-and-forth movement in this chapter. It seems that if the subject is something that has been addressed clearly by another, like money or urbanization, then Simmel is concerned with explicating the dynamics. On the other hand, Simmel appears more interpretive and philosophical when holding up a new idea for us to consider.

To think like Simmel, then, is to assume the integrity of the individual. Simmel does acknowledge that personalities are formed as people interact socially, but, unlike Mead, who makes a case for a completely social self, Simmel feels that there are psychological impulses that are part of the individual and only get modified as they come in contact with society. So, Simmel's big question concerns how society, especially changes that come with modernity, impacts individual psychology. As we will see, Simmel always comes back to the individual. In this way, he takes an ecological approach—he wants to know how changes in the environment change the human organism. Because of this, he doesn't spend a great deal of time telling us about how social changes come about—he assumes them in the face of modernity. Rather, Simmel explicates the impact those changes have on people.

Here are some specific ideas that Simmel would have us think about:

- What happens to culture as a society moves from traditional rural economies to modern industrial economies, and how do those changes influence the individual?
- How will a person's life be different if she or he lives in the city rather than the country? How might it affect their emotions and the way they think?
- How does living in a modern society influence the kind of friendship network that a person might have? How will this network affect the individual's emotions and moral values?
- What about money? Does it make a difference to a society or an individual whether or not the exchange system is based on barter or money? In what ways? If money does make a difference, what about the use of credit? Does the existence of credit affect the kind of person you are?
- We have seen how conflict can bring about upheaval and change in a society. But are there any functional consequences of conflict? Can conflict actually make a group or society more integrated?

The rest of this chapter is divided into three major sections: Social Forms; Objectifying Culture; and Two Objective Institutions: Religion and Gender. In the first section, we will be looking at how certain social forms like sociability, conflict, exchange, and group size shape the individual and create society. The second section examines how things like money and cities create objective culture and how that in turn influences the individual. In the final section, we will see how objective culture interacts with two very important social entities: religion and gender.

Social Forms

Sociability and flirtation: Human beings survive because we organize our behaviors. What we generally mean by society is this organization. Social structures are simply patterns of organized behaviors that persist through time and space. Human beings could not survive apart from society. Yet, underneath and essential for these practical (structured) interactions is a shadow world of encounters whose purpose goes no further than simple human contact. These encounters shadow the more concrete social world in that the more practical forms may be present but do not function as means to ends but as ends themselves.

> Sociability is, then, the play-form of association. It is related to content-determined concreteness of association as art is related to reality. (Simmel, 1971, p. 130)

For example, we had some people over for dinner awhile back and I got into an "argument" about religion with one of the women. However, the purpose of our argument was not to convince or persuade; the purpose was simply to interact. I don't care if she ever sees things as I do, nor does she care if she convinces me. In fact, we are both convinced that the other will never change her or his point of view. And, honestly, if she did change her perspective, it would change her and she wouldn't be the person who is now my friend.

So we didn't argue to convince; we argued for the simple pleasure of talking, of being with one another and feeling connected. In **sociability,** talking is an end in itself. The content, in this case the stuff we were arguing over, is simply the carrier of stimulation. For conversation to fulfill a purely social function, it must be interesting, engaging, dynamic, and challenging—but the real end of an argument, persuasion, cannot be present. The sociability function of talk is why the topic of conversations can change so quickly and frequently. There is no goal other than the act of talking.

Simmel's concern with sociability reveals him to be a sociologist of importance and insight. Now, why would I say that? He isn't talking about race or inequality or revolutions or structural interrelationships. So how does being intrigued by mundane, pointless interactions qualify Simmel to be a remarkable and creative sociologist? Simmel was one of the first people to understand that threads of insignificant interactions hold society together. Without the threads produced by pure sociability, the fabric of society would unravel. Probably the most famous person to see this feature of society is Erving Goffman, who notes the importance of maintaining "face": people in social interactions work very hard to not embarrass themselves or others. We will ignore or repair things that might otherwise cause awkwardness. That sounds like we are just being polite, but Goffman argues that in saving face, we save the encounter, and in saving the encounter, we create a stable sense of social reality. I don't want to spend more time on Goffman now, but I do want us to get a hint that the sociology of the mundane may be one of our most important areas of study. Mundane and chatty interactions may have important social functions that aren't immediately apparent.

In some ways, sociability is play: we play at society. In order to carry out this play, a particular social balance appears: objective features of society and the individual must be left outside and only certain subjective features are expressed. The objective features of the individual refer to those elements associated with social structure. Things such as riches and position, learning and fame, and exceptional capacities and merit must all be left unmentioned and unacknowledged. Personal qualities—the character, mood, and what Simmel calls "the light and shadow of one's inner life"—must be left outside as well. The moment any of this becomes an issue in an encounter, it takes on a specific social form and the sociability is lost. Both the objective and purely subjective must be kept at bay in order to bring about a pure interaction, what Goffman would later refer to as an "interaction ritual."

In sociability, we present and focus on the self and its non-objective connections to others. Let's think about my argument example again. There were some objective features that could have been brought into the conversation. For example, I hold a Ph.D. and one of my specialty areas is religion. There is thus an objective structure in which I am an expert in religion. I could have appealed to that to make my case. But it was never brought up in our conversation, because to do so would have destroyed the sociability of it.

Further, in sociability we focus on the self without the subjective being fully revealed. Remember the last time you were having a bad day/week/month? Were there times when people would ask "how's it going," and you *didn't* tell them about all your troubles? Why didn't you tell them? Well, you say, they weren't interested. Really? I bet that most of those people are caring individuals. I also imagine that they have spent many hours listening to a friend in trouble. Okay, so they aren't cold and disinterested, so why didn't you tell them? It wasn't the time or place, you say. Ah, why wasn't it the time and place? And what was it the time and place *for*? It was not the time or place for the social form that we might label "support." It was the time and place for sociability, and we only reveal certain elements of our subjective life when we are being sociable.

There are different forms of sociability. Or perhaps a better way of putting it is that there are different social forms wherein sociability is present. One such form that Simmel (1984) explores is flirtation. Before we get into our discussion, I want to note that part of the way Simmel sees flirtation is rather traditional. He sees flirting occurring between a man and a woman, with the woman being the object of gaze and the man as the responder to the image she presents. I'm going to maintain the way Simmel talks about this issue, but keep in mind that in today's society the roles can be reversed—men can flirt with women—and may involve homosexual flirtation, rather than the heterosexual model Simmel gives us. Because this is a social form (remember that a social form is more about structure than it is about content), Simmel's theory of flirtatious sociability will hold no matter the direction or the sexual orientation of the flirtation.

Simmel says that flirting is a "natural growth in the soil of sociability." Just as in many conversations we play at society, in flirtation we play at eroticism. Flirtation demands the "simultaneity of implicit consent and refusal." By definition, flirting can be flirting only when the individual hints at consent and denial without letting the interaction come to either. Simmel also rightly links this sociability function of flirtation with the history of clothing. For example, much of women's clothing contains semi-concealment: it simultaneously calls attention to and forbids.

Flirtation also requires a specific attitude on the part of the man. He cannot see himself as a "victim carried along by her vacillations from a half-yes to a half-no." Nor can he enter into flirtation with an eye to gratification. If the erotic element in flirtation ever becomes a true possibility, then it becomes a private affair played out in the realm of reality and it loses its sociability. The man must see himself as an equal player in a game. He must desire nothing more than the free movement of play in which the "erotic lurks only as a remote symbol."

We must be careful to not dismiss these apparently empty and non–goal-directed interactions. For Simmel, sociability is an important social form that has at least four distinct functions. First, humans have an impulse to be with people. This impulse to be with others, to be part of a social group, is one of our most deeply seated needs as human beings. Think about Durkheim's argument concerning the collective consciousness and the soul. Both come about and are maintained because individuals have a sense of belonging to a group. There is also Mead's notion of the self and consciousness. Human beings are conscious and have a self only because we participate in interactions. In contemporary theory, people like Anthony Giddens and Harold Garfinkel argue that feeling part of an interaction is important for preserving a sense of shared reality and mutual trust. In sociability, this impulse to be connected to others and to talk is met in a pure way. That is, sociability has no other purpose (such as exchange or conflict or sexual union) than to produce the feeling of being connected to other people. Thus, sociability's first function is to provide this sense of connectedness and group inclusion.

The second function of sociability is, as Simmel puts it, the "resonance" that is left over after the interaction. He doesn't go into detail about the importance of this reverberation, but some contemporary theorists do, most notably Randall Collins. Collins (1988) argues that society is made up of interaction ritual chains. Out of each interaction we come away with two important things: cultural capital and emotional energy. *Cultural capital* refers to the things we know and can talk about. In an interaction, I share some of what I know in the hopes of receiving more knowledge. This knowledge may be as mundane as the season statistics for the current Oakland quarterback. The importance here is that my level of cultural capital influences into which interactions I can go and how I will fare.

Emotional energy is the affective or emotional residue of the interaction—it also invigorates basic motivations. If the encounter goes well and we are able to establish a firm sense of belonging and identity, then we take away positive emotions: we feel good about ourselves, the others, and the groups involved in the interaction. And, most importantly, because of these positive emotions we are more likely to interact again with those same people. Thus, one of the important things that links interactions together into the whole that we call society is the emotional residue of successful, yet mundane, interactions.

The third thing that sociability provides is a delightful affirmation of social reality. In some sense, social reality is a burden—we have to work to make it happen. Human reality is a meaningful reality and meaning is always something other than the simple object itself. In sociability, we playfully and casually affirm social reality. As Simmel (1971) says, all the tasks and weight of life are expressed in artistic play, "in which the heavily freighted forces of reality are felt only as from a distance, their weight fleeting in a charm" (p. 140). Remember my example of the argument? My friend and I were *playing at arguing*. The goal of the interaction was not to convince or persuade. By playing at arguing we were not only able to experience a close connection, but we also reaffirmed our own realities. I didn't convince her, but I ended up feeling more certain of my own beliefs. Chatty conversations of sociability provide a backdrop of taken-for-grantedness about our reality. As Berger and Luckmann (1966) note, an exchange such as, "Honey, I'm going to work," and "Okay, see you when you get home" implies "an entire world within which these apparently simple propositions make sense" (p. 153).

Lastly, sociability resolves the problem of association. The problem of association or society is the measure of meaning and importance that belongs to the individual as compared to society. As social theorists have noted, there is a tension that exists between the individual and society. At any given moment, which is more important or meaningful? Do the needs of the many outweigh the needs of the one? Or is the individual the supreme entity? You've felt this tension yourself many times. For example, if you have ever been tempted to cheat on an exam or to buy a paper for a class, you've felt the tension. There are your needs as the individual to pass the course, but there are also the needs of society for legitimate evaluations. So, which is more important, the individual or society? This tension exists in all social forms except for one: sociability.

This problem of association is solved most purely in sociability in that there are no other motives or content. In all other forms of interaction, we are circumscribed by the form itself and by the desires of other individuals. Thus, if we want to trade something, say my Gibson guitar for your vintage tube amp, we are each held in check by the other person's desires and needs. But in sociability there is no utilitarian purpose and thus no limits other than the form itself, meaning that the individual becomes more important and meaningful than society.

In addition to the four functions Simmel gives us, sociability may have another important function. One of the concepts that contemporary theorists have become concerned about is the production of the self. Following the work of such people as Mead and Simmel, many see the self as dependent upon and produced in social interactions. Some contemporary folks, like Kenneth Gergen, have argued that since social structure and culture are no longer cohesive and unified, our patterns of interaction are no longer consistent and thus the self produced in such interactions is fragmented. Let me explain this a bit further. If we assume that the self is related to interactions, then the kind of self that we produce is not only connected to the content of those interactions but also to the structure. So if our interactions are fairly consistent, we will experience a consistent self. On the other hand, if our interactions are different and changing, then the self we experience will also be different and changing.

However, one of the things that we can see from Simmel's notion of sociability is that our sense of self may come from the non-directed, non-structured elements of our interactions. Just as we experience a sense of belonging to the group in what are seemingly frivolous encounters, we can establish a firm sense of self-affirmation. Let me give you an example. The other day when I was at the gym, a guy came in and said, "How's it going?" I said, "Fine, and you?" He said, "Good, except that today is leg day and that is just too much pain. I'm tired." "Yeah, I know what you mean," I sympathized. Now, there are two interesting things about this interaction. First, it serves no purpose other than sociability. Second—and here's where we can see a theory moment—I was able to see this interaction repeated two more times.

> Even though our life seems to be determined by the mechanism and objectivity of things, we cannot in fact take any step or conceive any thought without endowing the objects with values that direct our activities. These activities are carried out in accordance with the schema of exchange. (Simmel, 1971, p. 84)

As I continued to work out, another guy came in. The first fellow said, "Hey, what's up?" The new guy said, "Nothing, you?" And the first man said . . . you guessed it, "Good, except that today is leg day and that is just too much pain. I'm tired." A third man walked in and the same pattern repeated itself. Now, what was happening here? The first man was engaging in sociability (meaningless conversation) in order to have his subjective experience of self confirmed by others. There is a way, then, that our incidental encounters can provide a cohesive and unified sense of self in the face of late modern dynamics.

Exchange: As we saw with Marx, the concept of **value** is something that has fascinated social scientists and philosophers for many years. Simmel takes the issue a bit further than Marx, for Simmel sees that human beings cannot do or think anything without attaching value. In this he is like Weber. For Simmel, value is a sentiment, a subjective emotion. In order for me to perform any task, I must *feel* it is worth doing, even so mundane a job as taking out the trash. As Simmel notes, even our thoughts are influenced by values—our minds become occupied by those things that we feel are meaningful or valuable.

Thus, value is a part of the emotional life of people. It has little to do with the naturally appearing world. The world will reject as refuse something quite valuable to humans—the skeletal remains of a long extinct dinosaur are nature's trash but a paleontologist's life. Simmel's concern with value is its objectification, and the relationship between the subject and objective. While value governs all that humans do and it is subjectively experienced, value exists in exchanges and the social form of **exchange** transcends the individual.

Simmel argues that most relationships for people are governed by exchange principles. The primary evidence that Simmel has for this claim is that the majority of our relationships are reciprocal. That is, most of our activities are at least a two-way give-and-take. It's important for us to see this definition of exchange because many exchange theorists insist that it is governed by rationality, the evaluation of costs and benefits. This issue has opened exchange theory up to criticism, for there are other theories that insist humans are usually not rational in their behaviors—we are cognitive sufficers that often behave spontaneously or emotionally. While Simmel sees pure economic exchange as governed by cost and benefit analysis, it isn't necessarily true for exchange generally. Exchange, then, as understood through reciprocation, is the basic form of society: "value and exchange constitute the foundation of our practical life" (Simmel, 1971, p. 47).

Simmel argues that the form of exchange places sense data (the things we touch, see, smell, taste, and hear), the individual ego, other people, and other social and geographic spaces into a meaningful nexus. Our world is made up of various things. But what does it all mean? Simmel says that the way we experience our world is shaped by value. At a given moment, for example, I may value my sense data over other people. I may be hungry and smell hamburgers at Burger King and go to the drive-through, even though I'm giving a friend a ride to school. Or my sense data may tell me that lifting weights hurts, but my ego values an athletic physique so I continue to work out. Value orders our world, and exchange orders our values.

Value emerges through exchange and is determined by two factors: sacrifice and scarcity. For Simmel, the value of any object in an exchange is not an intrinsic feature of the object. In other words, nothing has a set value in and of itself. Nor is the value determined simply by raw materials and human capital, as Marx would argue. Rather, the value of an object in exchange is determined by the sacrifice needed to gain it and its availability. For example, there is a black, vintage, Guild Bluesbird electric guitar for sale on eBay. Yet my desire for the guitar alone does not constitute its value. This particular guitar is vintage, which means that generally there are not very many of them around, so the availability is low. But scarcity is also a contextual variable in that at any given moment there may be zero or 10 black Bluesbirds for sale on eBay. If there are 10 Guilds available from different sellers, then I have different sources and its exchange value at that moment is less (as is the other person's power in the exchange).

Value is also influenced by sacrifice; scarcity alone does not determine it. There are many scarce objects around, like a 1968 Silvertone guitar, for which most people will not sacrifice. There may be numerous reasons why people won't sacrifice for that Silvertone, but why people are or are not willing to make the sacrifice is not as important to Simmel as the *level* of sacrifice—because it is the level of sacrifice that influences exchange value. There are two kinds of sacrifice. We can think of absolute sacrifice in terms of actual labor: how much work did I have to put out in order to get the money to pay for the guitar? The physical, emotional, and mental discomfort that I endured during work is the measure of absolute sacrifice. The level of relative sacrifice, on the other hand, can be understood by thinking about the other objects I couldn't buy if I got the guitar (like a whirlpool for the back deck) or the other things I could have done with my time and effort.

Conflict: Conflict is also a social form that has patterned effects. Simmel argues that conflict is instinctual for us, so we find it everywhere in human society. There is the conflict of war, but there is also the conflict that we find in our daily lives and relationships. Simmel also argues that conflict is different for humans than for other animals in that our conflicts are goal related. There is generally something that we are trying to achieve through conflict and there are different possible ways of reaching our goal. The existence of the possibility of different paths opens up opportunities for negotiation and different types and levels of conflict.

Because Simmel sees conflict as a normal and functional part of human life, he can talk about its variation in ways that others missed, such as its functional consequences and variation. One of the more important ways that conflict can vary is by its level of violence. If people perceive conflict as a means to a clearly expressed goal, then conflict will tend to be less violent. A simple exchange is a good example. Because of the tension present in exchanges, conflict is likely, but it is a low-level conflict in terms of violence. People engage in an exchange to achieve a goal, and that desired end directs most other factors. Another example is a worker strike. Workers generally go on strike to achieve clearly articulated goals and the strikers usually do not want the struggle to become violent—the violence can detract from achieving their goals. The passive resistance movements of the sixties and early seventies are other examples. We can think of these kinds of encounters as the strategic use of conflict.

However, conflict can be violent, and Simmel gives us two factors that can produce violent conflict: emotional involvement and transcendent goals. In order to become violent, people must be emotionally engaged. Durkheim saw that group interaction could increase emotional involvements and create moral boundaries around group values and goals. He didn't apply this to conflict, but Simmel does. The more involved we are with a group, the greater is our emotional involvement and the greater the likelihood of violent conflict if our group is threatened.

Conflict will also tend to have greater levels of violence when the goals of a group are seen to be transcendent. As long as the efforts of a group are understood to be directed toward everyday concerns, people will tend to moderate their emotional involvement and thus keep conflict at a rational level. If, on the other hand, we see the goals of our group as being greater than the group and the concerns of daily life, then conflict is more likely to be violent. For example, when the United States goes to war, the reasons are never expressed by our government in mundane terms. We did not say that we fought the Gulf War in order to protect our oil interests; we fought the war in order to defeat oppression, preserve freedom, and protect human rights. Anytime violence is deemed necessary by a government, the reasons are couched in moral terms (we fight for individual freedoms; communists would say they fight for social responsibility and the dignity of the collective). The existence of transcendent goals is why members of the Right to Life side of the abortion conflict tend to be more violent—their goals are more easily linked to transcendent issues.

Simmel makes the case for two kinds of functional consequences of conflict: consequences for the larger social system within which conflict takes hold and consequences for the respective parties in a conflict. When considering consequences for the social system, Simmel is concerned with low-level and more frequent conflict. When explaining the consequences for the different groups involved, he is thinking about more violent conflict.

Conflict in the larger social system, as between different groups within the United States, releases hostilities, creates norms for dealing with conflict, and develops lines of authority and judiciary systems. Remember that Simmel sees conflict as instinctual for humans. So a society must always contend with the psychological need of individuals to engage in conflict. Simmel appears to argue that this need can build up over time and become explosive. Low-level, frequent conflict tends to release hostilities and thus keep conflict from building and becoming disintegrative for the system. It also creates pressures for society to produce norms for the expression of conflict. We can see this same dynamic operate at the dyad level. For example, when a married couple experiences repeated episodes of conflict, like arguing, they will come up with norms for handling the tension in a way that preserves the integrity of the relationship. The same is true for the social system, but the social system will go a step further and develop formal authorities and systems of judgment to handle conflict. Thus, frequent, low-level conflict creates moral and social structures that facilitate social integration.

The different groups involved in conflict also experience functional results, especially when the conflict is more violent. As a group experiences external conflict, the boundaries surrounding the group become stronger, the members of the group experience greater solidarity, power becomes centralized and more efficient, and the group tends to form coalitions with other groups (the more violent the conflict is, the more intensified are these effects). In order for any group to exist, it must include some people and exclude others. This inclusion/exclusion process involves producing and regulating different behaviors, ways of feeling and thinking, cultural symbols, and so forth. These differences constitute a boundary around the group, clearly demarcating those who belong from those who do not. As a group experiences conflict, the boundaries surrounding the group become stronger and better guarded. For example, during WWII the United States incarcerated those of Japanese descent. Today we may look back at that with shame, but at the time it made the United States stronger as a collective; it more clearly demarcated "us" from "them," which is a necessary function for any group to exist. Conflict makes this function more robust.

Along with stronger external boundaries, conflict enables the group to also experience higher levels of internal solidarity. When a group engages in conflict, the members will tend to feel a greater sense of camaraderie than during peaceful times. They will see themselves as more alike, more part of the same family, existing for the same reason. Group-specific behaviors and symbols will be more closely guarded and celebrated. Group rituals will be engaged in more often and with greater fervency, thus producing greater emotional ties between members and creating a sense of sacredness about the group.

In addition, the authority within a group experiencing conflict tends to become more centralized. A centralized government is more efficient in terms of response time to danger, regulating internal stresses and needs, negotiating external relations, and so on. Violent conflict also tends to produce coalitions with previously neutral parties. Again, WWII is a clear example. The story of WWII is the story of increasing violence with more and more parties being drawn in. Violent conflict produces alliances that would have previously been thought unlikely, such as the United States being allied with Russia.

Group size: As we've seen, social forms exist independently of their content and they can be applied to various kinds of actors. In the same ways, group size functions as a social form. So, for example, it matters if you are part of a dyad or triad (that is, a group with just two or three people). In a dyad, only another individual confronts each person, though to onlookers the dyad is seen as a group. In dyads, the group does not attain an objective existence above the individuals. There is no structure beyond the individual; there is only the other individual. What this means is that the dyad is always personal. You are always face-to-face in a dyad; there isn't an "objective other" to whom responsibility or blame can be shifted. This lack of objective other also implies that dyads have a greater natural tendency to exhibit subjective rather than objective culture.

Yet when one other person is added to the dyad, the individual in the group is confronted with a collective that exists beyond her or him. For example, I've been writing songs with a friend for about six years. The other day we decided to add a bass player to the group. Now there are three of us. Before Josh joined, if we had a disagreement, I only had to contend with Steve. But now if there is disagreement, I might have to contend with both Steve and Josh, a group within the group. In the move from dyad to triad, the possibility of social power appears for the first time, as well as the possibility of mediators and indirect relations. Thus, the size of a group brings its own dynamics regardless of the group's purpose or content.

Objectifying Culture

Objective and subjective cultures: We've seen that Simmel is concerned with the relationship between subjective and objective culture, and we know that the effects of objective culture on the individual trouble Simmel. Social forms are a necessary part of objective culture. They enable us to fulfill our individual desires and needs. Left alone, they would probably have little influence on the subject. However, there are certain social factors that tend to increase the distance between subjective and objective culture.

The following diagram (Figure 6.1) pictures the relationship between subjective and objective culture. What we see on the far left is probably what Simmel has in mind under ideal conditions. People need culture to interact with others, but in small, traditional communities the culture can be kept graspable and thus subjective. The double-headed arrows indicate reciprocal relations—subjective culture is made by and facilitates others, individuals, and interactions. Objective culture, on the other hand, stands apart from the individual psychology. Culture becomes more objective as interactions are extended to distant others and as certain features of modernity become more prominent. Objective culture is always produced out of subjective, individual culture, but it can take on a life of its own and can influence and dominate subjective experience. When this happens, a lag is produced between the individual and objective cultures. As the size and complexity of the objective culture increases, it becomes more and more difficult for individuals to embrace it as a whole. Individuals come to experience culture sporadically and in fragments. How individuals respond to this tension between subjective experience and culture is of utmost concern to Simmel.

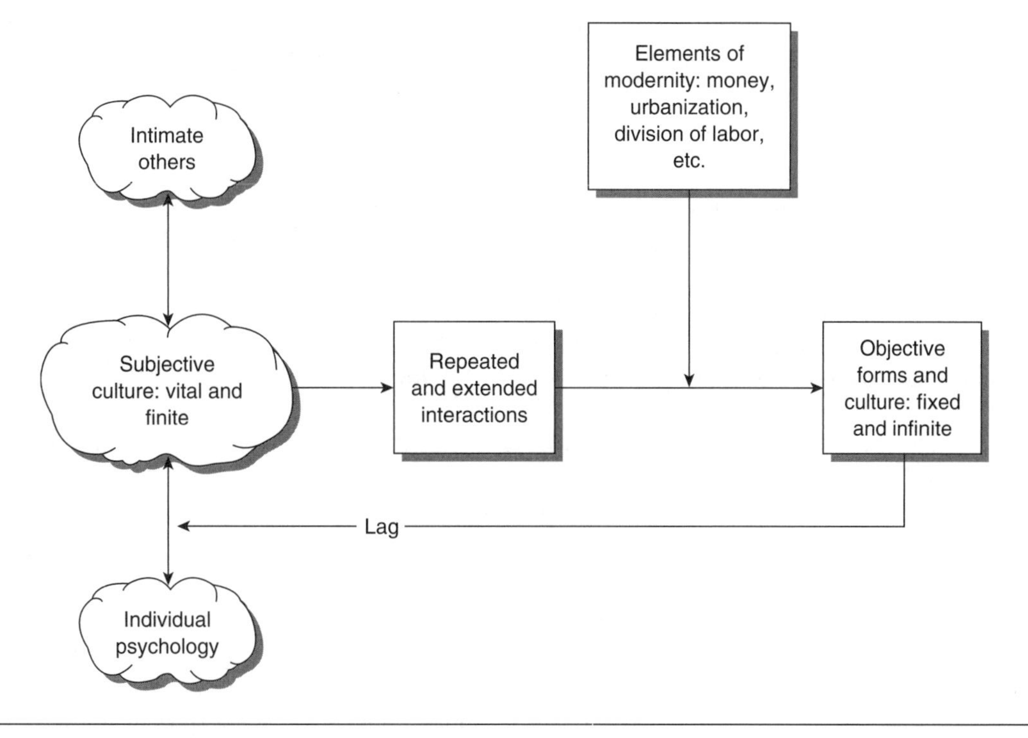

Figure 6.1 Subjective and Objective Culture

Simmel identifies three variables in objective culture, the first being that it can vary in its *absolute size.* The pure bulk of cultural material can increase or decrease. In modernity, the amount of objective culture increases continuously. For example, in the year 2000 the world produced approximately 1200 terabytes of scanned printed material. A terabyte contains over 1 trillion bytes. If we were to make a single book that contained just 1 billion characters, it would be almost 32 miles thick; a trillion is 1000 billion. It has been estimated that to count to 1 trillion would take over 190,000 years, if we counted 24/7/365. The human race created over 1200 trillion bytes of printed information in 2000. That's not counting the Internet. And that figure increases by 2 to 10% each year.

Objective culture can also vary by its *diversity of components.* Let's take fashion, for example. Not only are there simply more fashion items available (absolute size), there are also more fashion types or styles available—there are fashions for hip-hop, grunge, skater, hardcore, preppy, glam, raver, piercer, ad infinitum.

Finally, objective culture can vary by its *complexity.* Different types of culture can either be linked or unlinked. If different cultures become linked, then the overall complexity of the objective culture increases. For example, when this nation first started, there were only a few different kinds of religions (a couple of different Protestant denominations and Catholicism). Due to various social factors, the objective culture of religion has increased in its size and diversity, resulting in any number of different kinds of religion in America today. The culture of religion has also become more complex, especially in the last few years. Today we find people who are joining together in what was previously thought to be antithetical forms of religion. So, for example, we can find Christian-Pagans in North Carolina. As these different forms become linked together, the religious culture becomes increasingly more complex.

There are three interrelated forces in modernity that tend to increase objective culture in all three of its areas—urbanization, money, and the configuration of one's social network. Urbanization appears to be the principal dynamic. It increases the level of the division of labor, the extent that money and markets are used, and it changes one's web of affiliations from a dense, primary network to a loose, secondary one. As is typical with Simmel, we will find that these social processes bring some conflicting effects.

> The deepest problems of modern life derive from the claim of the individual to preserve the autonomy and individuality of his existence in the face of overwhelming social forces, of historical heritage, or external culture, and of the technique of life. (Simmel, 1959, p. 409)

Urbanization: Simmel's (1950) concern with objective culture is nowhere clearer than in his short paper, "The Metropolis and Mental Life": "The most profound reason . . . why the metropolis conduces to the urge of the most individual personal existence . . . appears to me to be the following: the development of modern culture is characterized by the preponderance of what one may call the 'objective spirit' over the 'subjective spirit'" (p. 421). This increase in objective spirit happens principally because of two interrelated dynamics: the division of labor and the use of money, both of which are spurred on through **urbanization**—the process that moves people from country to city living.

Historically, people generally moved from the country to the city because of industrialization. As a result of the Industrial Revolution, the economic base of the country changed and with it the means through which people made a living. As populations became increasingly concentrated in one place, more efficient means of providing for the necessities of life and for organizing labor were needed. This increase in the division of labor happened so that products could be made more quickly and the workforce could be more readily controlled. Simmel (1950) argues that the division of labor also increases because of worker-entrepreneur innovation: "The concentration of individuals and their struggle for customers compel the individual to specialize in a function from which he cannot be readily displaced by another" (p. 420).

The division of labor demands an "ever more one-sided accomplishment" and we thus become specialized and concerned with smaller and smaller elements of the production process. This one-sidedness creates objective culture: we are unable to grasp the whole of the product and the production process because we are only working on a small part. The worker in a highly specialized division of labor becomes "a mere cog in an enormous organization of things and powers which tear from his hands all progress, spirituality, and value in order to transform them from their subjective form into the form of purely objective life" (Simmel, 1950, p. 422).

Simmel claims that the consumption of products thus produced has an individuating and trivializing effect. It isn't difficult for us to see that we perform and create self out of things. I just recently bought a house and have spent a good bit of time decorating it. When I have friends over to see the new place, I glow when they say, "this place just fits you." We are all particular about the clothes we wear, the car we drive, the CDs we possess, the sound system we own, the perfume we use, and so on for the same reason: we are creating a sense of self. Notice a difference here between Simmel and Marx: for Simmel, there isn't anything intrinsically wrong with using commodities to create our self-concept and self-image.

On the other hand, Simmel also sees that the products we are using in modernity to produce and express self are changing. They are becoming more and more divorced from subjectivity due to the division of labor and market economy: there are too many of them and they are too easily replaced (it is thus difficult to get attached to them); they have become trivialized in order to meet the demands of mass markets (they thus have very little meaning in the first place); and products are subject to the dynamics of fashion and diversification of markets, which lead to inappropriate sign use (for example, dressing like a raver or a member of the upper class when you aren't a member of those groups).

Urbanization also increases the level of exchange in a society and thus the use of money-facilitated markets. The use of money increases the level of objective culture due to increasing demands for rational calculations. It also facilitates the processes we've just been talking about: the division of labor and trivialization of culture. In addition, we will see that money also has positive effects for the individual and society.

Money: When thinking about the presence of money in society, it is important to understand it in comparison to barter. Barter is the exchange of goods or services for other goods or services. In a barter system, there is no universal value scheme. Everything is equated on an item-by-item basis and every item is equally real. Thus, the bushel of corn that I grew may be exchanged for the 2 chairs that you made or for the 10 loaves of bread that Francis baked. The value of a product or service can vary tremendously, depending on local conditions or the subjective state of the trader.

> This web is held together by the all-pervasive money value, just as nature is held together by the energy that gives life to everything. (Simmel, 1978, p. 431)

Money creates a universal value system wherein every commodity can be understood. Of necessity, this value system is abstract; that is, it has no intrinsic worth. In order for it to stand for everything, it must have no value in itself. The universal and abstract nature of money frees it from constraint and facilitates exchanges. But it also has other effects; we will talk about four. These effects increase the more money (or even more abstract systems such as credit) is used.

One effect of money is that it increases individual freedom by allowing people to pursue diverse activities (paying to join a dance club or a pyramid sales organization) and by increasing the options for self-expression (we can buy the clothes and makeup to pass as a raver this week and a business professional next; we can even buy hormones and surgery to become a different sex).

Second, even though we are able to buy more things with which to express and experience our self, we are less attached to those things because of money. We tend to understand and experience our possessions less in terms of their intrinsic qualities and more in terms of their objective and abstract worth. So I understand the value of my guitar amplifier in terms of the money it cost me and how difficult it would be to replace (in terms of money). The more money I have, the less valuable my Laney amplifier will be, because I could afford a hand-wired, boutique amp. Thus, our connection to things becomes more tenuous and objective (rather than emotional) due to the use of money.

Third, money also discourages intimate ties with people. Part of this is due to the universal nature of money. Remember that money is a universal system. Because of this, it comes to stand in the place of almost everything, and this effect spreads. When money was first introduced, only certain goods and services were seen as equivalent to it. Today in America, we would be hard pressed to think of many things that cannot be purchased with or made equivalent to money, and that includes relationships. Much of this outcome is due to indirect consequences of money: the relationships we have are in large part determined by the school or neighborhood we can afford. Some of money's consequences are more direct: we buy our way into country clubs and exclusive organizations.

Money further discourages intimate ties by encouraging a culture of calculation. The increasing presence of calculative and objectifying culture, even though spawned in economic exchange, tends to make us calculating and objectifying in our relationships. All exchanges require a degree of calculation, even barter, but the use of money increases the number and speed of exchanges. As we participate in an increasing number of exchanges, we calculate more and we begin to understand the world more in terms of numbers and rational calculations. Money is the universal value system in modernity, and as it is used more and more to assess the world, the world becomes increasingly understood in terms of numbers. So, the use of money in exchanges multiplies the number and speed of exchanges and thus calculations; and money infiltrates our understanding of the world and relationships, which in turn quantifies the world and our relationships ("time is money" and we shouldn't "waste time just 'hanging out'").

Fourth, money also decreases moral constraints and increases anomie. Money is an amoral value system. What that means is that there are no morals implied in money. Money is simply a means of exchange, a way of making exchanges go easier. Money knows no good or evil: it can be equally used to buy a gun to kill school children as to buy food to feed the poor. So, as more and more of our lives are understood in terms of money, less and less of our lives have a moral basis. In addition, because moral constraints are produced only through group interactions, when money is used to facilitate group membership, it decreases the true social nature of the group and thus its ability to produce morality.

Thus, money has both positive and negative consequences for the individual. Money increases our options for self-expression and allows us to pursue diverse activities, but it also distances us from objects and people and it increases the possibility of anomie. In the same way, there are both positive and negative consequences for society. However, while it may seem that the negative consequences outweigh the positive for the individual, the consequences for society are mostly positive. We will look at a total of four social effects of the use of money.

The use of money actually creates exchange relationships that cover greater distances and last longer periods of time than would be possible without the use of money. Let's think of the employment relationship as an illustration. If a person holds a regular job, she or he has entered into a kind of contract. The worker agrees to work for the employer for a given number of hours per week at a certain pay rate. This agreement covers an extended period of time, which should only be terminated by a two-week notice or severance pay, depending on the contract. This relationship may cover a great deal of geographic space, as when the workplace is located on the West Coast of the United States and the corporate headquarters is on the East Coast. This kind of relationship was extremely difficult before the use of money as a generalized media of exchange, which is why many long-term work relationships were conceptualized in terms of familial obligations, such as the serf or apprentice. With today's more generalized forms of the money principle, such as credit and credit cards, social relations can span even larger geographic expanses and longer periods of time (for example, I am obligated to my mortgage company for the next 25 years, and I just completed an eBay transaction with a man living in Japan). What this extension of relations through space and time means is that the number of social ties increases. While we may not be connected as deeply or emotionally, we are connected to more diverse people more often, and that means that society has a broader base of connectedness. Think of society as a fabric: the greater the number and diversity of ties, the stronger is the weave.

Second, money also increases continuity among groups. Money flattens or generalizes the value system by making everything equivalent to itself. It also creates more objective culture, which overshadows or colonizes subjective culture. Together these forces tend to make group-specific culture more alike than different. What differences exist are trivial and based on shifting styles. Thus, while the fabric weave of society is more dense due to the effects of money, it is also less colorful, which tends to mitigate group conflicts.

Third, money strengthens the level of trust in a society. What is money really? In America it is nothing but green ink and nice paper. Yet, we would do and give almost anything in exchange for enough of these green pieces of paper. This exchange for relatively worthless paper occurs every day, without anyone so much as us blinking an eye. How can this be? The answer is that in back of money is the United States government and we have a certain level of trust in its stability. Without that trust, the money would be worthless (think of Confederate money). A barter system requires some level of trust, but generally speaking you know the person you are trading with and you can inspect the goods. On the other hand, money demands a trust in a very abstract social form—the state—and that trust helps bind us together as a collective.

Lastly, in back of this trust is the existence of a centralized state. In order for us to trust in money, we must trust in the authority of a single nation. When money was first introduced in Greek culture, its use was rather precarious. Different wealthy landowners or city-states would imprint their image on lumps or rods of metal. Because there was no central governing authority, deceit and counterfeiting were rampant. The images were easily mimicked and weights easily manipulated. Even though money helped facilitate exchanges, the lack of oversight dampened the effect. It wasn't until there was a centralized government that people could completely trust money. Thus, when markets started using money for exchanges, they were inadvertently pushing for the existence of a strong nation-state. Centralized authority is the structural component to a society's trust, and it binds us together.

Social networks: As a further result of urbanization, Simmel argues that social networks (what he calls the **web of group affiliations**) have changed. When we talk about social networks today, what we have in mind are the number and type of people with whom we associate, and the connections among and between those people. Network theory is an important part of contemporary sociology, and Simmel was one of the first to think in such terms. For his part, Simmel is mostly concerned with why people join groups. While this issue may seem minor, there are in fact important ramifications.

In small rural settings, there are relatively few groups for people to join, and most of those memberships are strongly influenced by family: we tend to join the same groups as members of our family do. In traditional, rural settings, the family is a primary structure for social organization, and families tend not to move around much. So there are likely to be multiple generations present. As a result, the associations of the family become the associations of the child. A child reared in such surroundings will generally attend the same church, school, and work as her or his parents, grandparents, cousins, and so on. For example, it is much more likely for a person living in Rural Hall, North Carolina, than for someone in Los Angeles, California, to live on property that has been handed down from grandparents, that is next to the houses of brothers and sisters and cousins, on a street that bears the family name; to attend large family dinners on Sunday; to marry her or his childhood sweetheart, and so on.

In rural settings, people tend to join groups because of **organic motivation**—because they are naturally or organically connected to the group. Many of the groups with which a person affiliates in this setting are primary groups. **Primary groups** are noteworthy because they are based on ties of affection and personal loyalty, endure over long periods of time, and involve multiple aspects of a person's life. In modern, urban settings, people join groups out of **rational motivation**—group membership due to freedom of choice. The interesting thing to note about this freedom is that it is forced on the individual—in other words, there aren't any organic connections. In large cities, people usually do not have much family around and the personal connections tend to be rather tenuous. In Southern California, for example, people move on average every 5 to 7 years. Most only know their neighbors by sight and the majority of interactions are work related (and they change jobs about as often as houses). What this means is that people join social groups out of choice (rational reasons) rather than out of some emotional and organic connectedness. These kinds of groups tend to have the characteristics of **secondary groups** (goal and utilitarian oriented, with a narrow range of activities, over limited time spans).

Not only are the motivations different, but the effects of one's web of group affiliations will have a dramatic impact on the subjective experience of the self. Consider the two pictures on the next page. Figure 6.2 depicts the group affiliations of someone who lives in a small, rural environment; Figure 6.3 depicts the affiliations of a person living in a city.

You will notice that the second configuration has a greater number of group affiliations. This is not only because the urban individual has a greater number of available groups, it is also because the motivation for joining is different (organic

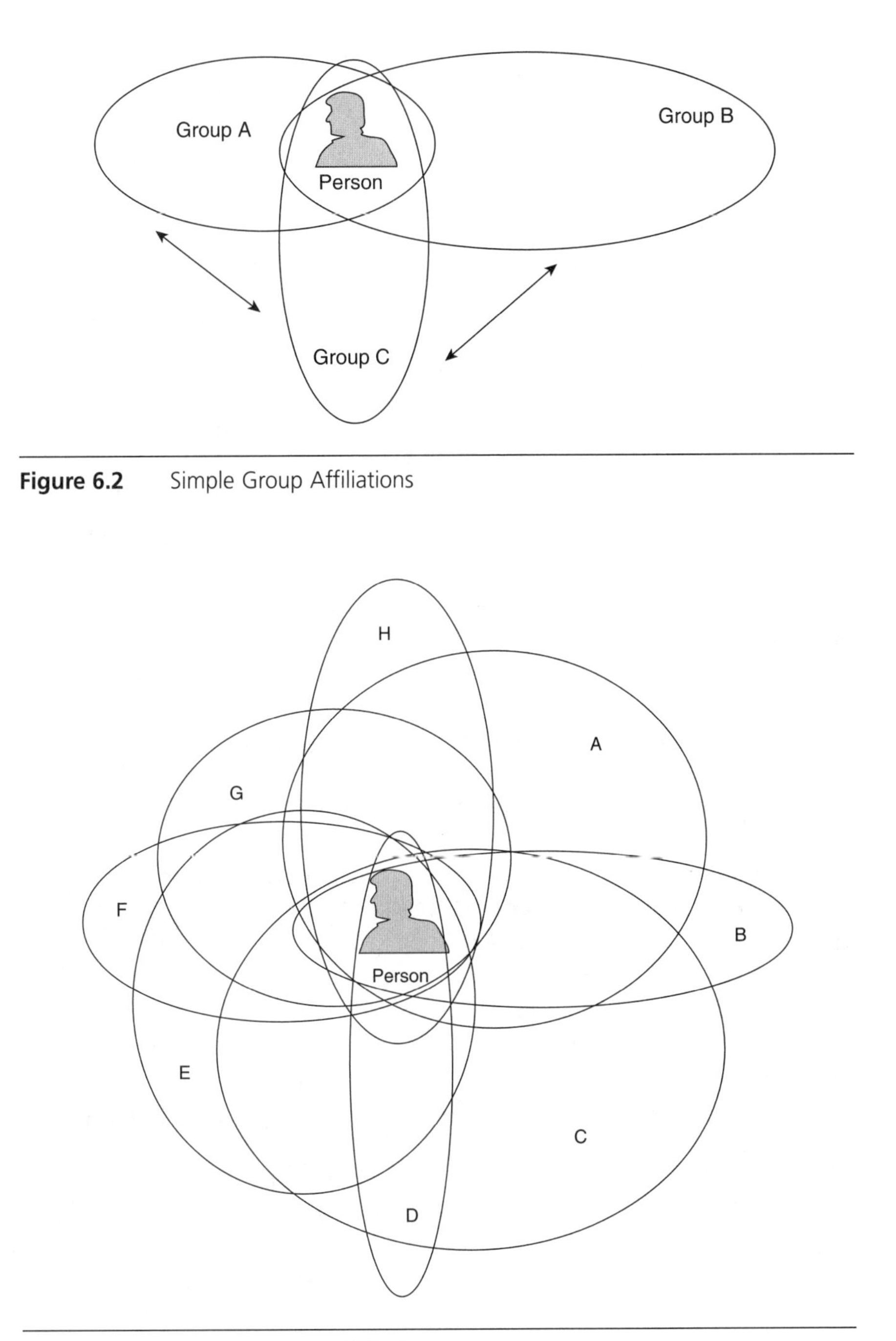

Figure 6.2 Simple Group Affiliations

Figure 6.3 Complex Group Affiliations

vs. rational). In the city, our choices tend to not be limited by family or primary relationships. You will also notice that the first picture has arrows connecting some of the groups, but the second one does not. This is to indicate that the groups in a small, rural social environment tend to be connected (so, for example, the people at your work either go to the same church or know people at your church). In the metropolis, there are few connections between groups—the bigger the city, the fewer the connections. So, for example, in New York City, the chances are quite slim that you would ever see someone from work at your dance group, or church, or your reading group, or the neighborhood housing coalition, or the Christmas play committee at your child's school, or in the yoga group you are going to join next month to replace the dance group with which you are growing tired.

As a result of rational group affiliations, it is far more likely that individuals will develop unique personalities. We reflect the kinds of groups with which we associate and the structure of our group associations. In the simple model in Figure 6.2, the person has few group influences and these groups have some association with one another. Thus, the community that is reflected in this model (simple, rural) will contain people who are very much alike. They will draw from the same basic group influences and culture and the groups will possess a compelling ability to sanction behavior and bring about conformity.

On the other hand, the person in the complex structure (Figure 6.3) has multiple and diverse influences and the groups' capacity to sanction is diminished. From Simmel's point of view, the group's ability to sanction is based on the individual's dependency upon the group. If there are few groups from which to choose, then individuals in a collective are more dependent upon those groups and the groups will be able to demand conformity. This power is crystal clear in traditional societies where being ostracized meant death. Of course, the inverse is also true: the greater the number of groups from which to choose and the more diverse the groups, the less the moral boundaries and **normative specificity**. In turn, this decrease in sanctioning power leads to greater individual freedom of expression. Many students, for example, are able to express themselves more freely after moving away from home to the university. This is especially true of students who move from rural to urban settings. Not only is the influence of the student's childhood groups diminished (family, peers, church), but there are also many, many more groups from which to choose. These groups often have little to do with one another. Thus, if one group becomes too demanding of time or emotion or behavior, you can simply switch groups. So, it may be the case that you experience yourself as a unique individual having choices, but it has little to do with you *per se:* it is a function of the structure of your network.

While decreased moral boundaries and normative specificity lead to greater freedom of expression, they can also produce **anomie**—also a concern of Durkheim's. For the individual, it speaks of a condition of confusion and meaninglessness. Unlike animals, humans are not instinctually driven or regulated. We can choose our behaviors. That also means that our emotions, thoughts, and behaviors must be ordered by group culture and social structure or they will be in chaos and will have little meaning. When group regulation is diminished or gone, it is easy for people to become confused and chaotic in their thoughts and emotions. Things in our life and life itself can become meaningless. So while we may think that personal freedom is a great idea, too much freedom can be disastrous.

Complex webs of group affiliations can have two more consequences: they can increase the level of role conflict a person experiences, and they contribute to the blasé attitude. **Role conflict** describes a situation in which the demands of two or more of the roles a person occupies clash with one another (such as when your friends want to go out on Thursday night but you have a test the next morning). The greater the number of groups with which one affiliates, the greater is the number of divergent roles and the possibility of role conflict. However, the tendency to keep groups spatially and temporally separate mitigates this potential. In other words, modern groups tend not to have the same members and they tend to gather at different times and locations. So we see the roles as separate and they rarely come in conflict.

Complex group structures also contribute to the **blasé attitude**, an attitude of absolute boredom and lack of concern. Every social group we belong to demands emotional work or commitment, but we only have limited emotional resources, and we can only give so much and care so much. There is, then, a kind of inverse relationship between our capacity to emotionally invest in our groups and the number of different groups of which we are members. As the number and diversity of social groups in our lives go up, our ability to emotionally invest goes down. This contributes to a blasé attitude, but it also makes conflict among groups less likely because the members care less about the group's goals and standards.

This blasé attitude is also produced by all that we have talked about so far, as well as overstimulation and rapid change. The city itself provides for multiple stimuli. As we walk down the street, we are faced with diverse people and circumstances that we must take in and evaluate and react to. In our pursuit of individuality, we also increase the level of stimulation in our lives. As we go from one group to another, from one concert or movie to another, from one mall to another, from one style of dress to another, or as we simply watch TV or listen to music, we are bombarding ourselves with emotional and intellectual stimulation. In the final analysis, all this stimulation proves to be too much for us and we emotionally withdraw. Further, this stimulation is in constant flux. Knowledge and culture are constantly changing. We could point to a variety of changes in medicine or style or "common knowledge" over the past few years, but I think that one of the most poignant examples of this constant change is the MTV show, "so 5 minutes ago." It is a show that looks at the changing culture of youth. The very existence of the show acknowledges and chronicles what Simmel is talking about for a specific segment of the culture. (I would invite you to check it out, but by the time you read this, the show itself will probably be "5 minutes ago.")

All that we have talked about in this section so far creates a social psychological need. Drowning in a sea of blasé and objectified culture, people feel the need to exaggerate their statements of difference. In order to sell them in mass markets, all truly personal or group-specific qualities have been bleached out of the products. And so, Simmel says, we must exaggerate any differences that do exist in order to stand out and experience our personal selves. This issue of exaggerated claims is very much in keeping with what some people are seeing happening in late modernity. As Zygmunt Bauman (1992) notes, "To catch the attention, displays must be ever more bizarre, condensed and (yes!) disturbing; perhaps ever more brutal, gory and threatening" (p. xx). Thus, modernity increases our freedom of expression, but it also forces us to express it more dramatically with trivialized culture:

> On the one hand, life is made infinitely easy for the personality in that stimulations, interests, uses of time and consciousness are offered to it from all

Table 6.1 Effects of Urbanization

The Division of Labor	↑ specialized culture ↑ production of trivialized products
Money Effects—Individual	↑ individual freedom ↓ attachment to products ↓ intimate ties with people ↑ anomie ↑ goal displacement
Money Effects—Societal	↑ number of social relationships ↑ continuity between groups ↑ social trust ↑ centralization of power
Rational Social Networks	↑ unique personality ↑ anomie ↑ role conflict ↑ blasé attitude

> sides. . . . On the other hand, however, life is composed more and more of these impersonal contents and offerings that tend to displace the genuine personal colorations and incomparabilities. This results in the individual's summoning the utmost in uniqueness and particularization, in order to preserve his most personal cores. He has to exaggerate this personal element in order to remain audible even to himself. (Simmel, 1950, p. 422)

We have covered a great deal of conceptual ground in this section. It's been made all the more complicated because each of the things we have talked about brings both functional and dysfunctional effects and the effects overlap and mutually reinforce one another. Normally, I would be tempted to draw all this out in a theoretical model; however, it would be way too complex for our purposes. Instead, I will finish this section with a list of the different effects (see Table 6.1). This list doesn't explain the theoretical dynamics, but it will provide you with a general and clear idea about how urbanization is influencing our lives. If in going over this list you find that you can't recall the theoretical explanation about why the effect is occurring, please go back and reread the text.

In general, then, modernity is characterized by a high level of urbanization. Urbanization brings with it three primary outcomes: increases in the division of labor, the use of money and markets, and rational group membership. Each of these has the effects (note that ↑ denotes an increasing and ↓ a decreasing tendency toward) noted in Table 6.1.

Two Objective Institutions: Religion and Gender

There are a number of different kinds social encounters and categories, and our subjective experience of each of these can be influenced by the presence of objective culture. Two experiences in particular give us insight into Simmel's thinking: religion and gender. With both, Simmel postulates a personality trait or essence that is unique to the individual. There is the "religious impulse" that is the fount of religion, yet is walled in and stifled by the objective culture of religion. There are also essential gender differences, according to Simmel, who makes the claim that objective culture is important because it is distinctly male. Women, then, are in a position where their subjective experience as women is denied or denigrated by the hegemonic culture.

The religious impulse: Simmel wants to trace the psychological origins of the historical and sociological developments of the objective form of religion. He begins with the assumption that there is within humans a "religious impulse," stronger in some than in others. Simmel argues that human beings have psychological needs for experiencing completeness in what is otherwise a fragmentary world, for having a fixed point in a world that is ever changing, for an overarching purpose for life, for a force that can reconcile conflicts between people, and for an external object toward which our human nature can be directed.

In some ways, Simmel expands on what Durkheim (1887/1993) claimed: "The fact is that we have a need to believe that our actions do not exhaust all their consequences in an instant . . . but that their consequences are extended for some distance in duration and scope. Otherwise they amount to very little . . . and they would hold no interest at all for us" (p. 119). Mead also touches on this notion that meaning and people do not exist as human in the present moment—the "I" and the "Me" exist in temporally different ways: the I is in the immediate social moment but the Me, the observing part of the self of which we are aware, exists outside of that context. The self must exist in this way so that we can observe our own behaviors. Alfred Schutz has this same issue in mind when he explains how humans impute meaning to experiences. It is always in a backward glance: experiences don't have meaning until we think about them, and we always think about them after they occur.

> Religion does not create religiosity, but religiosity creates religion. The workings of fate, as man experiences it in a certain subjective mood, consist of an interweaving of relationships, meanings, and feelings that do not yet make up religion in themselves and whose factual substance would never be linked with religion by souls in a different mood. Separated from these actualities, however . . . these elements constitute an objective realm of their own, and thus form "religion"—that is, the world of the objects of faith. (Simmel, 1997, p. 150)

"All of this," according to Simmel (1997), "nourishes man's ideas about the transcendent sphere" (p. 142). According to *Merriam-Webster's Unabridged Dictionary,* the first two meanings of transcendent are "going beyond or exceeding usual limits" and "proceeding beyond or lying outside of what is perceived or presented in experience." Meaning is just such a transcendent move; it doesn't exist in the object or the experience. Let's think about our example of the paper and green ink that mean money. If I were to hold up a $100 bill and ask you where the money is, you would probably point to the bill. But you would only be partially correct. The paper is a cue that tells us what meaning to give. However, the meaning isn't right there in the paper object. Actually, the meaning of money is the result of countless interactions that have taken place over history and are taking place at the very moment I show you the bill. The meaning of money isn't precisely in this place or time. In order to understand the meaning of the bill, we have to transcend this moment and the object. Here is the important point: this is true of every meaningful object or experience. It is, then, in the very nature of humans to exist transcendently and to dwell on transcendent issues.

Thus, transcendence and the religious impulse are normal for humans, because of the way in which we exist. In addition, there are certain feelings that are themselves religious. Among them are love, giving, and faith. All of these Simmel understands in their sociological context as well. Let's take faith, for example: "Practical faith is a fundamental quality of the soul that in essence is sociological, that is, it becomes concretized as a relationship with some being external to the self" (Simmel, 1997, p. 169). So faith is something that exists within the human soul, but it is expressed sociologically. Simmel argues that all social relationships require faith to one degree or another. If I use money at the grocery store, I am expressing faith in the social relationships that make money useful. On a more mundane level, every time I get into the elevator I am expressing faith in the civility of the other people, and every time I say "hi" to you in the hall I express faith in that same civility—because human behavior isn't circumscribed by instinct and thus anything is possible in face-to-face encounters.

Notice that faith and other kindred emotions don't become religious by being expressed in religious ways. They are by their very nature religious. This is an extremely important point in understanding Simmel's approach. There are elements in our nature and social relations that are by their character religious. However, when these impulses and facets of life become attached to and expressed through the objective form of religion, something else comes into existence. To understand what Simmel sees in religion, we have to think for a moment about categories.

Objective religion: It is the function of categories to include and exclude, to draw meaningful boundary lines. For example, we have categories of states that divide up the geography into meaningful chunks that we call the United States and Mexico. Are those states really *there* in the geography? No, but our categories function to create meaningful order out of what is otherwise simply a continuous flow. The same thing happens with race and gender. The social object "African American" that is standing in front of me comes to exist by the race categories through which we organize our sense perceptions. Thus, the contents of the empirical world are independent of our categories, but our *experience* of that content is dependent upon them. Collectively, categories can form conceptual schemes that order our existence according to some logic. For example, if a group of sociologists and a group of psychologists view the same phenomenon, they will see two distinct facts, according to the logic of their perspective. Such distinct systems include philosophy, science, art, politics, and religion.

In this way of thinking about things, there are empirical objects that our senses perceive. But our awareness of those objects is actually predetermined by the categories that we use to understand the world. As a result, those objects appear to us in the form of the category rather than the actual empirical object. All such categorical systems are born out of primary psychological impulses and sociological necessities, according to Simmel. In the case of religion, human beings have a religious impulse. It is tied to the transcendent nature of our existence and inner life. Religiosity is expressed in various emotions and relations, none of which are necessarily religion itself. However, there is a tendency for the religious impulse, perhaps because of its transcendent nature, to "overstep its bounds" and create for itself empirical categories. It does this by claiming exclusive right to such impulses as faith, love, and sacrifice, and by conceptualizing all available worlds through its scheme. When this occurs, religion as an objective cultural entity comes to exist. Religion, then, in turn determines experience just as any categorical scheme would.

Religion thus creates a new world. It transposes life into a "particular key" and "creates a new world alongside reality itself" (Simmel, 1997, p. 141). Keep in mind that Simmel isn't saying anything about the truth of any particular religion. The "reality" to which Simmel refers is the empirical world and our sense data of that world. And, as we've seen, our experience of that world is dependent upon the categorical schemes we use—in this case, religion.

In the end, religion, like all forms of objective culture, takes on a life of its own. The objective categories formed initially in response to the religious impulse in humans end up defining and ultimately changing the religious impulse. Objective religion can become an overarching scheme in which all other categorical systems are subsumed. It becomes a "self-evident context that imposes itself unquestioningly" upon other realities, like art, philosophy, and science. When this happens, religion becomes something other than what the religious impulse would have produced, for religion then acquires "elements as are subject to a different and nonreligious system of logic." This in turn creates tension for the subjective experience of people—for religiousness "derives from needs and impulses of the soul that do not have the least to do with the 'things' of the empirical world and with rational criteria" (Simmel, 1997, p. 142).

Essential versus constructed gender: There are a number of ways the inequality of women in Western European society can be viewed. One of the more pervasive points of view in sociology investigates the consequences of gender in society. So, for example, we know that the annual salary for women in the United States today is about 70% of what it is for men, all other things being equal. We also know that there exists a "glass ceiling" that prevents many women from occupying places of power.

Another approach tries to get behind the scenes and understand why or how the differences between the genders exist. One take in this approach argues that the differences between men and women are by and large socially constructed. For example, Randall Collins (1988) contends that a major factor in producing gender differences is warfare. In ancient human societies, there was a natural division of labor, with women generally staying close to the central dwellings due to lactation and childcare. In this model, women would move short distances from the camp and gather foodstuffs. Men, on the other hand, were more mobile and moved further away from the site in search of game. Men thus came to interact with other tribes and control the weapons.

War between collectives became more probable as humans moved from hunting and gathering to horticulture. Horticulture meant that land became the primary tool of survival and it thus had to be controlled and protected. Because men controlled the weapons and interacted with neighboring societies, they were the ones that went to war. One way to avoid war was to make the enemy part of the family. Familial relations allowed different tribes to see themselves as part of a larger whole and they set up land rights and responsibilities. Marriage was the key manner through which these family ties were created and women became the connecting tie. Thus, women began to be seen as sexual objects that could be traded to avoid war, and men began to be conceptualized as naturally aggressive.

There is a social/psychological component to Collins's theory: we internalize the structural conditions of our gender. So while there are some natural differences between men and women, according to this perspective the gendered differences that we observe today are due mostly to historical, structural processes. Other authors have argued for differences from the same viewpoint but with different elements. Nancy Chodorow (1989), for example, uses Freudian theory to contend that some of the differences between men and women are due to capitalism and the absent father.

The constructed differences theory approach is outlined in Figure 6.4. As you can see, this kind of theory begins with social factors that influence both the structural effects and the differences in gender. The effects and differences of gender mutually produce and reinforce one another. The cycle is thus self-reinforcing. Obviously, the cycle can be broken. From this perspective, the best approach would be to change the social factors, such as greater work-force participation by women and/or greater child-rearing participation for men. From Collins's theory, for example, we might argue that having women go to war as men do and giving them control over weapons would produce significant changes in gender roles and consequences.

My intent here isn't to give you a brief introduction to gender theory. Rather, I want you to see some of the different ways we can think about gender and the unequal position of women. When Simmel considers these issues, he is concerned with differences more than consequences. However, Simmel does not argue that the distinctions between men and women are produced through interactions with different social structures; he argues that men and women are *essentially* different.

Simmel's general model is pictured in Figure 6.5. We are going to explore the particulars of this model, but for now I would like you to see that this understanding of gender differences begins with an essentialist position. *Essentialism* argues that human action, feelings, and thoughts are the product of some fundamental,

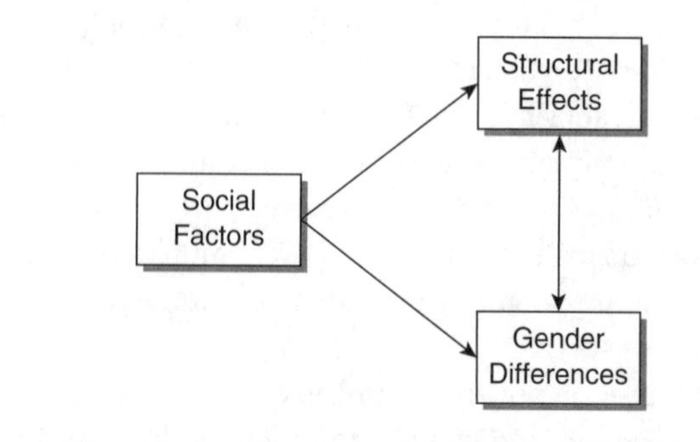

Figure 6.4 Structural Theory of Gender Differences

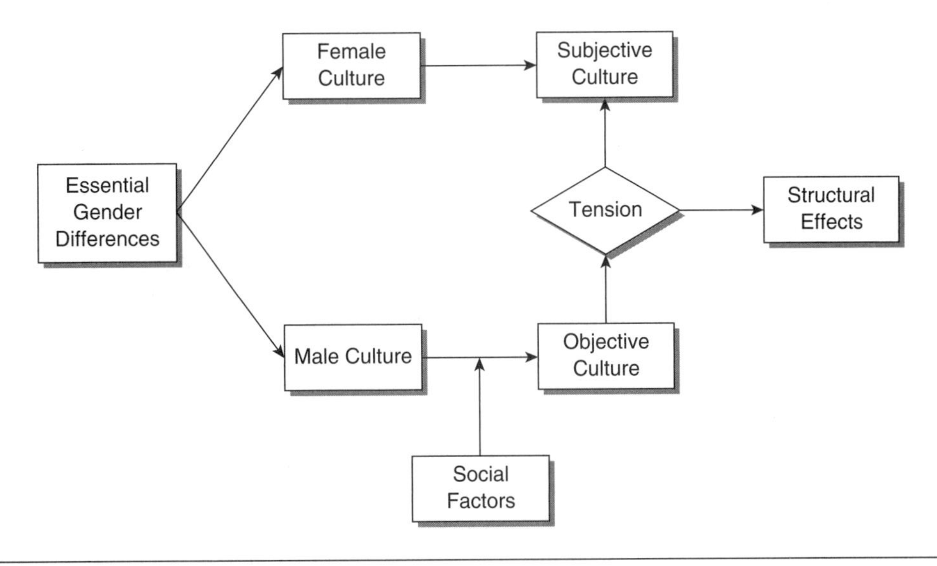

Figure 6.5 Simmel's Essentialist Theory of Gender Differences

internal, deep essence of the person. In other words, there is something intrinsic to being male or female, rather than the differences being socially based. Generally speaking, essentialist **perspectives** of gender come from either a religious or biological point of view. Simmel, on the other hand, describes gender as based on the human spirit, or the soul, but not necessarily in a religious sense. From a humanist perspective, he sees the human essence as unique. He argues that there is something about gender that is basic to society and being human: "The fundamental relativity in the life of our species lies in the relationship between masculinity and femininity" (Simmel, 1858–1918/1984, p. 102).

As you can see from the model, men and women produce different kinds of cultures, as would be expected if there were essential differences. Men's culture is objective, and, as we will see, it is part of their nature. Male culture becomes more objectified through the processes that we talked about earlier, and that is noted in the model by the exogenous variable "social factors." It's important to see the differences between the two approaches here—in the first, social factors are an integral part of the theory—they are what drive it; in Simmel's approach, they influence what is already going on but they don't drive the theory.

Simmel's sociological concern becomes how these essential differences affect how men and women relate to and are influenced by objective culture. Remember that culture, especially objective culture, is the most important social entity for Simmel, at least the most important in his analyses. Simmel argues that women are treated unequally and have the subjective experience of inequity because male culture is dominant and because women by their very nature cannot relate to objective culture. It is important to note that Simmel was one of the first to acknowledge that the general culture in Western society is male and that this presents a problem for women—there is little cultural space for women. He was also one of the first to talk about what has become known as the "slippery slope of sexism"—equating human with man.

Being male: Simmel assumes that men and women are different in the core of their being. The differences can be summarized in the statement "women are, men do." Simmel sees men as intrinsically motivated to produce objects. Man (and in this case we really do mean male) is driven to externalize himself. He becomes substantial in and through his accomplishments. Simmel says that the universal male norm is "the requirement that a man should be 'significant.'" Man finds this significance through building and creating. Though Simmel does not use this language, there is a way in which he sees Marx's species-being as peculiar to the male. In Simmel's scheme, men are supposed to produce; it is the fulfillment of their nature. They create objects, things external and separate from themselves.

> It is rather the case that, with the exception of a very few areas, our objective culture is thoroughly male. . . . The belief that there is a purely 'human' culture for which the difference between man and woman is irrelevant has its origin in the same premise from which it follows that such a culture does not exist—the naïve identification of the "human" with "man." (Simmel, 1858–1918/1984, p. 67)

As men produce, they create a culture that emphasizes individual accomplishments. For example, we have seen many TV shows or movies where someone is asked, "What have you done with your life?" (Your parents may be asking you this question in terms of your future.) The idea behind this question is that the value of life is understood though accomplishments. Achievements are objects that exist in and of themselves, apart from the individual. So, for example, I have published a book. That is an accomplishment. It stands outside of me. It has a Library of Congress number and as such it exists for others and will continue to exist long after I'm gone. Accomplishments are things that are differentiated "from the self with its emotive and affective centers." In other words, there is no real emotional connection between objects and the human soul—"there is something faithless in the separation of the person from the object" (Simmel, 1858–1918/1984, p. 71). According to Simmel, this emphasis on producing objects is distinctly male and results in professions, clusters of specific kinds of accomplishments.

> The deepest nature of one sex lies in an unceasing process of becoming and expansive activity. It is inextricably implicated in the temporal interplay of a most profound dualism. Its authentication, however, takes place in the domain of the objective, the continuous, the substantial. The other sex is concentric with itself. It reposes within its own meaning. As regards its authentication, however, it is devoted to the fleeting experience. It is not oriented to any result that could not be re-incorporated into this flux of actual interests and claims. (Simmel, 1858–1918/1858–1918/1984, p. 94)

As such, the division of labor is a distinctly male process. Men by their nature divide and objectify; it is the way they subdue the world. Thus, making work a thing that itself can be divided is in keeping with the male model. Marx argues that alienation comes from the division of labor, among other things. But for Simmel "alienation" is almost an intrinsic part of being male. The male nature is to produce, "to live creatively, and to prove itself." But this need to produce brings into being a desire to create beyond man's ability. Robert Browning once said, "Ah, but a man's reach should exceed his grasp." In this context, what that means is that men seek to accomplish more than they can ever hope. Inherent within the production process, then, is the frustration of the creative impulse. Man is driven to transcend himself. That's the effect of production: to become other, external, more than one's own self. If I design and build a skyscraper, I have transcended my own self. The building and its architectural statements make me part of the world writ large. But man is frustrated in his transcendence by the very limitations of production. Simmel calls this the "indigenous tragedy" of being male.

The male nature is "dualistic," "restless," and "the born transgressor of limits" (Simmel, 1984, pp. 92–93). Man is caught in the "uncertainty of becoming." He is always pushing the boundaries to create and to transcend himself, yet this vision is never complete, never at rest. However, the production of objectified culture provides a counterweight to this nature; it is at once part of the problem and part of the solution for man. Man produces "abiding, objective, trans-individual artifacts" that anchor his soul. While he is yet becoming, he has also arrived. So, for example, I am driven to write this book. It is, at this moment of writing, becoming. Because I am driven to do this work by my nature, according to Simmel, I am also in a state of being not-yet-finished and thus uncertain. Yet at the same time, I have already published. There are artifacts that exist that substantiate and anchor me in myself and in the world. This process produces a kind of equilibrium of existence for men.

Being female: Women, on the other hand, are not faced with this problem of becoming, for it is the nature of women to be. Simmel describes the female nature as one that is marked by close integration. Think of the inner makeup of human beings as a bicycle wheel. There is a center from which the spokes radiate to the periphery. Men are defined by distance and dis-integration from the core—it is the purpose of man to create objects that by definition are other than his inner being. Women are defined by close integration of the core with its activities. Women, according to Simmel, do not live apart from their emotive centers. There is no necessary detachment from their activities as with man; in fact, just the opposite is the case.

> The intimate relationship between all the aspects of her nature does not, as we are so often told, constitute the woman as a subjective being, but rather as a being for whom the distinction between the subjective and the objective really does not exist. (Simmel, 1858–1918/1984, p. 86)

Let's take an example of a stereotypical female behavior to help us understand what Simmel is arguing: women giving birth and rearing children. In this activity, women are not called upon to deny themselves or their emotion or their values. Neither are they called upon to objectify the recipient of the activity, the child. In fact, it is by virtue of emotional attachment and personal values that the activity is authenticated. In other words, we expect mothers to be emotionally attached to their children and to have strong sets of values about childcare. So the activity and the inner being of the woman are closely integrated. This is so, according to Simmel, because of the innate nature of woman.

Fidelity is thus the "fundamental structure of the female nature." This is not only true in the case of relationship fidelity, but also in the secondary meaning of the word—true reproduction. Though it isn't used much in advertising anymore, most of us are generally familiar with the term *hi-fi.* Kids from my generation would refer to their stereo systems as their hi-fi. The term comes from *high fidelity* and it means "to faithfully reproduce from a source." In this instance, the source may be a radio signal or a music album. When Simmel uses the term for woman, he has a similar meaning in mind. In the activities of women, there is a faithful reproduction of their center, because they do not disassociate from or objectify their activity. This drive for fidelity constitutes an obstacle to the "retreat of the ego" from its actions and embodiments that is characteristic of men.

Using Weber's terms, Simmel argues that instrumental rationality (means–ends rationality) is distinctly male and value rationality is particularly female. Life, and the living of it, has value in and of itself, not for what can be accomplished but for what it simply is. Woman doesn't have to transcend herself as man does. Men can live or die for an idea. In that passion, the idea is external and juxtaposed to man. For woman, her existence and the idea are the same. "She is always at home with herself. However, the man has his 'home' beyond himself" (Simmel, 1858–1918/1984, p. 116). There is, according to Simmel, no distance between a woman's inner being and the ideas she holds or the activities in which she engages.

Where man finds his tragedy in the mocking attraction of production, woman finds hers in the external and historical conditions of her life. The external conditions of her life require her to have relationships in a world filled with "the other." In this, she is required to transpose her existence from inner solitude to other people. These relationships then create a dualism that "shatters the state of pure repose in the inner center" (Simmel, 1858–1918/1984, p. 115). Historically, woman finds herself in a time where she is treated as an object or instrument. She is used to fulfill the life of the man, the child, and the home.

Because of the differences in the inner structures of men and women, the kind of knowledge they produce and hold are different. Here Simmel gives us a fundamental insight into the sociology of knowledge: different forms of knowledge come from different modes of existence. The knowledge that human beings hold is not something that is sterile, pure, and absolute. What we "know" is not a direct reflection of the universe as it is. Rather, what we know is intrinsically bound up with the way we live our lives. Thus, for example, what East Indians know about the way the universe functions can be quite different than what we in the United States know about that same universe. Here Simmel applies this principle to gender.

> The constitutive idea of the woman, on the other hand, is the unbroken character of her periphery, the organic finality in the harmony of the aspects of her nature, both in their relationship to one another and in the symmetry of their relationship to their center. This is precisely what epitomizes the beautiful. (Simmel, 1858–1918/1984, p. 88)

Men's knowledge is based on dualism, in keeping with their nature. The norm of logic is that truth lies in an autonomous realm. For example, the logic of the scientific method implies that the validity of any idea or theory is found by reference to the empirical world. Let's say we have a theory that states that capitalism will go through periodic cycles that produce greater levels of centralized capital (à la Marx). How would we verify that theoretical statement? We could create a database that contained the distribution of wealth in the United States by decade for the past 100 years. Then, by comparing the distributions at the decade points, we could see whether or not greater portions of the wealth are being held by the top 10% of the population. So, where does the truth of the idea lie? According to this logic, truth resides in a realm separate from the world of ideas—in the empirical world. This kind of logic constitutes a dualistic approach.

The way Simmel describes women's knowledge is more difficult to get at. This problem isn't unusual for those that posit an essential difference: the general knowledge and language that our society has is masculine in nature and thus insufficient to the task. Be that as it may, women's knowledge is not based on a dualistic approach, but rather, "is somehow situated in the things themselves." The best illustration of a type of women's knowledge is ethics. Ethics deal with what is good and bad, or right and wrong, or with moral duty and obligation. Ethical standards are true and right in and of themselves, without reference to a separate domain. For example, it is simply wrong to murder another human being. We don't have to refer to a world of proof; the ethic stands on its own.

What's important here is not whether or not we agree with the moral standard. For instance, we might disagree with the ethical statement, "It is always wrong to steal." What's important is the way in which these kinds of things come to exist as morals or ethics. They exist as ethical standards because we believe them to be true and right and good. We don't refer to a separate reference system to validate them, as we do in logical methods. In science, a theory is true if the theory and the empirical world correspond to one another. In ethics, it doesn't matter if there is a correspondence. A moral is held to be true simply because of its innate goodness. That kind of knowledge is in keeping with the nature of woman.

As we've seen, Simmel argues that the historical norm of masculinity is to be significant. For woman, the norm is beauty; however, it's important to realize that Simmel does not mean mere physical beauty. If you were to look up beauty in an exhaustive dictionary, you'd find a number of definitions. The first listed would undoubtedly be something like "extreme physical attractiveness that produces great sensory pleasure." This isn't what Simmel has in mind. The beauty that is characteristic of the innate qualities of womanhood has nothing to do with sensory pleasure; it isn't something that can be perceived through our optical nerves. If you were to keep reading the dictionary definitions, you would find something akin to "a particular grace; a perfection that calls forth delight for itself rather than its uses." This definition is much closer to what Simmel has in mind. The beauty that Simmel has in mind refers to a state of being where the inner and outer lives are in harmony with each other. The beauty that is the historical norm for women is the kind that is perceived through the soul.

Gendered social change: Simmel sees two possible roads to bring about social change for women. The first is concerned with the participation of women in the already existing world. Many feminists see this as the path to equality. Some argue from a Marxist perspective that women's workforce participation is key to parity between the genders. From this point of view, the key to power for anyone is class—money empowers. As women enter the workforce and are able to support themselves, they gain independence and begin to break down the patriarchic structure in back of the economy.

However, Simmel sees two problems with this approach. First, it is concerned with individual participation, wherein the number of women who are working and their levels of attainment are the gauge for success. Thus, this success is simply a sum of individuals. "It is not a struggle for something that in itself transcends everything individual" (Simmel, 1858–1918/1984, p. 66). The other problem with this tactic is that it requires women to internalize masculine culture. Interestingly, some contemporary research has found that one of the most significant predictors of women's success in the workplace is prior participation in sports. The argument is that sport socializes women into the competitive, masculine culture of corporate America. Simmel's critique here is that masculine culture is contrary to the female nature.

The second line of attack in bringing about gendered social change is to create and honor distinctly female contributions to society. Simmel sees this as ideal not only for women but also for society. The reason this is best for women should be apparent: it does not require them to internalize a culture that sets up inner conflicts.

Creating a sphere of feminized culture produces balance for society in a way that might surprise us. Simmel argues that humans grasp the meaning and value of something through relationships. That is, we never know a thing by itself. It is always understood in terms of its relationship to other things. Take a chair, for example. We understand "chair" as we are able to situate it in its relationships to other things, like tables and sofas. Thus, everything for humans derives its meaning relative to something else. Simmel sees relations as the fundamental structure of meaning. For Simmel, the most essential meaning-giving relationship for the human species is gender. However, as we've already noted, the dominant culture in society is masculine. This dominance means that the basic structure of meaning in our society is out of balance. We can think of this as a kind of false consciousness. If the basic organization of meaning is asymmetrical, then the entire system of meaning must be off balance. Nothing makes sense to people the way it should.

What a balanced system of knowledge would look like is difficult to talk about. As Simmel has already pointed out to us, the culture, knowledge, and language that we have are primarily masculine. Contemporary thinkers such as Dorothy Smith (1987) make the same observation. Nevertheless, Simmel does offer some examples and give some direction. A female culture would be one that is immediate to the person; there would be little distance between the subjective experience of the individual and the culture. The "classic" distinction between object and subject would disappear. The culture would be an intimate expression of the inner person. Society will value such a culture, and meaning will be balanced, when the perspectives of men and women are treated equally. And, ironically, masculine culture will only be truly honored when feminine culture is honored.

Thinking About Modernity and Postmodernity

Objective trivializing culture in modernity: Simmel argued that culture became increasingly objective (increasing size, diversity, and complexity) in response to social factors specific to modernity (urbanization, the division of labor, the use of money, and rationally motivated group memberships). As we've seen, there are complex and incongruous results from these forces: for example, greater individuality and freedom coupled with increased anomie and blasé attitude. Postmodernists suggest that factors specific to this age—mass media and advertising—have pushed Simmel's concerns to another level.

I'm going to give you a simplified explanation of one of the more important points of postmodern theory. In doing so, I will undoubtedly overstate and gloss over things, but I think it will help us get a handle on the idea of postmodernity. One of the important issues of postmodern theory concerns the way in which culture comes about. Durkheim is worried about the moral base of culture, and Simmel is concerned with the proportion of subjective versus objective culture. Both Durkheim's moral base and Simmel's subjective culture are created and firmly established in real social groups. For most of human history, culture was subjective and moral because it was used by the same social groups that created it. Most of us think that is still the case. I mean, Americans use American culture and Chinese use Chinese culture, right? Yes, but according to postmodernists, it isn't the same at all. Today in the United States, we have and use American culture, but a good portion of the culture we have isn't created by our social group. In fact, much of the culture we are surrounded by isn't created by a social group at all.

Postmodernists argue that a major shift has occurred between what might be called traditional society and postmodern society. Simmel was caught in the middle of this transition, which is one reason why his theory is so unique. In traditional social groups, the kind where Durkheim's moral base and Simmel's subjective culture prevail, culture is created and used in the same social context. For the sake of conversation, let's call this "grounded culture." In our society, however, much of our culture is created or modified by capitalists, advertising agencies, and mass media. We'll call this "commodified culture." Members in traditional social groups were surrounded by grounded culture; members in postmodern social groups are surrounded by commodified culture. There are vast differences in the reasons why grounded versus commodified culture is created. Grounded culture emerges out of face-to-face interaction and is intended to create meaning, moral boundaries, norms, values, believes, and so forth. Commodified culture is produced according to capitalist and mass media considerations and is intended to seduce the viewer to buy products. With grounded culture, people are moral actors; with commodified culture, people are consumers. Postmodernists argue that there are some pretty dramatic consequences for people in the shift to commodified culture.

Simulacrum and hyperreality in postmodernity: Jean Baudrillard (1981/1994) argues that this kind of trivialization can lead to some serious problems in postmodern culture. He puts forth that many of the things we do today in advanced capitalist societies are based largely on *images* from past lives. For example, most people in traditional and early industrial societies *worked with* their bodies. Today, people in postindustrial societies don't work with their bodies, they *work out* their bodies. In other words, the body has become a cultural object rather than a means to an end. In postindustrial society, the body no longer serves the purpose of production, rather, it has become the subject of image creation: we work out in order to alter our body to meet some cultural representation.

Further, humans have always had and used clothing, and there has always been a bond between the use of the body and the function of clothing. In previous eras, there was an explicit link between the real function of the body and the clothing worn—the clothing served the work being performed or was indicative of social status. For example, a farmer would wear sturdy clothing because of the labor performed, and his clothing indicated his work (if you saw him away from the field, you would still know what he did by his clothing). However, in the postmodern society, clothing itself has become the creator of image rather than something serviceable or directly linked to social status and function.

Let's review what we have established so far: in times past, the body was used to produce and reproduce; clothing was serviceable and was a direct sign of work and social function. In postmodern society, however, a greater proportion of bodies have been freed from the burden of physical labor than in previous eras and have instead become a conveyance of cultural image. The body's "condition" is itself an important symbol, and so is the decorative clothing placed upon the body. Now here is where it gets interesting: a fit body represented hard work and clothing in the past signed the body's work, but what do our body and clothing symbolize today? Today, we work out the body rather than work with the body. We work out so we can meet a cultural image—but what does this cultural image represent? In times past, if you saw someone with a lean, hard body, it meant the person lived a mean, hard life. There was a real connection between the sign and what it referenced. But what do our spa-conditioned bodies reference? Baudrillard's point is that there is no real objective or social reference for what we are doing with our bodies. The only reference to a real life is to that of the past—we used to have fit bodies because we worked. Thus, in terms of real social life, our worked out bodies represent the past image of working bodies. What does this imply about the clothes we wear? The clothes themselves are thus an image of an image that doesn't exist in any kind of reality. Baudrillard calls this **simulacrum**: an image of an image.

Tied in with all of this body image is, of course, mass media and advertising. The act of advertising itself reduces objects from their use-value to their sign-value (which is what we talked about with reference to the clothing—clothing isn't as *useful* as it is *symbolic* in postmodern society). For Baudrillard, the movement from use-value to exchange-value entails an abstraction of the former; in other words, the exchange-value of a commodity is based on a representation of its possible use. So when I buy a new truck, I have in my mind its use-value for me and that is what creates the exchange-value (I will pay $15,000 for it because I can imagine the use it will have). Advertising and mass media push this abstraction further. In advertising, the use-value of a commodity is replaced with a sign-value. What this means is that when I see an ad for the truck I want to get, chances are that what I observe are images of pretty girls, bottles of beer, and rock and roll. Advertising does not seek to convey information about a product's use-value; rather, advertising places a product in a field with unrelated signs in order to enhance its cultural appearance. Thus, there are certain signs and meaning contexts that get attached to commodities. As a result, we tend to relate to the sign rather than the use-value. People purchase commodities now more for the image than for the function they perform. Getting back to my truck: the use-value of a truck—transportation—is rarely seen among the advertised images of sex, sport, and status. (And I might pay $23,000 for a truck because I can imagine myself in its sign structure.)

Let's think again about clothing. The reality-based use-value of clothing is to cover us and keep warm. Yet right now I'm looking at an ad for clothing in *Rolling Stone* and it doesn't mention anything about protection from the elements or avoiding nudity. The ad is rather interesting in that it doesn't even present itself as an advertisement at all. It looks more like a picture of a rock band on tour. We could say, then, that even advertising is misrepresenting itself. On the sidebar of the "band" picture, it doesn't talk about the band. It says things like "jacket, $599, by Avirex; T-shirt, $69, by Energie"; and so on. We're not going to pay $599.00 for a jacket to keep us warm, but we will pay that for the status image that we think it projects. Baudrillard's point is that we aren't connected to the basic human reality of keeping warm and covered; we aren't even connected to the social reality of being in a rock band (which itself is a projected image of a life that never existed); we are simply attracted to the images. Thus, what we buy today aren't even commodities in the strictest sense. They are images of images, what Baudrillard calls a simulacrum of a simulacrum, or **hyperreality.**

A great deal of theoretical ink has been spilled in the debate concerning the relationship between the subject and culture in postmodernity. It is important to note that Simmel was aware of the link between culture and the individual's subjective experience long before postmodernism was even a word. It is ironic to think that the forces that produce individuality and personal freedom are the same forces that endanger the individual and her or his particular experience. Yet that is exactly Simmel's argument. Individual freedom is increasing as cultural cohesion and meaning is decreasing. Is the individual on the cusp between experiencing a core sense of individuality and a yawning abyss of meaninglessness where emotional depression (blasé attitude) is the norm? I don't know, and neither did Simmel. But he found that space between subjective experience and the state of objective culture to be a fascinating location for thinking about the social condition.

Summary

- There are two central ideas that form Simmel's perspective: social forms and the relationship between the subjective experience of the individual and objective culture. Simmel always begins and ends with the individual. He assumes that the individual is born with certain ways of thinking and feeling, and most social interactions are motivated by individual needs and desires. Encounters with others are molded to social forms in order to facilitate reciprocal exchanges. These forms constitute society for Simmel. Objective culture is that culture that is universal yet not alive to the individual's subjective experience. Thus, the person is unable to fully grasp, comprehend, or intimately know objective culture. The tension between the individual on the one hand and social forms and objective culture on the other is Simmel's focus of study.

- There are a variety of social forms; among them are sociability, exchange, conflict, and group size. Sociability occurs in interactions that have no other goal than chit-chat. Though the goal appears of no consequence, the functions of such interactions are important for the well-being of society and the person—they socially connect us to others, provide cultural and emotional capital, affirm social reality, and solve the problem of association. Reciprocal exchange is the foundation of society. Exchange is ordered by value and value is ordered by sacrifice and scarcity. Conflict is a social form that provides such functions as group solidarity, normative regulation, centralization of power, and coalition formation. The kind of function that conflict provides depends on whether the collective is a social system or a group. Group size also functions as a social form. Triads, for example, involve power relations among their members, whereas dyads do not. On the other hand, dyads are much more uncertain and thus perceived as unique due to their size.

- Urbanization increases the division of labor, the use of money, and changes the configuration of social networks. All of these have direct and indirect effects on the level of objective culture and its effects on the individual. The use of money increases personal freedom for the individual, yet at the same time it intensifies the possibility of anomie, diminishes the individual's attachment to objects, and increases goal displacement. People join groups for either rational or organic motivations. Rational motivations are prevalent in urban settings and imply greater personal freedom coupled with less emotional investment and possible anomie and role conflict; organic motivations imply less personal freedom and greater social conformity coupled with increased personal and social certainty.

- Both gender and religiosity are natural states for human beings. Each person has a religious impulse and a certain degree of religiosity. Religion, on the other hand, acts like a categorical scheme, and it overextends the valid limits of religiosity. Religion thus objectifies the world by claiming exclusive right to such impulses as faith, love, and sacrifice, and by conceptualizing all available worlds through its scheme. Gender is also an essential attribute of humans. Men naturally objectify themselves as they are motivated to produce. Women, on the other hand, are naturally integrated with all aspects of their being. Men operate through a dualistic knowledge system, seeking proofs for knowledge in the empirical world, while women's way of knowing is non-dualistic and centered. The problem of gender inequality is that men have dominated the social world and its culture. Culture, then, for the most part is objective and at odds with women's nature. True gendered social change will come as women's knowledge and way of being are valued as equal to men's.

- Simmel was concerned about the effect of objective culture on the individual's subjective existence. Postmodernists have taken that concern to another level. In times past, most of the culture was produced by people situated in real social groups that interacted over real issues. This grounded culture created real meanings and morally infused norms, values, and beliefs. In postmodernity, much of the culture is produced or colonized by capitalists using advertising and mass media. This historic shift implies that culture has changed from a representation of social reality to representations of commodified images, culture is produced rather than created, and people have changed from culture creators to culture consumers. In Baudrillard's words, culture in postmodern society is filled with a simulacrum that references hyperreality.

Building Your Theory Toolbox

Conversations With Georg Simmel—Web Research

Georg Simmel is one of the hardest classic theorists to get to know, particularly through Web research. This is generally due to his late discovery by contemporary American sociologists.

- Of particular interest is Simmel's position in the university. He was an outsider both because of anti-Semitism and his way of working. If your school has access to JSTOR, look up the *American Journal of Sociology*, Vol. 63, No. 6, May 1958. There you will find a number of interesting articles about Durkheim and Simmel. I'd like you to look at one in particular: "Georg Simmel's Style of Work," by Lewis Coser. Read the entire piece (it's short), and pay particular attention to the end quote from Dietrich Schaefer, a professor who was asked to evaluate Simmel's qualifications for a chair at the University of Heidelberg.
- There are a few additional questions you will be able to answer by using your Web search engine or by consulting one of the recommended Simmel pages: Why was Simmel's first dissertation rejected? What was Simmel's style of lecturing (type "Simmel" and "dentist" into your search engine)? Why would Simmel be considered an avant-garde philosopher (be sure to look up *avant-garde*)? What was Simmel's relationship with Max Weber?

Passionate Curiosity

Seeing the World (using the perspective)

Simmel's perspective is very unusual in sociology. He begins with the individual's innate creative energies and considers how objective culture constrains and enables the individual. Somewhat like Marx, Simmel considers the alienating aspects of modern society, but, unlike Marx, Simmel also sees exciting, creative potential coming from the interplay of objective culture and the individual.

- Remembering Simmel's definition and variables of objective culture, do you think we are experiencing more or less objective culture in the last 25 years? In what ways? In other words, which of Simmel's concepts have higher or lower rates of variation? If there has been change, how do you think it is affecting you? Theoretically explain what the proportion of subjective to objective culture will be like for your children. Theoretically explain the effects you would expect.
- Compare and contrast Simmel and Weber on the issue of culture. How has the centrality of culture made their theories distinct from the other theorists we've

covered? How do they approach culture differently? How has this made their theories different?

- How has Simmel's emphasis on the individual and inherent traits made his theory unique? How does Simmel's theory qualify as sociological, given the way he privileges the individual?

Engaging the World (using the theory)

- Simmel's idea of gender inequality is based on inherent differences between men and women. He also argues that gender equality will only truly come as society honors both male and female cultures. What, according to Simmel, is problematic about gender inequality in women's workforce participation? What structural changes would this society have to instigate in order to equally honor men's and women's cultures? What would this dual-honoring culture look like? What do you think of Simmel's understanding of gender and advice for social change? How would your life be different if Simmel's ideas were implemented?
- Keep a short diary of the times you interact with people during the day. Mark down the number of times the encounter is oriented toward purposeful or strategic considerations (when there are clear goals being achieved) and the number of times the interaction is simple talk or sociability. What are your results? Are you surprised? According to Simmel's theory, what does this imply is going on?
- We have heard lots of talk about globalization and the idea of the global village. Thinking of the entire world as if it were a large society, what would Simmel's theory predict will be the outcome of global conflict in the long run? Think through the issue considering both short, violent conflict and longer, less violent conflict.
- Perform a kind of network analysis on your web of group affiliations. How many of the groups of which you are a member are based on organic and rational motivations? In what kinds of groups do you spend most of your time? Over the next five years, how do you see your web of affiliations changing? Based on Simmel's theory, what effects can you expect from these changes?
- Both Durkheim and Simmel theorized about group membership and its effects on modern society. Bringing both of their theories together, how would you characterize the society in which you live? What kind of culture do you think is most prevalent in your society? Do you think there is regional variation to this general trend? If so, how would you explain it theoretically?

Weaving the Threads (synthesizing theory)

- Both Marx and Simmel are concerned with money. Compare and contrast Marx and Simmel on money. Are the dynamics for the increased use of money and its effects the same or different? In what ways can we say that the use of money is more present in our lives than 25 years ago (recall the more general definition of money)?

- How does Simmel's theory of gender compare to Marx and Engel's? How was the structure of gender inequality created? What is the core of gender inequality? In other words, what social issue or entity does gender inequality revolve around (Hint: for Simmel, it is differences in culture)? Do these theories complement each other in any way?
- Compare and contrast Durkheim and Simmel on the division of labor. What are the general conditions under which the division of labor will vary? What are the general effects of the division of labor in society? What does Marx add to our understanding of the division of labor?
- We have now had five theorists address the issue of religion. Using the material we have gathered thus far, answer the following questions: What is the definition of religion? What are the functions or effects of religion in society? How does religion vary? In other words, how do the forms of religion change, and how do the effects of religion vary? Considering these five theorists together, what can you say about the future of religion?
- Each of our theorists thus far has also been concerned about modernity. Using no more than a few words for each theorist, how would you characterize Spencer, Marx, Durkheim, Weber, and Simmel's theories of modernity? In other words, capture each of their concerns in a brief topic statement. Then write a short paragraph that explains modernity.

Further Explorations—Books

Featherstone, M. (Ed.). (1991). Georg Simmel [special issue]. *Theory Culture and Society, 8*(3). (Contemporary social thinkers show how Simmel's theory can be applied)

Frisby, D. (1984). *Georg Simmel.* New York: Tavistock. (The essential introduction to Simmel)

Ray, L. (Ed.). (1991). *Formal sociology: The sociology of Georg Simmel.* Brookfield, VT: Elgar. (Extensions of Simmel's theory of social forms)

Weinstein, D., & Weinstein, M. (1993). *Postmodern(ized) Simmel.* London: Routledge. (A post-modern reading of Simmel)

Further Explorations—Web Links

http://www.malaspina.com/site/person_1056.asp (Site maintained by Malaspina Great Books; contains brief introduction to Simmel's life and works)

http://socserv2.socsci.mcmaster.ca/~econ/ugcm/3113/simmel/society (Copy of Simmel's *American Journal of Sociology* article, 1910–1911, "How Is Society Possible?")

http://www2.fmg.uva.nl/sociosite/topics/sociologists.html#simmel (Site maintained by the University of Amsterdam, Netherlands; links to various informative sites)

Further Explorations—Web Links: Intellectual Influences on Simmel

Simmel, like Weber, had a number of different influences. Early in his career, he was influenced by Comte's positivism and Spencer's evolutionism. But Simmel was most influenced by German Idealism, and Immanuel Kant in particular.

German Idealism: http://www.iep.utm.edu/g/germidea.htm (Site maintained by The Internet Encyclopedia of Philosophy; this site contains a brief definition of and introduction to German Idealism, including Kant and Hegel's contributions)

Immanuel Kant: http://www.hkbu.edu.hk/~ppp/Kant.html (Site maintained by Steve Palmquist; contains a very thorough map of resources on the Web for Immanuel Kant)

CHAPTER 7

Self-Consciousness—George Herbert Mead (American, 1863–1931)

> *His grasp and learning were encyclopedic. When I first knew him he was reading and absorbing biological literature in its connection with mind and the self. If he had published more, his influence in giving a different turn to psychological theory would be universally recognized. I attribute to him the chief force in this country in turning psychology away from mere introspection and aligning it with biological and social facts and conceptions. Others drew freely upon his new insights and reaped the reward in reputation which he was too interested in subject-matter for its own sake to claim for himself. From biology he went on to sociology, history, the religious literature of the world, and physical science.* (John Dewey, spoken at Mead's funeral, April 30, 1931)

Dewey's quote indicates that George Herbert Mead is at least partially responsible for aligning psychology with the facts of biology and sociology. In doing so, Mead helped found an approach known as social psychology. Yet Mead's theory is much more social than psychological. As Mead (1934) says, "Social psychology has, as a rule, dealt with various phases of social experience from the psychological standpoint of individual experiences. The point of approach which I wish to suggest is that of dealing with experience from the standpoint of society" (p. 1).

Mead refers to his own method as *social behaviorism.* In using such a term, Mead is comparing himself to the kind of behaviorism that psychologist John B. Watson advocated. Watson focused on outward human behavior exclusively, arguing that investigating the internal workings and states of the human mind is unnecessary. Mead wants to maintain the emphasis on human action, but argues that understanding the mind is vital for understanding human behavior. In fact, it is the mind that makes human behavior unique. According to Mead, the mind itself is behavior, a kind of pre-behavior before actual action can take place. In other words, before people act, they think about possible personal and social consequences.

Mead's emphasis on *social* behaviorism also indicates that he is interested in the place social communication has in behavior. In other words, Mead sees social interaction as the chief motivating or organizing principle behind human behavior, rather than as operant conditioning. In a brilliantly creative move, Mead argues that the mind is itself internalized social communication—the internal thinking that precedes and makes unique human behavior is socially constructed.

As a result of this particular approach, all sociological social psychology begins with George Herbert Mead. Of course, there were other theories, such as those of Marx, Durkheim, and Weber, that contained some social psychological elements, and some theorists, such as Simmel and Cooley, who were explicitly concerned with the individual or self, but it is Mead who formed the solid foundation for a sociological perspective of what goes on inside a person. Simmel's work doesn't help us much with the internalization process, as he is mostly concerned with the intersections among the individual, subjective culture, and objective culture. Simmel's theory assumes an intrinsic individual psyche; Mead's theory assumes that the self is utterly social.

Mead gives us an entirely social way of understanding consciousness and thought. In fact, his theory indicates that we are not *born* human, we *become* human. Not only does Mead's theory indicate that we become human, it also implies a historically situated self. What I mean is that you not only think and feel different things than your parents and grandparents, you *are* different. The very essence of your self is potentially different because it is made up of different kinds of social ingredients. The difference between you and your grandparents may not be extreme, but using Mead's perspective we would say that the self of twenty-first-century America is fundamentally different than that of first-century Greece.

Mead's ideas form the base of the Chicago School of Symbolic Interaction. His work has influenced such people as Herbert Blumer, W. I. Thomas, Robert Park, Norman Denzin, and Gary Alan Fine. The Chicago School approaches human life as something that emerges through interaction. As a result, the interactionist approach, and thus Mead, has been the chief way through which we have become aware of the inner workings and experiences of diverse social worlds, such as Denzin's work on alcoholics. Symbolic interaction is also used as the principal perspective anytime culture and the individual meet. For example, Arlie Hochschild used an interactionist approach to understand how emotions are socially scripted through "feeling rules." John Hewitt has recently used Mead's approach to understand how self-esteem is culturally and socially created and then internalized by the individual. And Edwin Lemert and Howard Becker used Mead's ideas to understand deviance as an outcome of interaction and labeling.

The list of people and work that Mead has influenced over the years would probably fill more pages than this entire book. The short version is this: anytime a sociologist wants to consider the mind, self, socialization, the active intersection between self and society/culture, or the interactions between them, then the ideas of George Herbert Mead are salient.

Mead in Review

- Born on February 27, 1863, in South Hadley, Massachusetts. His father, Hiram Mead, was a Congregational minister who became professor of homiletics (the art of preaching) at Oberlin Theological Seminary when Mead was seven years old. Mead's father died in 1881. Mead's mother, Elizabeth Storrs Billings, taught at Oberlin after her husband's death, and was President of Mount Holyoke College from 1890 to 1900. Mead had one sibling, Alice, who was four years older than George Herbert.
- Mead began his college education at Oberlin when he was 16 years old and graduated in 1883. He began his career as a school teacher, but was discharged by the school board when he refused to teach rowdy students. For the next three years, he took work as a surveyor for the railroad and tutored during the winter months. Mead grew tired of railroad work and returned to school, pursuing graduate studies in philosophy at Harvard.
- At Harvard, Mead studied with Josiah Royce, and worked during the summer for William James as a tutor for his children. He received a scholarship to study advanced philosophy in Germany for a year. In Germany, Mead studied with Wilhelm Wundt, met G. Stanley Hall, and may have sat in on lectures given by Georg Simmel.
- In October of 1891, Mead married Helen Castle. The couple had one son, Henry. After the wedding, Mead and his wife went to the University of Michigan, where Mead had been appointed instructor of philosophy and psychology. While at Michigan, he became friends with Charles H. Cooley and John Dewey. It was also at Michigan that Mead began to draw out the implications of Wundt's and Hall's thinking.
- In 1893, Dewey asked Mead to join him to form the department of philosophy at the University of Chicago. Chicago contained the first department of sociology in the United States. Mead's major influence on sociologists came through his graduate course in social psychology, which he started teaching in 1900. Those lectures formed the basis for Mead's most famous work, *Mind, Self, and Society.* Among his students was Herbert Blumer, who was to later systematize Mead's thought into symbolic interactionism.
- While at Chicago, Mead became active in the educational reform movement headed by Dewey. Here, also, he became involved with the early feminist movement and marched for women's suffrage.
- Mead died on April 26, 1931. The bulk of his work, including *Mind, Self, and Society* (1934); *The Philosophy of the Present* (1932); and *The Philosophy of the Act* (1938), was published posthumously.

The Perspective: Society Inside

George Herbert Mead happened to live at the right place at the right time. I'm not taking anything away from his intellect; Mead was brilliant. But he also lived during a time and worked at a place where a lot of different ideas were coming to a head and where radical thought was encouraged. The ideas of evolution and psychology had been around long enough to have been accepted and to have inspired some new and radical ideas of their own. And a new, distinctly American school of philosophy had just been born, as well as a new university. The University of Chicago was created in 1892 by John D. Rockefeller and was intended to be one of the world's premier centers of knowledge. It was extravagant and radical and a place for new ideas.

As mentioned above, Mead was brought to Chicago by his friend John Dewey. Dewey was head professor in philosophy and wanted to create a department where the new philosophy could take root, away from the old schools of thought. The University of Chicago was very much on the cutting edge, and Dewey wanted his philosophy department to be in the lead. Mead was thus surrounded by new and exciting ideas and was working in an environment that encouraged new thought and risk taking. Because of this milieu of ideas and creativity, Mead ended up being a true synthesizer. He took elements from other theories and philosophies and creatively brought them together to form a new whole. Let's take a look at some of these ideas and philosophies.

Pragmatism: "What is truth?" That is an intense question and one that philosophers (and religions) have debated about for quite some time. Prior to the beginning of the twentieth century, there were two main philosophical schools of thought about truth: realism and coherentism. Realism holds that the truth validity of any statement is determined by its correspondence to reality. That seems like a straightforward method: something is true if it's real. But what is reality? The ultimate reality of the universe is a vastly different thing for the Yanomamö of South America than for the cosmopolitan in New York City. In addition, the question of "what is reality" is one that is up for debate among philosophers (that's where ontology comes in). So, this kind of correspondence theory, where truth corresponds to reality, while intuitively satisfying, runs into trouble defining what counts as reality and thus truth, at least for some philosophers.

Another school of philosophy, coherentism, holds that truth is a function of coherency. Every statement that may be true or false is a proposition. According to this perspective, a proposition is true if it is part of a wide-ranging and logically connected system of ideas about what is generally thought to be reality. This perspective is satisfying because it at least recognizes that human beings somehow affect truth and reality. But it runs into problems in that it is ultimately idealistic—reality ends up simply being a coherent set of ideas.

Pragmatism, on the other hand, argues that ideas are not meant to be representations of ultimate reality, as with realism; nor are ideas true because they match up with other ideas (coherentism). In pragmatism, ideas are organizational instruments. That is, humans organize their behaviors based on ideas. Ideas, then, are tools of action and have value as they facilitate behavior. Pragmatism (probably the only uniquely American school of philosophy) rejects the notion that there are any fundamental Truths; instead it proposes that truth is relative to time, place, and purpose. In other words, the "truth" of any idea or moral is not found in what people believe. Truth can only be found in the *practical actions* of the people. Thus, pragmatism argues that people hold onto what works; and what works is the only truth that endures.

Pragmatism developed out of the work of several men; among them are Charles Peirce, William James, and John Dewey. Peirce initially began thinking about how humans are able to practice self-control. He argued that the ability to control behaviors, to act rather than react, is one of the defining features of humankind and is the central focus of pragmatism. James and Dewey argued that the ability to control behaviors is what allows people to adapt to their environment (adaptation is an idea from Darwin and Spencer) and that this adaptation changes as the environment changes. In other words, truth is relative to practical concerns—as our environment changes, so do our truths, values, and beliefs.

Mead is also considered a founding thinker of pragmatism. He and Dewey, in particular, had similar concerns and developed a reciprocal working relationship. Mead specifically applies pragmatism to the problem of meaning: How does meaning come about and where is meaning found? Mead sees that meaning is not the direct result of words themselves. Even though we can look up a word in the dictionary and find a "meaning" for it, according to Mead that is not necessarily the word's true meaning. In true pragmatic form, Mead argues that meaning is found only in people's behaviors and it emerges out of social interaction in response to adaptive concerns and behaviors.

Emergence: The idea of emergence is extremely important in understanding Mead's perspective. In general, the word **emergence** refers to the process through which new entities are created from different particulars. Another way of putting this is that when something emerges, the whole is greater than the sum of its parts. Let's take a hammer as an example. People create social objects such as hammers in order to survive; and hammers only exist as such for humans (*hammers* don't exist for tigers, though they might sense the physical object). But the meaning of objects isn't set in stone, once and for all. While the hammer exists in its tool context, its meaning can vary by its use, and its use is determined in specific interactions. It can be an instrument of construction or destruction depending upon how it is used. It can also symbolize an individual's occupation or hobby. Or it could be used as a weapon to kill; and it is an instrument of murder or mercy depending upon the circumstances under which the killing takes place. Thus, the "true" meaning of an object cannot be unconditionally known; it is negotiated in interaction. The meaning pragmatically emerges.

Consciousness: William James was a pragmatist and also a friend of Mead's. James applied the pragmatic program to the ideas of self and consciousness in *The Principles of Psychology* (1890). According to James, the consciousness is not a structure as some argued (as with Freud and Wundt), but a process that can best be characterized as a stream. For James, **consciousness** emerges as the mind reflexively uses ideas as instruments in the struggle to live. All consciously held ideas, then, must be understood as the ongoing result of a series of unending experiences wherein the consequences of holding such ideas are known.

In other words, consciousness always implies a kind of double move. We are both consciously aware of an idea and are aware of *having* that idea—consciousness is being reflexively aware of our own ideas. And, according to James, consciousness contains a pragmatic motive: we consciously evaluate our ideas and hold onto the ones that prove useful. Let's think about this in terms of an example. I have a friend who has a cat, and on this particular day my friend was wearing a black suit. On the way to work, she used a lint roller to remove the cat hair. Not a bad idea. When she arrived at the office building, she got out of the car and continued to use the lint brush. After a couple of minutes, she noticed that there were other people in the parking lot. She then became reflexively aware of her own idea and saw it in terms of the pragmatic motive of the moment—it was not such a good idea to brush the lint in a public space.

Mead continues with this line of thinking and argues that consciousness or mind is a *behavior* that is constituted through the use of language based in social interactions. **Mind**, according to Mead, is an internalized conversation. In other words, we use language (a social entity) to think about becoming conscious of our own ideas—we use language to talk to our self about our ideas. This approach not only makes consciousness a matter of investigation (rejecting the black box approach of behaviorists), but it also makes consciousness immanently social rather than a property of the individual.

Self: William James also sparked Mead's thinking about the self. James posited that people recognize their self as an object in an interaction. According to James, the self isn't simply a concept without substance; people act toward the self as if it is an object with actual properties. James called this the "empirical self" or the "Me." Part of the empirical self is the social self, a diverse group of selves that people have as the result of interacting in different situations. Mead likes James's idea, but he feels that James did not explicate the mechanism through which these selves are formed. Mead supplies such a process through his idea of role-taking.

Charles Horton Cooley also influenced Mead's thinking about the self. In general, Cooley agreed with James that the mind is a social entity. But Cooley took this idea one step further and argued that society itself is a mental construct—society exists only in our ideas of it. More importantly for Mead, Cooley argued that people form an image of their self based on social interactions. Others in an interaction act as a sort of mirror in which we see the self. Cooley termed this the **looking glass self.** Cooley said that there are three steps to this mirroring process. First, we imagine what we look like to other people. It isn't enough for us to know what we look like in a physical mirror; we have to know what that image looks like to others. So, we imagine it. We put ourselves in their shoes and look back at our self. Second, we imagine how they feel about how we look. Do they like the way we look or not? Am I dressed and acting appropriately? Third, we have an emotional response to their feelings. That emotional response is either pride or shame, the most social of all emotions. Mead uses these ideas to form his theory of role-taking and self-formation. However, he disagrees with Cooley's emphasis on emotion. Mead argues that self-concept, the cognitions we hold about the self, takes precedence over self-esteem, the value judgments we have about the self—because, Mead argues, the concept must come before we can have any feelings about it at all.

Cooley also made the distinction between **primary** and **secondary groups**. These two groups vary (or differ) according to time, frequency of contact, and intimacy. The longer an individual associates with a group (time), and the more often the individual gets together with the group (frequency), and the more the individual reveals of her or his "true" self (intimacy), the more likely is the group to function as a primary group. These primary groups have a greater voice than do secondary groups in the construction of a person's self. Mead saw these primary groups (such as caregivers) as a chief link between society and the individual.

Thus, according to Mead, the self is a social object whose meaning emerges through successive role-taking experiences in interaction. Just like the hammer in our illustration, the meaning of one's self comes out of the way we act toward it, and its meaning changes as our action changes. Therefore, the self isn't a structure or core attribute of the individual. It is flexible and its meaning will tend to change as our interactions change. Further, the self is not so much an attribute of the individual as it is an object of interaction. In other words, the self is a social entity.

Society and the generalized other: Another man from whom Mead drew inspiration is Wilhelm Wundt. Wundt was the first experimental psychologist. He is generally attributed with moving psychology from the realm of philosophy into science. But Wundt is also credited with placing greater emphasis on totality and organic relations than others before him; and that is where his influence on Mead lies. Wundt believed in the reality of the general will (what Durkheim calls the "collective consciousness" and Mead terms the "generalized other") and the precedence of the social over the individual. Based on this latter idea, Mead argues that the particular quality of being human is only possible through social interaction and successive stages of role-taking—we aren't human by birth, we *become* human through social interaction. Wundt also influenced Mead through his theories of language and gesture. Wundt argued that gestures are signs that mark ongoing interactions and that they are elements of mental communities that take the place of instincts for humans.

Mead brings together Wundt's idea of mental communities and the notion of emergence to argue that society is not a thing that exists outside of human interaction, as Durkheim, Marx, or Spencer would posit. On the contrary, what we mean by society is produced through generalized sets of attitudes that we may or may not take in interaction. Let me ask you a question. Do you believe that Great Britain exists? Yes, of course you do. Since it exists, you should be able to point it out to me, right? So, where is it? Is it the land mass? Perhaps that is part of it, but we'd be hard pressed to think of Great Britain without the British people. So, is it the people? Not really; if those same people grew up in Canada, they would be Canadian, not British. Is the map or the flag Great Britain? No, they are just symbols of it. Well, does it exist in parliament or the royal family? Maybe part of it does. Hopefully you can see where I'm going with this illustration. The idea of Great Britain exists symbolically and in the attitudes we can assume about it in the interaction. The idea of the "British people" is particularly instructive here. There is nothing intrinsically British about them. They are British precisely because they grew up surrounded by interactions in which "Britain" was an important symbol and they could effectively claim the identity of "British." When stated like this, it may seem obvious to the extreme, but the ramifications are profound: society, like all meaningful objects, emerges out of the interaction.

Mead thus draws from various influences to form a new understanding of mind, self, and society. Mead is particularly concerned with how humans produce meaning on a continuous basis through pragmatic concerns in social interactions. People only pay attention to a limited number of things in interaction. These things are denoted symbolically and the meaning emerges through negotiated interaction. One of the most important things that emerges through interaction is the self. In fact, the self is the pivot point for society and individual thinking. Society is possible because of the self, and the same is true for thinking. What society is, the influence it can have and its meaning to people, can only be found in face-to-face interactions involving symbolic gestures and language.

Thus, to think like Mead is to pay attention to the ever-changing landscape of human meaning: it is to see interaction as the only acting unit in society, to understand consciousness as a social construction, to see the self as a multifaceted and supple entity that is utterly social, and to catch sight of the fact that society is only possible because of the self.

Our discussion of Mead will be broken up into three primary topics. In the *Basic Human Tools: Symbols and Minds* section, we will see that the chief survival tool for human beings is language. We use language to organize human behavior across time and space, and we use it to pass down our knowledge to other generations. Since humans use language to survive, individuals need a mind in order to use language. The mind, then, as we will see, is socially necessitated and produced. In *Living Outside the Moment—The Self,* we will also see that the self is required by and created by society; it, like the mind, doesn't originate or completely belong to the individual either. The self also has a strange quality: it doesn't actually live in immediate time or space. In order to work, the self must transcend our behaviors and this time-bound moment. Finally, in *The Emergent Big Brother,* we will see that society exists symbolically and is built up through various actions and interactions.

Basic Human Tools: Symbols and Minds

> All animals are confronted with the challenge of material subsistence, but only humans are saddled with the vexing question of its meaning. (Snow & Anderson, 1992, p. 228)

Symbols: Ernst Cassirer (1944) argues that human beings created a new way of existing in the environment through the use of signs and meaning. While this sounds a bit like what Marx might say, it is distinctly different. Marx argues that humans survive because of creative production, the work of our hands. This externalization of human nature then stands as a mirror through which ideas and consciousness come. Thus, ideas for Marx have a material base. Cassirer argues in the opposite direction: it is through ideas that we manipulate the environment.

Mead actually anticipated Cassirer's argument. According to Mead, language came about as the chief survival mechanism for humans. We use it to control our environment. And this sign system comes to stand in the place of physical reality. By and large, animals relate directly to the environment; they receive sensory input and react. Humans, on the other hand, generally need to decide what the input means before acting. This constitutes the difference between reacting and acting. Humans don't simply react; we can act on purpose because we put a sign system between the environment and ourselves. We suspend response, think about the world in symbolic terms, and decide on a line of action. But before we get too far into this issue, let's break it down to its most basic element: signs.

A sign is something that stands for something else, like smoke stands for fire. So, if we see smoke coming out of our room, we will call and report a fire, even though we may never actually see the fire. It appears that many animals can use signs. My dog Gypsy, for example, gets very excited and begins to salivate at the sound of her treat box being opened or the tone of my voice when I ask, "Wanna trrrreeeeettt?" But the ability of animals to use signs varies. For instance, a dog and a chicken will respond differently to the presence of a feed bowl on the other side of a fence. The chicken will simply pace back and forth in front of the fence in aggravation, but the dog will seek a break in the fence, go through the break, and run back to the bowl and eat. The chicken appears to only be able to respond directly to one stimulus, whereas the dog is able to hold his response to the food at bay while she seeks an alternative. This ability to hold responses at bay is important for higher-level thinking animals.

These signs that we've been talking about may be called **natural signs**. They are private and learned through the individual experience of each animal. So, if your dog also gets excited at the sound of the treat box, it is because of its individual experience with it—Gypsy didn't tell your dog about the treat box. There also tends to be a natural relationship between the sign and its object (smoke/fire; sound/treat) and these signs occur apart from the agency of the animal. In other words, Gypsy did not make the association between the sound of the box and her treats; I did. So, in the absolute sense, the relationship between the sound and the treat isn't a true natural sign. Natural signs come out of the natural experiences of the animal, and the meaning of these signs is determined by a structured relationship between the sign and its object.

Humans, on the other hand, have the ability to use what Mead calls **significant gestures** or **symbols**. This idea of the significant gesture is one of the most central concepts to Meadian theory. According to Mead, other animals have gestures but none have *significant* gestures. A gesture becomes significant when the idea behind the gesture arouses the same response (same idea and/or emotional attitudes) in the self as in others. So, when I call "fire" because I see smoke outside my room, you and I will both run. The symbol elicited the same response in both of us. Thus human language is intrinsically reflexive: the meaning of any significant gesture always calls back to the individual making the gesture.

In contrast to natural signs, symbols can also be abstract and arbitrary. With signs, the relationship between the sign and its referent is natural (as with smoke and fire). But symbolic meaning can be quite abstract and completely arbitrary (in terms of naturally given relations). For example, the year 2005 is completely arbitrary and is an abstract human creation. What year it is depends upon what calendar is used,

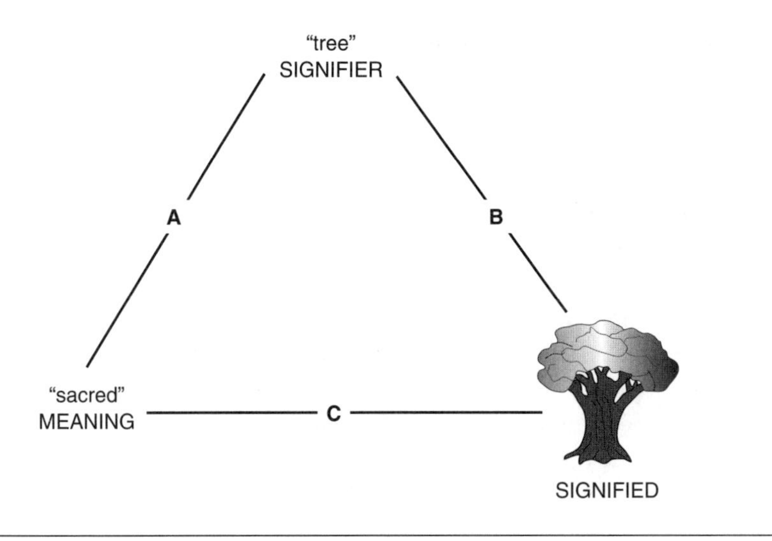

Figure 7.1 Three-Part Relation of the Sign

and the different calendars are associated with political and religious power issues, not nature. Because symbolic meaning is not tied to any object, the meaning can change over time. For example, in the United States there have been several meanings associated with the category of "people with dark skin."

What we mean by arbitrary can most clearly be brought out by looking at an idea that Charles S. Peirce proposed. Peirce explains that every word actually contains three elements and relationships: the signifier, the signified, and the meaning (the word *signifier* is being used in the same way we are using symbol). I've depicted these relationships in Figure 7.1. The important point to see is that none of these elements or relationships is necessary. There is no necessary reason why we call the object with green leaves and brown bark a "tree." We could have just as easily called it a stone. There isn't anything inside the object that compels us to name it as we do. Nor is there any obligatory reason why we attach the meaning of sacred to the sign and the signified. All of the relationships and points are made and remade every day in interaction.

According to Mead (1934), the meaning of a significant gesture, or symbol, is its "set of organized sets of responses" (p. 71). Symbolic meaning is not the image of a thing seen at a distance, nor does it exactly correspond to the dictionary definition; rather, the meaning of a word is the action that it calls out or elicits. For example, the meaning of a chair is the different kinds of things we can do with it. Picture a wooden object with four legs, a seat, and a slatted back. If I sit down on this object, then the meaning of it is "chair." On the other hand, if I take that same object and break it into small pieces and use it to start a fire, it's no longer a chair—it's firewood. So the meaning of an object is defined in terms of its uses, or legitimated lines of behavior. This definition of action is created through social interaction, both past and present.

Because the meaning—legitimated actions—and objective availability (they are objects because we can point them out as foci for interaction) of symbols are produced in social interactions, they are **social objects**. Any idea or thing can be a social object. A piece of string can be a social object, as can the self or the idea of equality. There is nothing about the thing itself that makes it a social object; an entity becomes an object to us through our interactions around it. Through interaction we call attention to it, name it, and attach legitimate lines of behavior to it. For example, because of certain kinds of interactions, a Coke bottle here in the United States is a specific kind of social object. But to Xi, a bushman from the Kalahari Desert (in the film, *The Gods Must Be Crazy*), the Coke bottle became something utterly different as a result of his interactions around it. For Xi, the Coke bottle dropped from the sky—an obvious gift from the gods. But when he brought it to his village, this playful gift from the gods became a curse, because there was only one and everybody wanted it. It became a scarce resource that brought conflict. Eventually, Xi had to go on a religious quest because of this gift from the gods (to us, a Coke bottle).

The signs, symbols, and social objects that are reciprocally held by any group are called language. **Language** is the repository of *social* experiences. Humans use language to express social and cultural events, experiences, and pragmatic meanings. Language therefore exists outside of the individual; it is a social entity. Thus, when we use language to understand our own experiences, those experiences become social. Let me explain this a bit. Have you ever had an experience that you couldn't grasp or understand? For me, one such experience was watching my son being born. It was amazing. I was literally awestruck. For days afterward I tried telling my friends about it but couldn't. Words failed to convey my experience. I was speechless.

That last line is extremely important. I couldn't grasp my experience, nor could I tell it to others, because I lacked a language for it. And so, that experience remains mine and mine alone. However, most of our experiences aren't like that. We can usually understand them and tell others about them, and, of course, the way that we do that is through language. But notice, we explain our experiences not only to others but also to our own self through language. Because language is a social thing, it renders those understandable experiences social as well.

Moreover, language transcends the here and now. Symbolic language is the only way in which the past and future exist for humans. Language isn't stuck in the present; it can go outside of the right here/right now. Because of language, we can talk to others and ourselves about the past, and we can symbolically plan for a future that does not yet actually exist. Due to this quality of language, we can construct symbolic worlds and meanings, such as the idea of freedom or the world of religion. This transcendent quality of language becomes very important when we consider how the self is formed.

Taken together, we have seen that language is abstract, arbitrary, reflexive, and transcends time and space. We have also learned that symbols are social objects because we are able to point out and refer to them in interaction. This fact of language also socially demands that each individual has a mind. It is an interesting little equation: human survival depends upon language; language depends upon individuals having minds; human survival, then, is dependent upon the mind.

Minds: Ellen Langer (1989) makes the distinction between mindfulness and mindlessness. She conducted an experiment at a nursing home in Connecticut. There, one group of elderly residents was allowed to make small decisions about their lives and environments, while others were not. One year later, that group was not only happier than groups who were not given such choices, but the death rate was significantly lower in the choice group as well. Langer concluded that the difference could be attributed to mindfulness—being actively engaged in one's life, rather than living by routine (mindlessness).

Langer discovered what Mead had talked about some years previously: the mind is behavior. For Mead, the mind is not something that resides in the physical brain or in the nervous system, nor is it something that is unavailable for sociological investigation. The mind is a kind of behavior, according to Mead, that involves at least five different abilities. It has the ability

- to use symbols to denote objects,
- to use symbols as its own stimulus (it can talk to itself),
- to read and interpret another's gestures and use them as further stimuli,
- to suspend response (not act out of impulse), and
- to imaginatively rehearse one's own behaviors before actually behaving.

Let me give you an example that encompasses all these behaviors. A few years ago, our school paper ran a cartoon. In it was a picture of three people: a man and a woman arm-in-arm, and another man. The woman was introducing the men to one another. Both men were reaching out to shake one another's hands. But above the single man was a balloon of his thoughts. In it he was picturing himself violently punching the other man. He wanted to hit the man, but he shook his hand instead and said, "Glad to meet you."

> Mind arises in the social process only when that process as a whole enters into, or is present in, the experience of any one of the given individuals involved in that process. When this occurs, the individual becomes self-conscious and has a mind. (Mead, 1934, p. 134)

There are a lot of things we can pull out of this cartoon, but the issue we want to focus on is the disparity between what the man felt and what the man did. He had an impulse to hit the other man, perhaps because he was jealous. But he didn't. Why didn't he? Actually, that isn't as good a question as, *how* didn't he? He was able to not hit the other man because of his mind. His mind was able to block his initial impulse, to understand the situation symbolically, to point out to his self the symbols and possible meanings, to entertain alternative lines of behavior, and choose the behavior that best fit the situation. The man used symbols to stimulate his own behavior rather than going with his impulse or the actual world.

The kind of interaction that we've been talking about, upon which society depends, is predicated upon each of the participants having a mind. The mind stops people from acting on impulse (perhaps a more instinctual, animalistic response) and allows them to consider what would be socially responsible. Without the mind, society would be impossible; we would simply have chaos. It's important to note that since the mind is seen as behavior rather than a thing, people can just as easily be mindless as mindful. And the mind, because it is behavior, is something that can be observed.

Mead also argues that the mind is a result of social interaction. Things in the environment become objects to humans as they are invested with meaning—sets of responses—and are symbolized. According to Mead (1934), these symbolic objects represent not only our organized responses,"but they also represent what exists for us in the world" (p. 128). Thinking is the symbolic manipulation of these socially defined and invested objects in relation to a self. Mead argues that the mind appears when the social process is taken in through language acquisition: "Reflexiveness, then, is the essential condition, within the social process, for the development of mind" (p. 134).

Infant dependency and society are the preconditions for the formation of the mind. Human babies are completely dependent upon their parents for survival, more so and for longer periods of time than other species. They are thus forced to interact with others in already organized social environments. When babies are hungry or tired or wet, they send out what Mead would call "unconventional gestures," gestures that do not mean the same to the sender and hearer. In other words, they cry. The caregivers must figure out what the baby needs. When they do, parents tend to vocalize their behaviors ("Oh, did Susie need a ba-ba?"). Babies eventually discover that if they mimic the parents and send out a significant gesture ("ba-ba") that they get their needs met sooner. This is the beginning of language acquisition; babies begin to understand that their environment is symbolic—the object that satisfies hunger is "ba-ba" and the object that brings it is "da-da." Eventually, a baby will understand that she has a symbol as well: "Susie." Thus, language acquisition allows the child to symbolize and eventually to symbolically manipulate her environment, including self and others. The use of reflexive language also allows the child to begin to role-take, which is the primary mechanism through which the mind and self are formed.

Living Outside the Moment—The Self

Defining the self: Have you ever watched someone doing something? Of course you have. Maybe you watched a worker planting a tree on campus, or maybe you watched a band play last Friday night. And while you watched, you understood people and their behaviors in terms of the identity they claimed and the roles they played. In short, when you watch someone, you understand the person as a social object. After watching someone, have you ever called someone else's attention to that actor? Of course you have, and it's easy to do. All you have to say is something like, "Whoa, check him/her/it out." And the other person will look and usually understand immediately what it is you are pointing out, because we understand one another in terms of being a social object. Sometimes we will notice someone and be embarrassed for her or him, sometimes we laugh at or with the person, and other times we might feel a sense of admiration. When we point out the other person to our friend, she or he will probably feel the same way—because the fact that the person observed is a social object calls out the same responses in both of us.

> The self has the characteristic that it is an object to itself. . . . How can an individual get outside himself (experientially) in such a way as to become an object to himself? This is the essential psychological problem of selfhood or of self-consciousness. (Mead, 1934, pp. 136, 138)

People-watching is a pretty common experience and we all do it. We can do it because we understand the other in terms of being a social object. But let me ask you something. Have you ever watched yourself? Have you ever felt embarrassed or laughed at yourself? How is that different than watching other people? Actually, it isn't. ("Whoa, check it out.") If I could, I'd send you home right now and have you think through this deceptively simple fact: watching your self is the same as watching others. We relate to and understand others because they are social objects; we relate to and understand the self because it, too, is a social object.

But there is something decidedly odd about this statement. It's easy to watch someone else, and it is easy to understand *how* we watch someone else. If I am watching a band play, I can watch the band because they are on stage and I am in the audience. We can observe the other because we are standing outside of them. We can point to them because they are *there* in the world around us. But how can we point to our self, call our own attention to our self, and understand our self as a social object? Do you see the problem? We must somehow divorce our self *from* our self so that we can call attention to our self, so that we can understand our self as meaningfully relevant as a social object. Whew. So, how is that done? That, as Mead says, is the most significant problem in social psychology.

I think the self is the most amazing of all human productions. One of the things that I find so fascinating about it is that it doesn't live in the here and now. By definition, the self must exist outside of our direct experience. It is this ability to not be in the moment that allows us to monitor our behaviors and produce society. The mind considers immediate behaviors with reference to a mental object (the self) that is made up of selected images from the past and hoped-for images from the future. Thus, our conscious experience is predicated upon our being able to achieve a perspective that is divorced from the constraints of time. What this implies is that the **self** is a perspective—it is a symbolic platform on which to stand and view our own behaviors as if someone else were performing them.

You have no doubt had the experience of doing something while saying to yourself, "I can't believe I'm doing this." At that moment, you are looking at your own behaviors as if "the inner you" wasn't doing them, as if someone else was acting. You are both the actor and the audience—part of you is living in the moment but part of you is living outside of the moment. The self is thus constituted by the perspective and the reflexive dialog between the two parts of you.

Our self exists as an object outside of our direct experience, and there is a sense in which it exists as something other than the body as well (see Mead, 1934, p. 357). The body may be used to experience the world, but the self is something that is conceived of as being greater than the body that is situated in time and place—the self is an understanding of the person that has a past and future, while the body is physically only in the here and now. The self is therefore a separate thing from the body and from immediate experience; the self is a socially produced mental concept (social object), differentiated from the body and separated from time and space. But how is it that we can exist in this way, divorced from our direct experiences? This is the question that drove the bulk of Mead's work.

> He becomes a self in so far as he can take the attitude of another and act toward himself as others act. (Mead, 1934, p. 171)

Forming the self: Mead argues that the essence of the self is cognitive and lies in the internalized conversation of gestures. Mead notes that both Cooley and James argued that the self is found in the reflexive emotional experiences that make up self-feeling. But, Mead maintains, there must already be a self present to develop feelings about it. The process whereby the self originates, then, is the cognitive/linguistic development of the individual through role-taking, whereby the individual learns to distinguish self as one of the objects in the environment that must be taken into consideration for action. The genesis of self occurs through the play, game, and generalized other stages of role-taking.

Role-taking is the key mechanism through which people develop a self and the capacity to be social, and it has a very specific definition: **Role-taking** is the process through which we place our self in the position (or role) of another in order to see our own self. Students often confuse role-taking with what might be called role-*making.* In every social situation, we make a role for ourselves. Erving Goffman wrote at length about this process and called it *impression management.* Role-taking is a precursor to effective role-making—we put ourselves in the position of the other in order to see how they want us to act. For example, when going to a job interview, you put yourself in the position or role of the interviewer in order to see how she or he will view you—you then dress or act in the "appropriate" manner. But role-taking is distinct from impression management, and it is the major mechanism through which we are able to form a perspective outside of ourselves.

A perspective is always a meaning-creating position. We stand in a particular point of view and attribute meaning to something. Remember that meaning is not a function of the object itself. Meaning has to be given, and that's why there can be so many different meanings for one symbol. Let's take the Confederate flag, for example. What does it mean? To some it means Southern heritage, to others white supremacy, and to still others it signifies a rebellious way of life (akin to flying the pirate's flag). Thus, what the "Stars and Bars" means depends on where you stand. The flag itself is almost immaterial. Its only real function is to cue the different perspectives so that people can interact around it as a social object.

The self is just such a perspective. It is a viewpoint from which to consider our behaviors and give them meaning. It is the way in which we become a social object to our own self. Social objects become significant as people create perspectives and attach sets of organized responses to a thing. The process of self-formation is the process of being defined in social interaction as a social object. Keep in mind two things as we consider the formative stages of the self: they all explain one thing—how we can get outside our own behaviors; and, they move from concrete to abstract role-taking episodes.

Before the self begins to form, there is a preparatory stage, which I mentioned above when discussing the mind. While Mead did not explicitly name this stage, he implied it in several writings. The preparatory stage is the stage of imitation. The child imitates the adults in its environment. She imitates words, gestures, or sounds but without understanding the meaning. The child uses the terms of the symbolic world without grasping it meaningfully. Social objects, including self, are not invested with clear sets of responses at this point.

The second stage in the process of self-formation is the **play stage**. During this stage, the child can take the role, or assume the perspective, of certain significant others. Significant others are those upon whom we depend for emotional and often material support. These are the people with whom we have long-term relations and intimate (self-revealing) ties. Mead calls this stage the play stage because children must literally play at being some significant other in order to see themselves. At this point, they haven't progressed much in terms of being able to think abstractly, so they must act out the role to get the perspective. This is important: *a child literally gets outside of him- herself* in order to see the self.

Children play at being Mommy or being Teacher. The child will hold a doll or stuffed bear and talk to it as if she were the parent. Ask any parent; it's a frightening experience because what you are faced with is an almost exact imitation of your own behaviors, words, and even tone of voice. But remember the purpose of role-taking (notice this isn't role-*playing*): it is to see one's own self. So, as the child is playing Mommy or Daddy with a teddy bear, who is the bear? The child herself. She is seeing herself from the point of view of the parent, literally. This act is the origin of the divorced perspective that we call self. This is the genesis of the objective stance: being able to get outside of the self so that we can watch the self as if on stage.

As a child acts toward herself as others act; the child begins to understand self as a set of organized responses and becomes an object to herself. Children begin by taking the role of significant others, and during this time the child can only take the role of a single person at a time. The individual is incapable of seeing herself from the perspective of many others at the same time. The significant others are segmented, therefore the view of self is a segmented one. So, there is not yet a sense of a whole self.

> This organized reaction becomes what I have called the 'generalized other' that accompanies and controls his conduct. And it is this generalized other in his experience which provides him with a self. (Mead, 1925, p. 269)

The next stage in the development of self is the **game stage**. During this stage, the child can take the perspective of several others and can take into account the rules (sets of responses that different attitudes bring out) of society. But the role-taking at this stage is still not very abstract. In the play stage, the child could only take the perspective of a single significant other; in the game stage, the child can take on the role several others, but they all remain individuals. Mead's example is that of a baseball game. The batter can role-take with each individual player in the field and determine how to bat based on their behaviors. The batter is also aware of all the rules of the game. Children at this stage can role-take with several people and are very concerned with social rules. But they still don't have a fully formed self. That doesn't happen until they can take the perspective of the generalized other.

The **generalized other** refers to sets of attitudes that an individual may take toward him- or herself—it is the general attitude or perspective of a community. The generalized other allows the individual to have a less segmented self as the perspectives of many others are generalized into a single view. It is through the generalized other that the community exercises control over the conduct of its individual members.

Up until this point, the child has only been able to role-take with specific others. As the individual progresses in the ability to use abstract language and concepts, she or he is also able to think about general or abstract others. So, for example, a man in the privacy of his home may eat a bag of potato chips and think, "If the guys at the gym could see me now, they'd have a fit." Or a woman may look in the mirror and judge the reflection by the general image that has been given to her by the media about how a woman should look. There are no specific other people involved, but these individuals are able to see the self through the eyes of some generalized other.

An insightful example of how the generalized other works is given to us by George Orwell (1946) in his account of "Shooting an Elephant." Orwell was at the time a police officer in Burma. He was at odds with the job and felt that imperialism was an evil thing. Yet at the same time, the local populace despised him precisely because he represented imperialistic control; he tells tales of being tripped and ridiculed by people in the town. One day an elephant was reported stomping through a village. He was called to attend to it. On the way there, he obtained a rifle, only for scaring the animal or defending himself if need be. When he found the animal, he knew immediately that there was no longer any danger. The elephant was calmly eating grass in a field. And he knew that the right thing to do was to wait until the elephant simply wandered off. But at the same time, he knew he had to shoot the animal.

It had never been his intention to kill the elephant, and it was apparent that it was not needed. However, as Orwell (1946) relates, "I realized that I should have to shoot the elephant after all. The people expected it of me and I had got to do it; I could feel their two thousand wills pressing me forward, irresistibly" (p. 152). He felt the expectations of a generalized other. Though contrary to his own will, and after much personal anguish, he shot the elephant. As Orwell (1946) puts it, at that time and in that place, the white man "wears a mask, and his face grows to fit it. . . . A sahib has got to act like a sahib" (pp. 152–153). Not shooting was impossible, for "the crowd would laugh at me. And my whole life, every white man's life in the East, was one long struggle not to be laughed at" (p. 153). We may criticize Orwell for his decision (it's always easy from a distance); still, each one of us has felt the pressure of a generalized other.

The I and the Me: Thus far it would appear that the self is to be conceived of as simply a reflection of the society around it. This concept could not account for change or deviance. But for Mead, the self isn't simply this socialized entity. Remember that part of what we mean by the self is an internalized conversation; there are at least two people involved (the observer and the actor). Mead actually postulates the existence of two interactive facets of the self: the "I" and the "Me." The Me is the self that results from the progressive stages of role-taking and is treated as a social object. The Me doesn't fully come into existence until we are able to role-take with the generalized other: "only in so far as he takes the attitudes of the organized social group to which he belongs . . . does he develop a complete self" (Mead, 1934, p. 155). Further, it is role-taking with the generalized other that gives us a unified self. Up until that point, our self perspectives are based on the points of view of individual others. We see our self as would Mom, or Dad, or sister, or teacher, or our friend Joe. The generalized other unifies all those perspectives into a community. It is through the Me that "the community exercises control over the conduct of its individual members" (Mead, 1934, p. 155). The Me is the perspective that we assume to view and analyze our own behaviors; the "I" is that part of the self that is unsocialized and spontaneous. The I is the seat of the impulse, and it is the actor. Part of what we mean by the self, then, is this reflexive, internal dialog between these two parts. Notice the way in which this dialog occurs—the I is the subject and the Me is the object: what I do, I do to Me.

> Human society as we know it could not exist without minds and selves, since all its most characteristic features presuppose the possession of minds and selves by its individual members. (Mead, 1934, p. 227)

We have all experienced this conversation and the tension it creates. The cartoon that I referenced earlier is a good example. We have all felt the impulse of anger or irritability or, on a more positive note, spontaneous playful behavior. There is this part of us that wants to react immediately to stimuli. We may want to jump for joy or shout in anger or punch someone we're angry at or kiss a stranger or run naked. And sometimes we immediately act upon our impulses, but not very often. To the extent that we do act on these impulses, we are not acting fully human. A less extreme way of putting it is that we are not acting social (but remember that, for Mead, to be human is to be social).

Here is a key to understanding Mead: the self is essential for society. We don't have a self because there is a psychological drive or need for one. We have a self because society demands it. Remember our conversations about Spencer and Durkheim and their ideas of requisite functions? The self is a requisite function as well; society cannot exist without it. Think about it this way: what would we call someone who could not engage in this internal dialog between the I and the Me? We would call such a person a sociopath, or what psychologists are now referring to as someone with an antisocial personality disorder (APD). A society filled with sociopaths would be impossible. Further, the self is not an individual or psychological construct. The self, as we have seen, is constructed through language acquisition and role-taking in social interactions.

As I said, we have all experienced the conversation between the I and the Me (sometimes I would be much happier if I didn't think so much about Me). The Me presents to us the perspectives of society, the meanings and repercussions of our actions. The I presents our impulses and drives. These two elements of our self converse until we decide on a course of action. But here is the important part: the I can always act before the conversation begins or even in the middle of it. In true pragmatic form, the individual cannot know the action of the I until it is experienced, until it is past. The I can thus take action that the Me would never think of; it can act different than the community. These two elements of the self are reflexively aware and continually converse back and forth: "The self is essentially a social process going on with these two distinguishable phases. If it did not have these two phases, there could not be conscious responsibility and there would be nothing novel in experience" (Mead, 1934, p. 178).

It should be fairly clear at this point that the mind and the self are mutually constituted. That is, they both create and reinforce one another. Through role-taking, the individual is able to take the perspective of the generalized other, and it is at that point that the individual truly has a self. Mead also argues that it is the ability to take into consideration the generalized other that makes abstract thinking possible.

Abstract thinking: We can conceive of our thoughts as laying on a continuum between concrete particulars and abstract universals. If I think about my dog Gypsy, then I'm thinking fairly concretely. If, however, I think about the category of dog, then I'm using an abstract universal to think. There has been quite a bit of philosophical ink spilled in debating whether or not abstract universals exist. The category of dog, or even more abstractly, mammal, doesn't exist *there* in the same way Gypsy does. But Mead isn't concerned with the ontology of universal categories; he is presenting us with an understanding of how the mind is able to think abstractly, and abstract thought is socially based. We are able to think abstractly because we are able to role-take with the generalized other.

Think about it this way: What we mean by human thought is always tied up with *awareness* of our thoughts. If we are thinking, we are thinking *about* something and not simply experiencing it. For Marx, ideas and conscious thought came about for us because of the process of externalization—as we are faced with our own products, we can be reflexive about our own nature. The ability to think, then, always involves a move away from immediate, sensual experience.

For Mead, that movement is accomplished through role-taking. If we are able to role-take only with significant others, then our thought will be concrete and specific, because our thought will be merely with reference to the particulars of another individual. In this case, the mental movement away from sensual experience extends only so far as the given situation or act and the particular attitudes of the individuals involved. On the other hand, if we role-take with a generalized other, we develop the ability to think outside of the situation and its particulars. As we develop the ability to role-take with a generalized other, our mind is pushed to stand in a very abstract position. This process obviously varies. The further removed is the generalized other from the particulars of the situation, the more abstract is the internalized conversation that we call thinking (Mead, 1934, pp. 155–156n).

The Emergent Big Brother

Without the self and mind, society could not exist. Society is dependent upon our being able to suspend responses, to take our own behaviors as an object of consideration from the perspectives of significant and generalized others, and to choose from alternative lines of behavior. Mind and self are thus social, not individual, entities. They are not only produced socially, they are necessary because of society. Mind, self, and society are intrinsically bound together through mutual constitution and need. This fact is why the question of individuality is so intriguing.

Given this mutuality, how is it the person is able to experience the self as a psychological rather than social entity? Both Durkheim and Simmel give us some direction on this. Basically, as society differentiates and the moral boundaries connecting the groups become thin and meager, the person is freed in varying degrees from the constraints of groups. Mead gives us the internalization mechanism. As the individual role-takes with different groups, she or he forms divergent "Me's." The person can see her or his behavior from the perspective of many groups. As these groups become less connected and more different from one another, the person develops a sense of individuality. She or he is different from all the groups and others encountered because the person is the focal point of a unique matrix of group relations. The challenge then, of course, is to develop and keep a single narrative about the self going—which is part of the postmodern problem.

But we are getting ahead of ourselves. Because of the ways in which mind, self, and society are intertwined, it is important for us to consider Mead's theory of society. We are actually going to have to build a context within which to understand the Meadian idea of society. We'll begin with the act.

The act: Humans act; they don't *re*act. One of the things that make us unique is that we have free will. We make decisions about our behaviors. But we don't do that all the time, and that is part of what Weber is getting at when he implies that traditional and affective behaviors are less than fully human. But often we do make a conscious decision to act, and Mead begins his understanding of society with the individual acts of human beings. In his conceptualization, we'll see how it is that humans act rather than react.

As Mead characterizes it, the **act** contains four distinct elements: impulse, perception, manipulation, and consumption. Generally speaking, the impulse originates in the "I" (I want to eat or I want to say something). For most animals, the route from impulse to behavior is rather direct—they react to the stimulus using instincts or behavioristically imprinted patterns. But for humans, it is a circuitous route. After we feel the initial impulse to act, we perceive our environment. This perception entails the recognition of the pertinent symbolic elements—other people, absent reference groups (generalized others) and so on—as well as alternatives to satisfying the impulse. After we symbolically take in our environment, we manipulate the different elements. This is the all-important pause before action; this is where society becomes possible. This manipulation takes place in the mind and considers the possible ramifications of using different behaviors to satisfy the impulse. We role-take with significant present and generalized others ("Man, this band is great. I wonder what they would say if I asked to sit in on a couple of songs?" "I wonder if there are any students here. . . . What would they think if they saw me up there playing guitar with just a little bit too much to drink?" "Oh, jeeze, what if word got back to the department?!?!?" "Ok, I and Me are just going to sit here and have another beer!"). And we think about the elements available to complete the task. After we manipulate the situation symbolically, we are in a position to act. This final phase is what Mead termed *consumption.*

Interaction: But humans don't simply act, we interact. **Interaction** is the ongoing negotiation and melding together of individual actions. There are three distinct steps that must occur before we can say that there is interaction. Mead refers to them as the *triadic relation.* First, there is an initial cue given. But, as we've seen, the cue itself doesn't carry any specific meaning. Let's say we see a friend crying in the halls at school. What does it mean? It could mean lots of things. In order to determine (or more properly, create) the meaning, we have to respond to that cue: "Is everything alright?" But we still don't have meaning yet. There must be a response to our response. After the three phases, a meaning emerges: "Nothing's wrong—my boyfriend just asked me to marry him."

> The logical structure of meaning . . . is to be found in the threefold relationship of gesture to adjustive response and to the resultant of the given social act. . . . This threefold or triadic relation between gesture, adjustive response, and resultant of the social act which the gesture initiates is the basis of meaning. (Mead, 1934, p. 80)

But we probably still aren't done, because her response has become yet another cue. What does that cue mean? We can't tell until you respond to her cue and she responds to your response. This process is how meaning emerges from the interaction. Meaning results from the back-and-forth negotiation over a symbolic object. Thus, we cannot tell until after the interaction is completed what the different social objects meant to the people involved. And we may not even be done then. Sometimes we take the meaning of one interaction and make it a cue in another interaction.

Let's suppose you told your friend who was crying in the hall that marrying this guy was a bad idea. You saw him out with another woman last Friday night. At this point, the social object, marriage, which was a cue that caused her to cry in happiness, has become an object of anger. So the meaning that emerges is now betrayal and anger. She then takes that meaning and interacts with her fiancé. In that interaction, she presents a cue (maybe she's crying again, but it has a different meaning), and he responds, and she responds to his response, and so on. Maybe she comes to find out that you misread the cues that Friday night and the "other woman" was just a friend. So, she comes back to you and presents a cue, ad infinitum.

Thus, meaning isn't predetermined at all and can't be predicted. This lack of predictability is what Mead is trying to get at when he says that meaning is emergent. Meaning is something that is achieved in and through interactions; it doesn't exist in the object or in the culture in any determined way. Let's take a more serious example. Can the meaning of "female" change from one interaction to another? Sure. A woman may have an interaction wherein she is treated with respect as a professional colleague, but in the very next interaction be treated as a sexual plaything, an object of gaze. But it isn't simply a one-way street. All meanings are subject to negotiation within each and every interaction by all the participants. So the professional woman may see a man that she finds attractive and tries to get his attention so that she can be an object of gaze for him. The meaning of any social object thus depends on the interaction and how we use symbols.

Emerging society: But this doesn't imply that meaning is absolutely free-floating. Mead indicates that there are **institutions**, but these institutions aren't like the structures that we see in functionalism or conflict theory. Those institutions are macro-level entities that by and large determine human behavior from the top down. And, as we've seen, the theorists who talk about such things assume society is objective. Mead (1934), on the other hand, argues that "an institution is, after all, nothing but an organization of attitudes which we all carry in us"; they are "organized forms of group or social activity—forms so organized that the individual members of society can act adequately and socially by taking the attitudes of others toward these activities" (pp. 211, 261–262). So, in any interaction there may be several different attitudes or perspectives that we can take and through which we can organize our behaviors. Society, then, doesn't exist objectively outside the concrete interactions of people. Herbert Blumer, the man who formalized symbolic interaction, tells us that the interaction is the only acting social unit. In other words, things like "institutions" or "societies" can't act because they don't exist objectively. They exist only as sets of attitudes, symbols, and imaginations that people may or may not use and modify in an interaction. In other words, society exists only as sets of potentials (ideas that people could possibly use).

In general, these perspectives or potentials are called the **definition of the situation**, which is the reigning characterization of any interaction. In other words, most encounters can be defined in some broad manner—as a university classroom, for example. This is an important point because it tells us that all situations for humans are by their very nature symbolic. It also implies that the roles that we play are prescribed for us by the definition of the situation. Roles are behavior clusters or scripts that are characteristic of some position in a community. So, in a university classroom, we would expect to find the roles of professor and student.

But notice that the definition of the situation isn't as clear-cut as the above example would indicate. I may be at the grocery store and the role of professor doesn't live in that definition of the situation. However, the definition of the situation changes as soon as a student calls me Dr. or Professor Allan. It is no longer simply a grocery store. By invoking a different definition, the student presents the possibility of overlaying the definition of store with that of the university. I say that the student presents the possibility because, as with every meaning, it is unfixed. I may respond by simply saying "hello," in which case I allow the world the student proposed, or I may ask the individual to please call me Ken. Of course we wouldn't be done yet, as the student may tell me that she or he isn't comfortable calling me by my first name. And we are off and running again.

The same is true for our above example of the word *female*. The role of "potential date partner" doesn't exist in her professional situation, yet she may suggest a different definition by simply winking her eye or by allowing a touch to linger longer than expected, thus making available a new set of scripts or roles. We can never tell in advance what definitions, roles, or meanings are going to come out of an interaction—they all emerge as the interactions go along.

Society is thus an emergent quality of the interaction. Society is *achieved* as people role-take with institutionalized perspectives within a given definition of the situation. All of what we've talked about in this section so far can be summarized by stating that human action is not determined or released in response to institutionalized pressures (structures), but it is built up and emerges through the negotiated definition of the situation. Further, Mead's theory implies that human society must be studied in terms of acting units and research methods, and theory must be geared toward understanding the meaning to the actors. The only acting unit is real people in real interaction, because that is where meaning is achieved and where people produce roles and selves. And, according to Mead's perspective, theory can't be general, other than a few orienting concepts, because human nature is symbolic (not empirical like Durkheim and Marx argue) and human behavior cannot be predicted.

Another important point for Mead is that the self is seen as one of many social objects in the environment, the level of importance and definition of which must be determined by participants. So, the self isn't something that has an essential meaning either. Like all social objects, it must be symbolically denoted and then given meaning within the definition of the situation. And like all social-symbolic objects, the meaning of the self is flexible and emergent.

A theory moment happened for me not too long ago that might help to illustrate this point. My brother-in-law (Don) and my sister (Patti) both told me about an incident in back-to-back telephone conversations. Don is a distance runner and was training for the Pike's Peak run. During one of his training sessions at the Peak, he experienced heart fibrillation on the way down. He didn't tell Patti at that time, and he ran the race a week later. He did not experience any further problems. After the race, he told Patti about the training incident over dinner and margaritas. Patti got upset. She then told me about it. She said that Don is too much into machismo posturing and doesn't deal with reality (her definition of Don's self). The "reality" that Patti referred to is that he is 65 years old and had previously experienced a five-hour fibrillation problem at the doctor's office. In this interaction with me, Patti referred to herself as a "caregiver and an organizer." That's her self as she sees it vis-à-vis Don. Don told the same story, as far as the actual events are concerned. But the meaning of all the events changed, as did the definitions of the selves involved. Don defined the "condition" as not life threatening, as one that is normal for athletes, and he said that lots of doctors say it's okay for an athlete in his condition to continue training. Don defined himself as a "competitor and an optimist" and Patti as a "worrier that mothers too much."

The important thing to see in this example is *not* that every individual has an interpretation of an event—that would be more psychological than sociological. What's important to note is what is happening socially: these two people are negotiating with a third party over the meaning of an event as well as the kinds of selves that that event indicates that they have. Out of this interaction and negotiation emerge a definition and a sense of self for each of the participants. The process through which that occurs is the focus of a Meadian way of perceiving the social world.

Thinking About Modernity and Postmodernity

Reflexivity and the fragmenting of the self: Mead never saw his theory as particularly modern. His argument implies that the self is essentially a reflexive construction involving the production of the self as social object, regardless of the historical context. However, there are some elements of postmodern theory that seem to challenge Mead's notion of an integrated self. Recall that Mead argues that the self becomes integrated through role-taking with a generalized other. When role-taking with only significant others, the self will seem segmented, divided as it is among the different points of view. The generalized other is able to link all those individual perspectives into one abstract whole, thus giving the self a sense of integration. But what would happen if the generalized other was itself fragmented or constructed from vacuous images?

Some postmodernists argue that just such a state exists. Rather than a self that is created through role-taking with actual social others, the postmodern self is created through role-taking with media images. These images are not connected to one another in any real way. Through the media, individuals are inundated with disconnected images of different groups, different people, different values, and different modes of expression—all of which has increased the amount of reflexivity and self-monitoring in which individuals engage. The problem is that this increased reflexivity is directed toward a cultural self that is socially saturated and filled with constant change and doubt. The result is what Gergen (1991) refers to as *multiphernia*—vertigo of self-values filled with expressions of inadequacy—and the pastiche personality. Jameson (1984) goes so far as to argue that the culture of the postmodern era "stands as something like an imperative to grow new organs, to expand our sensorium and our body to some new, as yet unimaginable, perhaps ultimately impossible, dimensions" (p. 80).

As we've seen, Mead is certainly aware of the reflexive dimension of creating a self. In fact, according to Mead, both the mind and the self are created through reflexive thought. The self is found in the social object (Me) as well as the reflexive conversation an individual has with her- or himself from the perspectives of specific and generalized others. Though Mead does not go into much detail, he is aware that this reflexiveness, though necessary for social life, is not altogether comfortable for the individual. It is perhaps problematic in postmodernity.

Being too reflexive about things can actually deconstruct them (Allan, 1998, pp. 84–85). Sometimes this can be good. Racial and sexual inequalities can only continue if people believe that the social categories themselves are real. When we realize, for example, that race is not an essential quality, but rather a symbolic meaning that has been created through historical, social processes and is defined in interactions, then we can deconstruct race and it will no longer influence our behaviors. There is, however, a limit. If subjective reflexivity pushes deconstruction too far, we are then faced with nothing but a yawning abyss of meaninglessness—a placeless surface that is incapable of holding personal identity, self, or society. Meaning is set loose from its social and material moorings and becomes completely free-floating.

This is part of what postmodernists are getting at. Reflexivity has increased in postmodernity because of a number of social factors, like the legitimation of doubt, constantly changing knowledge, disconnected media images, rapidly shifting markets, and so on. This increased reflexivity is particularly important for the self. Too much reflexivity can destroy the meaning that we give the self. However, Mead gives us a theory through which we can understand how the destructive tendencies of postmodern reflexivity can be lessened.

Fusing the I and the Me: Let me give you a practical example of how reflexivity works. Let's pretend you are at a bar and see someone that you think is attractive. What do you do? Well, if you're like most of us, you think about it and you role-take. You put yourself in that person's shoes and see how you would look to her or him. You might try out different scenarios in your mind. Depending on how these reflexive thoughts come out, and depending on who that person is with and what that other person may think of you, you may or may not approach the person. If you are like most of us, that process can be emotionally draining.

But let me add one small detail to our story. In this scenario, you are at the bar and you see someone attractive and you are drunk. Now what do you do? Chances are good that you will simply walk right up to that person without so as much as a second thought. Why? Because you're drunk, right? Yes, but let's think like theorists and go below the surface. What is it that being drunk does? Being drunken short circuits the reflexive loop inside you. And because of that, you exist more in the immediate moment than you generally do. In other words, the impulses of the I aren't subject to the scrutiny of the Me (at least we can't hear it talking). This is a shadow example, because this dampening of the reflexive loop is in some ways artificial since it is due to alcohol, but it helps us see a bit of what Mead is talking about.

Mead recognizes an important phase in the interaction between the I and the Me. There are times in which the I and the Me can be fused and the reflexive loop stopped. During those times, things feel more real and we sense the true "meaning of life." The individual experiences those times when we aren't reflexive as essentially real. In other words, we feel that we are not attributing meaning; rather, it simply is what it is (direct sensual experience).

Mead argues that those times when the I and the Me are fused are "intense emotional experiences." This emotional response is all the more powerful when experienced with a group. "The wider the social process in which this is involved, the greater is the exaltation, the emotional response, which results" (Mead, 1934, p. 274). This description is reminiscent of Durkheim's explanation of ritual. Mead supplies what is missing from Durkheim: the *internal* mechanism through which increased emotional energy is produced.

Mead says that fusing takes place in circumstances where there is complete identification of the person with the expectations of the situation. This can most easily happen in group activities. Mead gives the example of saving someone from drowning. Everybody who sees the person drowning will respond in the same manner. There is a sense of common effort and identification with all. Other situations that give rise to this phenomenon are religious and patriotic events. Durkheim, of course, sensitizes us to another kind of practice that brings about reduced reflexivity, that being ritual. There is yet another source for this kind of experience. It's called flow.

Mihaly Csikszentmihalyi (1990) uses the term **flow** to capture what he calls optimal experience: "the state in which people are so involved in an activity that nothing else seems to matter" (p. 4); it is due to being so involved in one's activity that "self consciousness disappears, and the sense of time becomes distorted" (p. 71). In a state of flow, people are not reflexive; they are completely immersed in the situation. Based on a psychology of optimal experience, Csikszentmihalyi argues that flow comes about in rule-bound situations where there is a clearly known goal, where the individual is pushed to the functional limits of her or his skill level, and where there are unambiguous signals about how well she or he is doing. In other words, the situation must enable people to concentrate all their attention on the task at hand. And "whenever one does stop to think about oneself, the evidence [of good performance] is encouraging . . . and more attention is freed to deal with the outer and the inner environment" (Csikszentmihalyi, 1990, p. 39).

If the postmodernists are correct, Mead's theory would predict that people who participate in more activities where the reflexive loop is dampened and the I and the Me fused will have more core-self experiences. In particular, from just our conversation we can see how increasing participation in emotion-producing rituals (like rock concerts) and activities that focus one's cognitive energy (like extreme sports) could create an array of experiences that feel essentially real and produce a firm sense of self.

Summary

- To think like Mead is to see that mind, self, and society are ongoing social productions that emerge from interactions.

- There are basic elements, or tools, that go into making us human. Among the most important of these are symbolic meaning and the mind. We use symbols and social objects to denote and manipulate the environment. Each symbol or social object is understood in terms of legitimated behaviors and pragmatic motives. The mind uses symbolic-social objects in order to block initial responses and consider alternative lines of behavior. It is thus necessary for society to exist. The mind is formed in childhood through necessary social interaction.

- The self is a perspective from which to view our own behaviors. This perspective is formed through successive stages of role-taking and becomes a social object for our own thoughts. The self has a dynamic quality as well—it is the internalized conversation between the I and the Me. The Me is the social object, and the I is the seat of the impulses. When the self is able to role-take with generalized others, society can exist as well as an integrated self. Role-taking with generalized others also allows us to think in abstract terms.

- Society emerges through social interaction; it is not a determinative structure. In general, humans act more than react. Action is predicated on the ability of the mind to delay response and consider alternative lines of behavior with respect to the social environment and a pertinent self. Thus mind, self, and society mutually constitute one another. Interaction is the process of knitting together different lines of action. Meaning is produced in interaction through the triadic relation of cue, response, and response to response. What we mean by society emerges from this negotiated meaning as interactants role-take within specific definitions of the situation and organized attitudes (institutions).

- Mead's theory along with postmodern theory implies that the self can become fragmented through role-taking with multiple and disconnected images and through radical reflexivity. An extension of Mead's theory indicates that a sense of core self can be produced through the practice of Durkheimian rituals and intense personal experiences. In such experiences, the I and the Me fuse, reflexivity is dampened, and the self is experienced as essentially real.

Building Your Theory Toolbox

Conversations With George Herbert Mead—Web Research

Every once in awhile we are fortunate to find something personally enlightening about one of our theorists on the Web; with Mead, we have a portion of John Dewey's speech given at Mead's funeral. Read Dewey's tribute to Mead at http://spartan.ac.brocku.ca/~lward/Dewey/Dewey_1931.html. Based on your reading, what five words do you think characterize George Herbert Mead? What does Dewey say about Mead's work? How does Dewey characterize Mead's influence on his thinking? Using the Dewey site as well as the sites given below for Mead, answer the following questions?

- In what way can Mead be considered part of an "oral tradition"? How was his most famous book written? How did Mead feel about writing?
- When and how did Mead create a nationwide disturbance?
- What kind of lecturer was Mead?
- What was the City Club of Chicago? What was Mead's association with the City Club?
- What did Mead think about poetry?
- Search for links concerning Oberlin College. What is the history of Oberlin? Mead's father taught at Oberlin, and George received his undergraduate education there. How do you think Oberlin influenced Mead's thinking?
- Mead was one of the first professors at the University of Chicago. Research the history of the University of Chicago. What kind of university was it? Why and how was it founded? What did it mean that Mead was invited to Chicago?
- Who was Jane Addams and what was the settlement house movement? What was Mead's relationship to Addams? How did Addams influence Mead?

Passionate Curiosity

Seeing the World (using the perspective)

Using Mead's theory ought to give you an entirely different outlook. So far, most of the people we have dealt with have assumed that society is at least somewhat objective and they have by and large been concerned with historical or structural processes. Even Simmel assumes that society has objective forms, though they aren't at the macro level that Spencer, Marx, or Durkheim would talk about. Mead, on the other hand, basically says that society only exists symbolically and its meaning emerges out of interaction. How would Mead talk about and understand race and gender? According to Mead's theory, where does racial or gender inequality exist? From a Meadian point of view, where does responsibility lie for inequality? How could we understand class using Mead's theory? From Mead's perspective, how and why are things like race, class, gender, and heterosexism perpetuated (contrast Mead's point of view with that of a structuralist)?

Engaging the World (using the theory)

- More and more people are going to counselors or psychotherapists. Most counseling is done from a psychological point of view. Knowing what you know now about how the self is constructed, how do you think sociological counseling would be different? What things might a clinical sociologist emphasize?
- Using Google or your favorite search engine, enter "clinical sociology." What is clinical sociology? What is the current state of clinical sociology?
- Mead very clearly claims that our self is dependent upon the social groups with which we affiliate. Using Mead's theory, explain how the self of a person in a disenfranchised group might be different than one associated with a majority position. Think about the different kinds of generalized others and the relationship between interactions with generalized others and internalized Me's. (Remember, Mead himself doesn't talk about how we feel about the self.)

Weaving the Threads (synthesizing theory)

- We did something different in this chapter. Rather than pointing out the social factors and effects of postmodernity, I offered what might be viewed as a remedy. Now is a good time to gather our thoughts about postmodernity. Look back through what I've said thus far about postmodernism. You should find that with each theorist, I concentrated on one particular aspect. For example, with Durkheim it was the institutionalization of doubt and with Weber it was the rise of a new class. Write out each of the main issues, the social factors that bring that issue about, and the effects. You can use what we talked about in Spencer as a framework. Now, you should have the beginnings of a fairly decent theory of postmodernity. How does the extension of Mead's theory mitigate or change some of those effects?
- Mead gives us our clearest alternative thus far to positivism. Using Mead's theory, critique the assumptions of scientific theory. From this critique and using Mead's perspective, explain why the symbolic interactionist approach demands an entirely different theory than science requires.

Further Explorations—Books

Baldwin, J. D. (1986). *George Herbert Mead: A unifying theory for sociology.* Beverly Hills, CA: Sage. (Part of the *Masters of Social Thought* series; excellent short introduction to Mead's life, thought, and continuing relevance)

Blumer, H. (1969). *Symbolic interactionism: Perspective and method.* Englewood Cliffs, NJ: Prentice Hall. (Blumer's classic text; he systematized Mead's thought and formally founded symbolic interaction)

Blumer, H. (2004). *George Herbert Mead and human conduct* (T. J. Morrione, Ed.). Walnut Creek, CA: AltaMira Press. (Series of articles by Blumer on Mead's theory)

Cook, G. A. (1993). *George Herbert Mead: The making of a social pragmatist.* Urbana, IL: University of Chicago Press. (Definitive biography)

Plummer, K. (Ed.) (1991). *Symbolic interactionism.* Brookfield, VT: Elgar. (Part of the *Schools of Thought in Sociology* series; brings together the most important writings in symbolic interaction)

Further Explorations—Web Links

http://www.lib.uchicago.edu/projects/centcat/centcats/fac/facch12_01.html (Site maintained by the University of Chicago; short background of Mead's work at Chicago, includes handwritten notes from Mead)

http://www.iep.utm.edu/m/mead.htm (Site maintained by The Internet Encyclopedia of Philosophy; excellent information about Mead's life and theory, including extensive quotes)

http://spartan.ac.brocku.ca/%7Elward (Site maintained by the Mead Project, Department of Sociology, Brock University, St. Catharine's, Canada; extensive information on Mead, his influences and contemporaries; links to many of Mead's published works)

http://www.cla.sc.edu/phil/faculty/burket/g-h-mead.html (Site maintained by the Department of Philosophy at the University of South Carolina; link to open discussion group about Mead's work)

Further Explorations—Web Links: Intellectual Influences on Mead

More than any other of our theorists so far, Mead was influenced by and synthesized specific schools of thought. Using your Internet search engine, define the following: utilitarianism, pragmatism, and behaviorism. In addition, Mead was directly influenced by the following thinkers:

Charles Darwin: http://www2.lucidcafe.com/lucidcafe/library/96feb/darwin.html (Site maintained by Lucidcafe, a commercial concern; contains short introduction to Darwin and various helpful Web links)

John Dewey: http://www.iep.utm.edu/d/dewey.htm (Site maintained by The Internet Encyclopedia of Philosophy; excellent source for Dewey's life and philosophy)

Wilhelm Wundt and William James: http://www.ship.edu/~cgboeree/wundtjames.html (Site maintained by Dr. C. George Boeree, professor of psychology at Shippensburg University; combines Wundt and James and their influence on psychology; excellent information about both men and their influence)

Charles H. Cooley: http://sobek.colorado.edu/SOC/SI/si-cooley-bio.htm (Site maintained by sociology graduate students at the University of Colorado, Boulder; part of a fuller site dedicated to symbolic interactionism—I encourage you to check out the entire site)

CHAPTER 8

A Society of Difference

Harriet Martineau, Charlotte Perkins Gilman, and W.E.B. Du Bois

An Open Letter

Steven Seidman, a contemporary social theorist, has written a book called *Contested Knowledge: Social Theory Today* (I highly recommend it). The title of Seidman's work indicates that theory is not something about which everyone agrees. We've talked a bit about this in our discussions of the nature of society (objective or symbolic) and the purpose of theory (descriptive or critical). Part of the discussion that sociologists and other social thinkers have about this concerns the definition of the *canon,* or the group of people that we should reference in theorizing today. For quite a bit of the twentieth century, American sociologists thought they had a fairly good idea whom to reference—people like Marx, Weber, and Durkheim.

Today, and for about the last 20 years or so, the issue of the founding thinkers has been contested knowledge. I think this shift has been very, very good for sociology and social thinking. We should be uncomfortable with our list of most important theorists, for in limiting the list we are excluding voices and perspectives. For much of the twentieth century, sociology systematically excluded voices of race and gender. Even though most of us are aware of this problem, many theory books still do not include some of the early thinkers who theorized about gender and race.

For this book, I have consciously included not only the theorists we have in this chapter, but also the contributions that the other thinkers in this book made to the idea of difference. So, the explicit gender theories of Marx and Simmel have been brought out, and the activism or involvement in this issue of people like Weber and Mead is made clear. However, I'm not satisfied nor am I completely comfortable with what I've done. There are many other ways this book could have been put together and written, particularly as it revolves around Martineau, Gilman, and

Du Bois, and there are other theorists of race and gender that I could have included instead of these three.

There are, of course, reasons for writing the book the way I have, and I'm going to take a moment to explain them to you. But before I do, I want you to know that social theory is contested terrain, and it should be. Critique and reevaluation can only make us and our theories better. I also want you to know that I view my representation of theory (this book) not as an end, or even as a beginning, but simply as a tentative statement in an ongoing dialog. As Ritzer and Goodman (2000) say in *The Blackwell Companion to Major Social Theorists*, "although any list of theorists covered in a collection such as this one can be read as an official canon, this book is intended to be used as 'cannon fodder' in an open, contestable process of theory construction and reconstruction" (p. 2).

In the other chapters of this book, I've considered one social theorist and his perspective and theories at a time. My major reason for focusing on theorists instead of paradigms or major concepts is that these men each present us with unique views of the social world. If we were to organize our talk in terms of functionalism, conflict theory, and so on, we would miss the insights that we can glean from considering them one at a time. The work of these men is regarded as classic for good cause: their ideas and questions are amazingly provocative, and getting inside their heads can change the way we see the social world. Another reason that I've focused on individuals is that a student of sociology should be able to talk about each of these theorists separately. Professional sociologists talk to each other in terms of Durkheimian, Weberian, or Meadian theories (despite how it may appear, we rarely talk in terms of functionalism, interactionism, or conflict theory). These thinkers are part of our cultural capital and you'll need to know them to carry on a conversation in sociology. Not knowing about them individually would be like going to a sports bar and not knowing the difference between baseball and hockey. In addition, these men and their theories still hold sway. You can't really understand contemporary theory nor can you do your own theoretical work without knowing about their ideas and theories.

This chapter is unique in that I am putting three different theorists together and I'm focusing on a general idea. Why these theorists and why this idea? There are two basic reasons why I've chosen to put these folks together. The first rationale concerns the way theory has been built in sociology—we haven't built it around or included the idea of difference. Hence, Martineau, Gilman, and Du Bois don't have as clear a theoretical lineage nor do they form a major part of our cultural capital. Sad to say, but you could get through your entire graduate career in sociology and never need to know the name Harriet Martineau; but you would fall flat on your face in your first graduate class if you didn't know about Max Weber. Hopefully, books like this one will help change this issue of cultural capital. Martineau, Gilman, and Du Bois were chosen for this book because they seem to be the most often referenced people when the issue of exclusion is brought up.

For most of the people in this book, we can draw clear lines between them and contemporary theory; that isn't necessarily true for Martineau, Gilman, and Du Bois. The reason for this discrepancy isn't because these three people didn't say anything profound or inspiring; quite the opposite is true: they have a lot to say to us. But social theory isn't simply built around the power of explanation: values are always at the core of what we do. Thus, *what* is being explained is sometimes just as important, if not more, than the efficacy of the theory. The questions we ask, and thus the answers we hold onto, are very much guided by the cultural values that we have.

The end result is that there isn't a clear line of heritage in contemporary race and gender theory that reaches back to the eighteenth and nineteenth centuries. All three of our present authors were enormously popular in their day. Martineau's first serialized piece, for example, outsold Charles Dickens. By the time of her death, Martineau had published over 70 books and around 1500 newspaper articles. Yet, despite this popularity, the works of Martineau, Gilman, and Du Bois were never adopted into the canon of social theory. As I've mentioned, in sociological and social theorizing, it is standard practice to cite Marx, Weber, Durkheim, and the others. A long lineage of academic work can be traced back to them. There's a clear "paper trail." This isn't the case for Martineau, Gilman, or Du Bois. Even contemporary race and gender theory isn't explicitly built upon them. It's rare that a gender theorist says that she or he is constructing a theory on or through modifying Gilman or Martineau. The same is true in race theory for Du Bois, though he is more often cited because of his political influence.

The other reason why I'm combining these three in one chapter is that the volume of actual theory in their work isn't as large as that of the other people in this book. Don't get me wrong, there is quite a bit to say about each of our present theorists—many books have been written about each of them. However, a great deal of what has been written concerns their lives, political influence, and/or their consciousness-raising activities. When Martineau, Gilman, and Du Bois wrote, they were among the disenfranchised, and most people of their time were blissfully unaware of or unconcerned about the plight of blacks and women. Thus, much of their work is in the form of popular readings like novels, short stories, poems, newspaper articles, and the like. Part of this was due to the political structure of the academy, but the larger part of it had to do with the need to speak out to the general public. As Marx's theory implies, there must be a high level of critical consciousness before things can change. Writing for academics, then, hasn't been a major concern for the oppressed. Further, most members of disenfranchised groups aren't concerned with producing a general theory of society; they are rightfully concerned with one specific aspect: the social oppression of their group.

The most important reason for including these three in this book is their general theoretical concern: Martineau, Gilman, and Du Bois theorize about difference, and difference is extremely important in contemporary theory. There are two kinds of differences that are highly important to people: the differences of gender and of race; both are fundamental in the way we understand ourselves and the way we relate to others. (There is in contemporary theory another difference that is extremely important: sexual orientation.) In fact, in normal face-to-face interactions, we have a more pronounced awareness of our gender and our race than of our class. The reason for this is that class impacts the structure of our relationships and encounters more than race or gender does. What I mean is that most of the people we interact with are members of our class position. If you're middle class, you usually don't hang around with the corporate president of Disney, nor do you generally go out to lunch with a homeless person (unless you're being particularly benevolent at the time). Without thought or action on our part, class structures our lives; but race and gender cut across class (race less so, but it's catching up) and we are becoming increasingly aware of them. This distinction between class and race/gender is one reason why we talk about race and gender as theories of difference.

Harriet Martineau (British, 1802–1876)

Harriet Martineau was born in Norwich, England, on June 12, 1802, the sixth of eight children. Martineau was a sickly child, and by the age of 12, she had lost most of her hearing. Many of the adult pictures of Martineau show her with her hearing trumpet (an early hearing aid), which she acquired when she was 18. Martineau grew up in a Unitarian household, and was generally educated at home (most women at this time were barred from the university). Unitarians believe in one god, in contrast to the doctrine of the trinity, and are committed to the use of reason in religion, universal brotherhood, and a creedless church. Unitarians were part of the religious dissenters' movement in England, which fought against the dominance of the Anglican Church. Though she eventually became agnostic, Martineau's religious background clearly informs her thinking, as we will see when we get to her typology of religion. Her father was a modest capitalist, whose business eventually

Photo 8.1 Harriet Martineau

collapsed due to competition from larger, more industrialized companies. This business failure forced Harriet to seek gainful employment, which she found, writing for the *Monthly Repository,* a Unitarian journal. Martineau began by writing a series of short stories with ethical and political points. She saw writing as a way of teaching and illustrating the ideas of political economy for the general public. Her talent for writing soon became obvious: in only her second year of full-time writing, she published a novel, a book on religious history, a variety of essays, and 52 articles for the *Monthly Repository.*

Among her best-known works are *How to Observe Morals and Manners* (a book of sociological methods), *Society in America* (a three-volume sociological study of America), *Deerbrook* (a novel about English domestic life), and the translation of Comte's *Positive Philosophy* into English. She wrote *Morals and Manners* on the ship traveling to America, and spent two years in the United States gathering data for *Society in America.* Martineau was a prolific writer, traveled extensively, and was well respected by the intellectuals of her day. Harriet Martineau died of natural causes at 74 years of age. She passed away at The Knoll, a home she designed and had built in the Ambleside Lake District in England; and she left her own obituary.

Martineau's Perspective: Natural Law and the Hope of Happiness

Positivistic social theory: Harriet Martineau is one of the earliest sociologists. She is best known for her translation of Auguste Comte's *Positive Philosophy,* one of the major foundation stones of sociology. Martineau, however, did more than translate. She revised Comte's work down from six volumes to two. So taken was Comte with Martineau's version, that he adopted it rather than his own and had the revision translated back into French. If you've read Comte, chances are you've read Martineau's version. In praise of Martineau, Comte wrote, "And looking at it from the point of view of future generations, I feel sure that your name will be linked to mine, for you have executed the only one of those works that will survive amongst all those which my fundamental treatise has called forth" (as quoted in Harrison, 1913, p. xviii).

Martineau is obviously a positivist. We spent some time talking about positivism when we discussed Spencer. In brief, positivism is one of the philosophical roots of scientific sociology. Positivism asserts that the universe is organized by invariant laws, and human beings through reason and observation can discover those laws and use them to understand, predict, and control every area of their lives.

I want to take this opportunity to make the distinction between positivism and neo-positivism. **Neo-positivism** began to be formulated around the turn of the twentieth century and found one of its strongest sociological spokespersons in George A. Lundberg, past president of the American Sociological Society (later renamed the American Sociological Association). Lundberg agrees with Comte that the defining feature of positivism is its method. All academic disciplines are created and based on unique methods through which each of these forms of knowledge is created. One of the major factors that makes history, philosophy, and biology different from each other is their methods of creating knowledge. Positivism is based on the scientific method of empirical observation. As Comte points out, it is this method that provides an inherent unity among the sciences.

It is primarily the method that makes neo-positivism different than historic positivism. Lundberg argues that the three elements of the neo-positivistic method are quantitivism, the position that enumeration and measurement are essential in any scientific investigation; behaviorism, the assertion that consciousness is objectively unknowable; and positivistic epistemology, that knowledge is dependent on sense impressions. Comte would agree with the third position, but the second has by and large fallen out of the picture, at least in terms of its influence in sociology. The first position, however, which exalts quantitative analysis as the primary method of positivism, still holds sway in contemporary sociology and is something with which Comte would disagree.

Comte gave us four methods for accumulating scientific knowledge about society: observation, experimentation, comparison, and historical analysis. I don't want to go into a long discussion about each of these methods, but I do want to point out what is missing in Comte's methods and is primary in neo-positivism: statistics. Comte, though a mathematician himself, did not want the positivistic method identified with statistics and quantification. He held that the use of probabilities was a delusive theory. Comte's methodology can be generally classified as historical-comparative. In neo-positivism, Comte's historical model has been replaced with statistical methods and the thesis of social progress has been replaced by "the aspiration of cumulative growth within scientific sociological theory itself" (Timasheff, 1967, p. 211).

The last part of the quote by Timasheff is particularly important. Comte proposed positivistic sociology as a way to fulfill the project of the Enlightenment. The **Enlightenment**, as you'll remember, was a European intellectual movement of the seventeenth and eighteenth centuries. The thinkers of the Enlightenment exalted empirical observation, the use of reason, and systematic doubt. Empirical experience was considered the foundation of human truth—more important than authority, whether it came from the state, church, or even science. Reason, observation, and doubt were to be used to discover how human life, both social and individual, can be controlled in the same way as the natural world in order to bring about progress in human existence. Neo-positivism not only replaced historical-comparative methods with statistics, it also reduced the Enlightenment's hope of creating a better society to the refinement of theory. In other words, the social theory of the Enlightenment had an external purpose: to improve the human condition. Neo-positivistic theory, on the other hand, has no external purpose. Here theory is statistically tested so that the theory itself can be improved. It is knowledge for the sake of pure knowledge.

In the chapter on Talcott Parsons, I mention the difference between sociological and social theory. The central distinction I make there is that *social theory* is designed to incite social discourse with an eye to change and *sociological theory* is designed to describe things as they exist. As you can see, we can talk about the distinction between positivism and neo-positivism in those same terms. Along these lines, Ben Aggar (2004) has recently said,

> Since about 1970, theorizing has been displaced from sociology, which is increasingly methods-driven, into neighboring disciplines such as English, comparative literature, philosophy, anthropology, history, cultural studies and women's studies. What remains as theory in sociology is largely middle-range explanation that, through literature reviews introducing empirical journal articles, collects cumulating research literatures into generalizations about variables.

I am not arguing for or against either approach; I am just trying to make some clear lines where the distinctions are pretty fuzzy. Some contemporary theorists, for example, are opposed to middle-range theory and the limitation of statistical analysis, yet prefer to be named sociological theorists because they are not interested in taking a critical stand in public sociology. Nevertheless, I believe it is important to recognize the shift in theory for sociology, particularly because many of the theorists we are talking about in this book practice what could now be called social theory, at least in the terms I am using. The specific reason, of course, to go into this detail now about Comte, positivism, and neo-positivism is that it is important for our consideration of Martineau. Martineau is a positivist in the Comtean tradition. As such, she sees the social world as predictable and best known through empirical methods. However, Martineau's methods are comparative rather than statistical and she has a critical purpose in her work.

> [A]ll men entertain one common conviction, that what makes people happy is good and right, and what makes them miserable is evil and wrong. (Martineau, 1838/2003, p. 4)

Natural law and happiness: Martineau is also a thinker in the tradition of the Enlightenment, particularly in her belief in natural rights and law. Natural law is the idea that, apart from human institutions, there are laws and rights that apply to every human being. The United States Declaration of Independence contains this idea in the phrase, "We hold these truths to be self-evident, that all men are created equal, that they are endowed by their Creator with certain unalienable Rights." Martineau (1838/2003) believes that "every element of social life derives its importance from this great consideration—the relative amount of human happiness" (p. 25).

Happiness is thus the touchstone for sociological analysis, and it's this emphasis that qualifies Martineau as a critical theorist—her belief that every society must be held up to this one standard. However, Martineau isn't dogmatic or self-righteous in her approach. She is very careful to understand that her own view of what is right and wrong is influenced by the society in which she lives and was brought up. In her analysis of *Society in America,* Martineau (1837/1966) tells us how to go about making such judgments: "to compare the existing state of society in America with the principles on which it is professedly founded; thus testing Institutions, Morals, and Manners by an indisputable, instead of an arbitrary standard" (p. viii).

In the rest of this section on Martineau, we will be looking at her methodological recommendations, *Observing Morals and Manners: Methods of Social Research,* as well as her take on *Institutions.* Martineau's methods are important because they constitute one of the first statements of social methodology, which predates both

Weber's and Durkheim's books. More important, however, is the reason why Martineau wrote her book on methodology: she felt that every responsible citizen ought to make informed, scientific observations of society. Only in all of us being involved can we hope to make a better society. Her recommendations, then, are directed toward all of us. We will see that Martineau's recommendations not only cover a huge expanse of society, but she also believes that the observer her or himself must be morally prepared to observe. Martineau's focus in social observation is happiness, and she measures a society's progress toward that focus by comparing the cultural, moral code of a society and its behaviors. In other words, she is interested in the distance between what we believe and what we do. The other topic for our consideration will be Martineau's take on *Institutions.* We will specifically look at gender and family, religion and society, and education and freedom, and we will note two exciting ways in which Martineau was ahead of her time (the sociology of knowledge and the sociology of space). We will also see that Martineau's hope for progress lies in public education.

Observing Morals and Manners: Methods of Social Research

Comparing moral codes and real life: In her comparison of **morals and manners,** Martineau prefigures the contemporary work of Robert Wuthnow (1987). Wuthnow argues that the moral foundation of any society is its cultural code. The content *per se* isn't really a concern for Wuthnow, rather, he is interested in the distinctions between the code and the way people relate to the code. The first distinction is between the moral object, the object of commitment, and the real program, how people observably act. This is the same distinction Martineau makes between morals and manners: **Morals** are the beliefs of a society and **manners** are the behaviors that are supposedly motivated by the morals. Wuthnow adds two more distinctions: the difference between one's true inner self and the roles that that self plays; and a distinction composed of "some particularly powerful symbols . . . [that] have powerful symbolic value because they happen infrequently, are unusual or divine, and sometimes influence circumstances or represent changes in one's life" (Wuthnow, 1987, p. 73). The distinction made here is between the things that people see as inevitable and the behaviors that are intentional.

> To test the morals and manners of a nation by a reference to the essentials of human happiness is to strike at once to the centre, and to see things as they are. (Martineau, 1838/2003, p. 26)

The important thing for us to notice is that the difference between all the pairs of symbols is that one may be empirically known (manners) and the other can't be seen (morals). Thus, the moral object of commitment, for example, God, is unknowable. It exists as a symbolic code. But the action that is connected to the moral object (Bible reading in the example of God) may be known. Wuthnow sees a dynamic aspect to the distance between morals and manners. Individuals must maintain a close but distant relationship between the ideal and the real. If the distance gets too close, people will tend to become either discouraged, because no one can uphold the ideal, or self-righteous and judgmental in their supposed adherence. On the other hand, if the boundary between the two is blurred and the individual ceases to be able to distinguish between the ideal and the real, then either cynicism or loss of justification for action will result. Thus, Wuthnow elaborates Martineau's view by including social psychological issues; the structural core of this theoretical perspective is found in her methodology, though Wuthnow did not cite or explicitly incorporate Martineau's work.

Things: In general, Martineau advocates the observation of *things* in order to discern the cultural morals and manners of a collective, and using discourse with people as a commentary on them. Like Durkheim, she sees that morals are part of the "common mind" of a people and that this public consciousness can be understood through social facts. However, Martineau's approach is much broader than Durkheim's. I've listed the social facts that Martineau uses to measure morals and manners in her work in America in Table 8.1. I'm not going to be describing each element of her methodology because they are of varying degrees of theoretical

Table 8.1 Martineau's Measurements of Morals

Structure	*Observance*
Religion	General type: licentious, ascetic, or moderate Physical places of worship Condition of the clergy Superstitions Suicide
General Moral Notions	Cemetery (age, and cause of death; visions of future) Degree of attachment to birthplace and family Representative characters—two kinds: dead (deepest beliefs) and living (currents of belief) National celebrations Treatment of guilty Songs Literature
Domestic State	Physical environment Distribution of resources Method and ownership of land Extent of commerce Physical health Condition of marriage
Idea of Liberty	Amount of feudal arrangements still present Kind of police (military, private, public) Urbanization Conditions of servants Freedom of press Education Persecution of opinion

importance. I do, however, want to point out the variety and objectivity of the *things* she has on her list. Martineau uses typologies, buildings, the physical environment, social practices such as suicide, cultural artifacts such as cemeteries, social networks such as attachment to birthplace and family, land ownership, health, marriage, education, police, urbanization, newspapers, songs, literature, and so on. One is almost overwhelmed at the breadth by simply reading the list. The really interesting point for me is that almost all of these items are actual things and facts that can be observed. One more issue I'd like to point out is that this list, along with her follow-up interviews with people, represents a very sophisticated and contemporary triangulation of methodologies. She used observational, interview, archival, and content analysis methods.

Philosophical and moral preparation: Further, Martineau sets out requirements for observation. In addition to knowing what to observe, the investigator must be prepared. Martineau points out that if someone asked you or me to explain the physics of the universe, or the biology of the human body, or the geographic strata of the Grand Canyon, we would respond by saying the request is silly, unless we had some basis to claim expertise. Yet people do not hesitate for a moment to "imagine that he can understand men at a glance; he supposes that it is enough to be among them to know what they are doing; he thinks that eyes, ears, and memory are enough for morals, though would not qualify him for botanical or statistical observation" (Martineau, 1838/2003, p. 14).

For Martineau, there are two main areas of preparation: philosophical and moral. The first thing to notice about her preparations is that they pertain to the inner person. Martineau points out that the instrument of observation is the human mind. Here we can really see the difference between old-school positivism and neo-positivism. Of course, neo-positivism requires training, but the real instrument of measure is the survey or statistical analysis. For the old school, it's the inner person that must be prepared.

The philosophical preparation of which Martineau speaks is very close to an emphasis I have used in this book: what you see depends on the perspective you take. As we will see, Martineau takes this general point to its deepest level. The first philosophical preparation you have to make is to know what it is you want to know. For Martineau, what she wants to know is the condition of the society with regard to happiness. I want you to notice the depth of what Martineau is saying here. Happiness for her is the essential human condition. It is the natural standard against which all human institutions must be gauged. Essentially, what Martineau is saying is that before going out to do research, you must settle in your own mind what is essential about being human. You must decide in advance what the fundamental qualities of humanness are and how they affect your study and the people you observe. The benefit of this preparation is that it focuses your view on the most important issues.

The observer must also have in mind the principles that will "serve as a rallying point and a test of the facts." Martineau is telling us that we must have in mind the basic categories of distinction that we will use to analyze our data. Again, this categorical scheme must refer to the essential human condition. Martineau's scheme is determined by the natural law of happiness and by the distinctions between morals and manners.

The third element of philosophical preparation involves understanding that human beings create social institutions and their beliefs and practices. Martineau lived at the beginning of the age of modernity; the period of time when the world began to shrink. Communication and transportation technology had shown the civilized nations that there are many truths and many ways of organizing human behavior. Martineau observes that this information should give the researcher the ability to step out of her or his own culture. Martineau sees two important results from this open point of view. The prepared observer of human morals will "escape the affliction of seeing sin wherever he sees difference" and will be freed from emotional responses that can cloud observation. Rather than experiencing shock and alarm, she is able to perceive and understand the morals and manners of a people as they exist for that society.

In addition to philosophic preparation, the observer of the human condition must make moral readiness. In this, Martineau pulls no punches, for she tells us that to make accurate observations the observer "should be himself perfect." Here we can see the depth of relationship between the observer and observed. Martineau truly believes that the inner condition of the researcher is the prime concern and preparation for making social observations. Martineau also recognizes that no one is perfect and if we should wait for perfection, no observations would be made at all. Thus, we are to take stock and understand the issues that would most easily beset our observations and make us prejudiced. Knowing what can pervert our vision, we can "carry with us restoratives of temper and spirits which may be of essential service to us in our task" (Martineau, 1838/2003, p. 52). In other words, we should know enough about the human spirit, and our own psychological and emotional development in particular, to be able to plan ahead to meet the issues that may cause us to be prejudiced.

> Natural philosophers do not dream of generalizing with any such speed as that used by the observers of men; yet they might do it with more safety, at the risk of an incalculably smaller mischief. (Martineau, 1838/2003, p. 18)

In addition, and more practically, the researcher must have unreserved sympathy. Martineau says that a geologist can have a heart of stone and a statistician can be as abstract as a column of numbers and still be successful. But the observer of society will be subject to deception at every turn if she cannot find her way into the hearts and minds of people. The observer of humanity must herself be most human. She must be able to feel her way through the labyrinth of social institutions and interactions and find what is truly meaningful to the subjective actor. She must be able to empathetically understand what people do and why they do it, what people believe and why they believe it, and what people feel and why they feel it.

Safeguards in observation: Martineau adds safeguards to the philosophic and moralistic preparation of the investigator. She advises that no preemptory decision be made, not only in public when she returns to her home culture but also in the private, daily journal she keeps. The researcher, in other words, must deny herself or himself the luxury of coming to any conclusions until massive amounts of data are gathered. This advice stands in stark contrast to the American system of academia that honors the number of publications over the profoundness and depth of thought and insight. Further, the observer must guard against making generalizations too quickly. The connections we see between different situations and belief may in fact be only skin deep. The data we gather must, then, be complex and nuanced. Only by knowingly guarding ourselves against premature generalization can we hope to do justice to the variety of subjective worlds that we observe. Obviously, Martineau isn't telling us not to make generalizations; remember that she is a positivist. Rather, she is telling us to make them slowly and only after compiling enough detail to respect the social worlds of others. Of course, one way to check our observations of others is to actually ask them, a practice Martineau recommends.

The final safeguard that the traveler must keep is hope. The path that Martineau advises is long and arduous. The researcher must make philosophical and moral preparations. Her heart and mind must be ready to observe human behavior. The observations must be made in great detail and breadth. She must keep a journal of her observations and considerations, being careful to not generalize or draw conclusions too quickly. This kind of research takes time, patience, and part of our soul. With such costs, it's easy to become discouraged, particularly when we have to be so careful and take so much time. But Martineau (1838/2003) tells us to keep hope: "Every observer and recorder is fulfilling a function; and no one observer or recorder ought to feel discouragement, as long as he desires to be useful rather than shining; to be the servant rather than the lord of science; and a friend to the home-stayers rather than their dictator" (pp. 20–21).

Institutions

Several institutions that Martineau looks at give us clear indications of her theoretical framework. Specifically, we want to talk about gender and family, religion, and education. All through her discussions of these social institutions, she gives us ideas pregnant with theoretical ramifications, yet little in the way of fully developed theories. Thus, I will be picking out specific ideas for fuller consideration.

> His more philosophical belief, derived from all fair evidence and just reflection, is, that everyman's feelings of right and wrong, instead of being born with him, grow up in him from the influences to which he is subjected. (Martineau, 1838/2003, pp. 34–35)

Gender and family: I want to begin our discussion of Martineau's understanding of family and gender with a particularly significant idea. One of the important areas of debate surrounding and within feminist literature is the idea of women's knowledge. The most sociological of these arguments is that women's knowledge is different and more insightful than men's due to their relationship to social structures. I'll consider this idea in a bit more detail when we get to Du Bois, but for now it's interesting to note that much of Martineau's work contains this idea of the **sociology of knowledge**, that what we know is formed by the social, political structures wherein we live and by the kinds of social practices in which we engage. Of course, Marx had a sociology of knowledge position as well. His, however, was based on his brand of humanist materialism and the production of ideology by the elite. It wasn't until the work of Karl Mannheim in 1928 that Marx's theory was generalized. Durkheim also produced a sociology of knowledge in his work on religion. Part of his argument was that the essential categories of existence were produced through religious rituals and the structured distribution of populations in time and space. Like Marx, Durkheim doesn't have a general concern with the sociology of knowledge. While Martineau doesn't explicitly explain the theories behind this portion of her work, she nevertheless demonstrates a clear and general sociology of knowledge.

Martineau does not, however, privilege the female point of view. Later we will see that Du Bois feels the black position in terms of truth in knowledge is superior to that of whites. This perspective is based on the idea that systems of oppression can only be understood from the outside—an idea that came from Marx. This notion of outsider knowledge can be found in the generally accepted methods of sociology, albeit as a cautionary note. You have probably been exposed to it already. The general idea is that it's difficult to study one's own culture because we will tend to be biased toward it or fail to see things that might be apparent to someone looking from the outside. Critical theory simply takes this a logical step further and claims that there is a special kind of knowledge associated with being the subject of a structure of oppression.

But Martineau doesn't make such a claim. In fact, Martineau very clearly thinks that every person, not simply the oppressed person nor the professionally trained sociologist, could and should make rigorous and insightful observations of society. The only assertion that Martineau makes with regard to women's point of view is in response to the criticism that being a woman is a disadvantage in doing sociological research. In response, she points out that women have particular insights into some places that men don't, most particularly to the home.

The home is important for several reasons in Martineau's scheme. It is the primary part of what she calls the domestic state. As you can see in Table 8.1, the domestic sphere of society includes economic and distributive structures (manufacturing, markets, roadways, and so forth), commerce in general (Martineau includes dealings between individuals or groups in society; and interchanges of ideas and opinions, or sentiments), the general level of health, as well as marriage. Among all these domestic institutions, family (marriage in particular) stands out. It's the place where the primary socialization of children takes place, and a place where the morals and manners of a people can best be observed. Further, the treatment of women in marriage is one of the most important keys for understanding how a society measures up to the universal law of happiness. Keep in mind that happiness is the basic human right and that marriage is the fundamental social institution. Thus, unhappy marriages are a strong indicator of the moral state of an entire society. Martineau also says that the treatment of blacks is a measuring rod for happiness. It's the position of the oppressed that is particularly important for critiquing a society's overall standing with regard to happiness. The elite in any society are going to be happy; but whether or not human happiness is generally valued can only be gauged by how universal it is in a collective, and that is determined by the experience of society's minority. The reason for this goes back to the issue of natural rights. If human beings have natural rights, then the success of government is judged by the groups it excludes from those rights and the degree of that exclusion.

Let's return to the central place that marriage has in Martineau's research and society. The state of marriage in particular provides insight into the moral fabric and practices of a people. Insightfully, Martineau finds a link between marriage and the practice of gender. While the institution of marriage and the structure of gender are two distinct things, they also have reciprocal effects on one another. Thus, the unhappiness in marriage that is prompted by gender inequality is given strength by our cultural concepts of gender. Specifically, Martineau finds clear links between the condition of marriage and the beliefs that men should have courage and women should be chaste. The connection between gender on the one hand (masculinity ↔ courage; femininity ↔ chastity) and marital unhappiness on the other is a cultural link.

What's wrong with men being identified with bravery? Martineau argues that as long as the chief defining feature of masculinity is courage, then men's primary obligation is to society. Bravery is a social virtue. It is played out on the stage of society for the benefit of society: Men are brave so that society can exist. They demonstrate their courage most clearly on the battlefield and are given medals, parades, and sacred burials in return. Being brave in war is the pinnacle toward which all other forms of masculine bravery is oriented, whether it is bravery in the face of pain in sports or emotional strength in the face of sad stories. Little boys must be brave and not give in to pain or emotion because someday they may be called to battle.

The cultural logic of this gets played out in gender in a couple of ways. First, and probably most insidiously, the principal male quality is expressed in obligation to society, not to the home or to individual people, most specifically wives. There is, then, a clear separation between public and private spheres in this culture of courage and chastity. The logic of the culture dictates that men are responsible for public society and women for the private fields of home and sexual purity. Thus, we see that men are players and women sluts when sex is for recreation; the culture of gender makes them responsible for different social spheres. This culture also dictates that men are more present in public office and the houses of power and women are the caretakers of sexual morality and the home.

The other no less significant result of this cultural logic is that women are subject to the whims of men. Martineau tells us that honor associated with courage is shallow, because it comes to be "viewed as regarding only equals," which in this case would only be other men. This kind of honor is not a personal virtue. It is associated with courage, which we have seen is a public sentiment, and is therefore a public issue. A public authenticity is at stake in this kind of honor, not personal authenticity. And its honorability is dependent upon its base. What I mean by that is that this kind of honor is defined by courage. Therefore,

> The requisitions of honour come to be viewed as regarding only equals, or those who are hedged about with honour, and they are neglected with regard to the helpless. Men of honour use treachery with women,—with those to whom they promise marriage, and with those to whom, in marrying, they promised fidelity, love, and care; and yet their honour is, in the eyes of society, unstained. (Martineau, 1838/2003, pp. 173–174)

Let me reemphasize that Martineau does not see these as intrinsic features of gender. Men are not naturally aggressive and women are not naturally moral. The sentiments and values associated with gender are found in specific kinds of societies—structure and culture go together in Martineau's scheme.

Martineau also argues that the level and kind of workforce participation women have influences the moral codes of a society. In this, of course, she foreshadows a great deal of feminist theory, as well as the work of Charlotte Perkins Gilman. There is a relationship between the way we think of marriage and the workforce participation of women. Martineau gives us two ways that domestic morals are influenced. First, to the degree that women are left out of the workforce, they will be seen as "helpless" and men "vicious." There is a relationship between personal power or efficacy and money. This relationship makes perfect sense, of course, in a society where one's survival is linked to work and making money. In societies where a living wage is guaranteed to all, or women are given equal opportunity, this association would be very weak. Women would not be dependent upon men for their survival, and the power between the genders would be equal. However, in a society where living is dependent upon working, as it is in capitalist countries, and women are denied equal access to jobs, then men will have greater power not only in society but also in personal relationships.

The second result of unequal workforce participation is that the institution of marriage is "debased." The contemporary ideal of marriage in our culture is that it is a union freely entered into by two people in love. In a society where marriage is the chief way that women get ahead or survive, it can never be the case that marriage is secure "against the influence of some of these motives even in the simplest and purest cases of attachment" (Martineau, 1838/2003, p. 183). Martineau actually frames this issue more pointedly. She talks about it in terms of marriage being the principal or only aim of women in society. When the chief goal of women is to marry and have children, then the values associated with that goal are enculturated in the very young. Young girls are socialized to value being attractive to men (yet chaste), dating, marriage, and having children. The toys, games, and images provided for girls generally have these themes. Thus, when grown, women sense that they are unfulfilled and incomplete unless they are married and have children. One's femininity becomes synonymous with marriage and child-bearing. In such a society, "the taint is in the mind before the [marital] attachment begins, before the objects meet; and the evil effects upon the marriage state are incalculable" (Martineau, p. 183).

Martineau makes an interesting observation about the condition of women and the legitimacy of the state. She says that every civilized nation claims legitimacy by their treatment of women. In modern society, the rights of women were probably an issue before racial inequality. No matter what race or society or culture in which you find yourself, there are men and women and gender relations. As I've mentioned, one of the earmarks of modernity is the idea of equality. The idea began first in religion and governments as relations between clergy and laity (the Protestant removal of the priesthood), and royalty and masses. In most cases, the latter was formalized by the idea that all men (white-Anglo-Saxon-Protestant-heterosexual-property-owners) are created equal, without regard to status positions in religion or government. Eventually, this cultural frame of equality was applied to other groups, and gender is the ubiquitous other.

Thus, nation-states claiming to be modern point to their treatment of women as proof of modernity. However, as Martineau notes, all the proofs are different and in the end only serve to legitimate the exclusion of women. One society may claim that women are honored in their country because women are "cherished domestic companions" and afforded high respect. This respect might be shown by allowing women to go first through a door, by standing when they enter or leave a room, or by covering them from head to toe in a veil of sacredness. Another society will claim honor because their religion safeguards women by not requiring them to deal with the harsh aspects of life; another will declare the honor of women because all of society knows that behind every great man is a great woman. The list of legitimations goes on, but Martineau claims that only in societies where women are guaranteed equal workforce participation is there true honoring of women.

However, it would be a grievous error for us to conclude that Martineau only concentrated on women. She is much more generally concerned with society. Undoubtedly, through her own experience as a woman she became sensitive to women's issues, but she is really concerned with happiness as a general feature of society. The institution of family and the position of women and other disenfranchised groups are key indicators of the progress of society toward the universal good of humanity.

Religion and society: Religion is another important social structure for measuring society's morals and manners. As we've seen, religion is consistently theorized as an important social institution. For example, Durkheim argued that religion forms the base of society. Out of it sprang the primary categories of thought and the moral basis of society. Durkheim noted that the two functions of religion eventually split, with science taking over speculation about the origins and workings of the universe and religion the issues of faith. Religion, in Durkheim's way of thinking, will evolve to a very general form that embraces all humankind under universal values.

Martineau isn't concerned with the origins of religion, as Durkheim is, but she does seem to say that religion evolves toward more inclusive, less judgmental forms. The *forms* of religion are of chief interest for Martineau. She gives us three types of religion: licentious, ascetic, and moderate. **Licentious religions** are those where the gods are seen as part of nature and having human passions, as in the early Greek pantheons. During the early stages of this type of religion, licentious or immoral conduct is encouraged. This form changes when people are seen as less than the gods, and the power for doing good or harm is attributed to the gods. When this happens, the idea of propitiation (an act or gift given to appease the wrath of a god) enters and ritual worship begins. Even in its advanced form, this kind of religion is still preoccupied with licentious behavior, albeit in a negative way: the rituals performed for an almighty god counter the ill done by sin.

Ascetic religions are highly ritualized as well. Emphasis here is on rituals that provide proofs of holiness. Martineau argues that spiritual vices rather than immoral acts of the flesh are the center of this type of religion. The spiritual vices of "pride, vanity, and hypocrisy,—are as fatal to high morals under this state of religious sentiment as sensual indulgence under the other" (Martineau, 1838/2003, p. 79). Martineau sees religious ritual as misplaced emphasis. In this way of thinking, rituals are an empty form that takes our attention away from "self-perfection, sought through the free but disciplined exercise of their whole nature" (p. 80). **Moderate religions** are the least ritualized of the three, and focus on self-actualization through broad education that involves the entire being.

Martineau doesn't argue that the different forms are replaced as religion evolves. On the contrary, each of these forms existed in her day and are still with us today. For example, traditional Christianity is based on propitiation and the sins of the flesh. The older forms of Christianity, like Catholicism, still emphasize ritual as the vehicle through which come propitiation and forgiveness. The big question for the social researcher is which form dominates. The importance of the different religious forms is that religion plays a key role in forming the morals of any society. If licentious religious forms dominate, then the political and domestic morals will be very low or contain mostly ritual. Where asceticism governs, the natural behaviors and emotions of humanity are judged to be sin. Martineau (1838/2003) prefigures symbolic interaction and labeling theory by arguing that once a behavior or emotion is called licentiousness, "it soon becomes licentiousness in fact, according to the general rule that a bad name changes that to which it is affixed into a bad quality" (p. 81). Moderate religion encourages a culture of freedom and acceptance. It's a "temper" rather than a pursuit, a settled state of mind focused on inner peace and strength rather than behaviors.

The things of religion that Martineau advises us to look at include the places of worship, the condition of the clergy, generally held superstitions, holy days, and suicide. In talking about the places of worship, Martineau is giving us a sociology of space. Sociology of space looks at how the designs and placements of buildings influence human sentiments and behaviors. Martineau tells us to look at where the religious buildings are located, how they are built, their number, their diversity, and whether or not the physical structure is conducive to ritual or spiritual service.

Sociology has been concerned with place for quite some time, particularly in the field of urban sociology. Obviously, Martineau's theories predate this, but her conceptions of space are somewhat different than those of urban sociology. The way she thinks about space and its influence on people is much more in keeping with the ideas that postmodernism has brought into sociology. A postmodernist approach to space argues that in the world of architecture, modernism is defined by function: a denial of everything idiosyncratic about humanity, coupled with the rationalization of space and the imposition of a strict and systematic order on daily life. Modern architecture thus tends to be cold and functional. On the other hand, postmodern architecture uses "double-coding" (Jencks, 1977) to incorporate elements of both modern and historical forms. The reason for this is that postmodern architecture acknowledges how indeterminate our lives are, living in a postmodern world. In postmodern architecture, space becomes more personalized through the blending of styles and the inclusion of historical references. There is also another vision of postmodern architecture, one that sees the expression of the placelessness of a postindustrial society as primary. Here buildings are used to signify and reinforce a person's frustration with space by discouraging independent, creative mobility and by providing only isolated and disjointed spaces in which to exist (Jameson, 1984). In both Martineau's and postmodernism's accounts of space, there is a reciprocal relationship between the buildings that structure space and the subjective experience of the individual. The relationship is reciprocal in that the placement of buildings and the way they are formed both influence the psychology of the individual and reflect it.

Martineau's sociology of space takes in much more than architecture. The first thing she recommends in studying domestic institutions is to look at a geological map. Martineau recognizes that how a society is situated geographically, that is, its physical environment, is important for how social structures develop. For example, nations whose territory is isolated but filled with natural resources will have a different set of structural arrangements than a society that has boundaries on all sides and is dependent upon other nations for raw materials. The natural environment also strongly influences a society's division of labor, and the psychological makeup of the workers. Farming needs large expanses of fertile soil and mining requires deposits of valued ore—farmers and miners are different kinds of people because of the different work they perform. Areas that are conducive to city dwelling and trade will contain artisans and craftsmen, who are both different than miners and farmers. Martineau tells us that the psychological dispositions of a society are significantly dependent upon these jobs and political realities, all of which are influenced by the environment.

In addition to architecture and building placement, the clergy are important for the study of religion because they are the first learned people in a society. Education begins with them, so it is important to ask, what kind of knowledge do they espouse and how are they esteemed in society? Are they respected because of their perceived holiness or their great learning? Suicide, of course, gives us some idea of religious sentiment as well. Why do people commit suicide? Is it from shame, from devotion, or to shrink from suffering? What kind of religious sanctions or motivations are there regarding suicide?

Martineau places her thinking about religion under a research agenda she calls the "idea of liberty." The way a society is structured around the idea of liberty is extremely important to Martineau. It is clearly related to the idea of happiness. The way in which the structures of a society are related to happiness is gauged by the level and kind of oppression. As we've seen, her key cases are those of race and gender. The opposite of oppression is liberty. Thus, liberty can be seen as the positive counterpart to oppression in measuring the structured potential of happiness in any society. Martineau mentions several factors that provide measures of liberty: the level of class divisions in society (the greater the bifurcation, the more likely the use of ideology and suppression); the kind of police force (national and military versus local and custodial); the level of urbanization and the influence city culture has on the nation (urbanized culture will tend to be more diverse and accepting of difference); the level of democratic governance; the level of direct or indirect governmental control of the press; and the degree of persecution of opinion. However, Martineau argues that in many ways the most important indicator of the idea of liberty is the condition of the education system in society.

Education and freedom: As a modernist thinker, Martineau believes in the progress of humanity. According to Martineau, there are two great powers—force and knowledge—and the story of humanity is the movement away from one and toward the other. Social relations began through physical force and domination and the idea that "might makes right." In such societies, "the past is everything." The kind of knowledge that is of most value, then, is traditional knowledge. The tastes, ideas, sentiments, and ambitions in such a society are those that come from antiquity. This kind of knowledge is nonreflexive and maintains the status quo; it "falls back upon precedent, and reposes there" (Martineau, 1838/2003, p. 45).

Martineau (1838/2003) argues that society moves away from force and toward knowledge as the basis for moral power in response to increases in commerce, professionalization, communication, and urbanization (p. 46). Under these conditions, knowledge cannot stay fixed in the past. It is opened up to new horizons through increases in communication and commerce with other societies. Urbanization and commerce tend to expand the limits of knowledge as well as push for constant change. City living also brings people more in contact with diversity, which in turn opens up the boundaries of knowledge. Professionalization creates expert knowledge systems and, as more and more jobs and trades become professions, people in general begin to value expert knowledge over traditional information.

Concerning the education system itself, Martineau gives us both interesting (like the use of uniforms, which she says is "always suspicious") and obvious measures (such as the distribution of social types—gender, race, religion, class—and the general level of educational attainment). However, two of the most important indicators of education and its connection to the idea of freedom are its financial base and the position of the university. The influence of education's financial base should be apparent, but it bears mentioning. The first issue is the extent of free education. The level to which a society supports public education is an unmistakable measure of the support of the ideas of freedom and democracy. In civilized countries, education is the way to socially advance; it's the legitimate way to get ahead. A high level of support for free, public education indicates a high level of support for equal opportunity. The level of supported equality is linked to the kinds of jobs or careers one can get with a given level of education. In other words, all we have to do to understand the level of equality that society wants to support is look at the kinds of job opportunities free and public education provides. For example, if a society supports public education only through high school, it indicates that the level of equality that the state is interested in providing is equal to the jobs that require a high school education.

> The extent of popular education is a fact of the deepest significance. Under despotisms there will be the smallest amount of it; and in proportion to the national idea of the dignity and importance of man,—idea of liberty in short,—will be its extent, both in regard to the number it comprehends, and to the enlargement of their studies. (Martineau, 1838/2003, p. 203)

The second indicator of the place of education in society is the position of the university. Martineau (1838/2003) claims that "in countries where there is any popular Idea of Liberty, the universities are considered its stronghold" (p. 203). The reason for this link between liberty and universities might surprise you. Weber tells us that in societies where bureaucracy is the principal form of social organization, education will serve to credential a workforce. Martineau's function for the university is different than Weber's. Of it she says, "It would be an interesting inquiry how many revolutions warlike and bloodless, have issued from seats of learning" (p. 203). According to Martineau, the function of the university in its relationship to liberty is to inspire critical thinking. A university education should make you question authority and society. Freedom must be won anew by every generation. Almost by definition, democracy is alive and expansive. A university that doesn't challenge its students with new ideas and doesn't provide an atmosphere wherein new principles can be founded, is "sure to retain the antique notions in accordance with which they were instituted, and to fall into the rear of society in morals and manners" (Martineau, p. 203).

Not only is the purpose, content, and environment of the university important, so are its students, in particular their motivation for study. To the degree that students are motivated to obtain a university education for a job, education for freedom is dead in a society. Martineau (1838/2003) makes a comparison between students in Germany and those in America (recall that she is British and thus has no vested interest). German students are noted as having a quest for knowledge: the student may "remain within the walls of his college till time silvers his hairs." The young American student, on the other hand, "satisfied at the end of three years that he knows as much as his neighbours . . . plunges into what alone he considers the business of life" (p. 205). These words were of course written almost 170 years ago. And Martineau (1838/2003) had great hope for the American system of education: "a literature will grow up within her, and study will assume its place among the chief objects of life" (p. 206). I will let the reader judge the fulfillment of her hope.

> The great importance of the fact lies in this,—that increase of knowledge is necessary to the secure enlargement of freedom. (Martineau, 1838/2003, p. 206)

Summary

- Martineau's perspective is based on positivism and the Enlightenment. She believes that all humans have an inalienable right to happiness. This notion is the touchstone for all empirical and theoretical work. Because happiness is a natural right for humans, governments must be measured in terms of their facilitation or denial of this right. In particular, Martineau uses the way society treats gender and race as key indictors of happiness. Empirically, societies are measured by comparing their stated morals and practiced manners.

- Martineau feels that every citizen should be part of the enlightened observation of society. Society is generally moving forward, but its true progress toward providing for the natural rights of human beings must be measured and societies must be held accountable to these measures. However, because the human mind is the principal tool of observation, quite a bit of preparation must take place: observers must not judge the behavior of others simply because it is different than their own; the researcher must also be able to empathize with others.

- Martineau argues that there is a link between the culture and practice of marriage, and gender. Courage as a particularly masculine trait will obligate men to society, rather than to family or relationships; and motivated by courage, men will give honor to socially honored heroic people, rather than women and children. The feminine culture of chastity also has practical effects. Equating chastity with femininity makes women responsible for intimate relations. Together, these cultural equations help to create separate spheres of influence and concerns for men and women.

- Marriage and gender are also negatively affected as women are denied workforce participation. Not only are women seen as helpless and men enabled to be vicious, but marriage is always to some degree motivated by economic concerns rather than more enlightened incentives, when women are given limited workforce opportunities. Children are socialized into this gendered culture and economy. Thus, when grown, people feel that their gender is essentially tied to expressions of courage and work, chastity and relationships.

- In order for societies to progress toward providing happiness for their citizenry, education must be freely provided. The level of equal opportunity in a society is measured by the level of free education. Education must also be sought after for its own sake, and not simply economic gain. In addition, the universities in a society must provide a critical, reflexive education. The kind of religion most prevalent in a society will also influence how far a society goes in providing for and safeguarding happiness. Overly ritualized religions, or religions that focus on holiness or redemption, will stand in the way of the self-actualization of the people. Moderate religions provide a mindset that is focused on inner peace, strength, and authenticity. Societies will tend to move toward greater levels of free and open education and moderate religious forms due to the influences of commerce, professionalization, communication, and urbanization. However, the true level is always determined through empirical research.

Charlotte Perkins Gilman (American, 1860–1935)

Charlotte Stetson Perkins Gilman was born Charlotte Perkins on July 3, 1860, in Hartford, Connecticut. Her father was related to the Beecher family, one of the most important American families of the nineteenth century. Gilman's great-uncle was Henry Ward Beecher, a powerful abolitionist and clergyman, and her great-aunt was Harriet Beecher Stowe, who wrote *Uncle Tom's Cabin,* which focused the nation's attention on slavery. Gilman's parents divorced early, leaving Charlotte and her mother to live as poor relations to the Beecher family, moving from house to house. Gilman grew up poor and was poorly schooled.

Gilman's adult life was marked by tumultuous personal relationships. Gilman's first marriage in 1884, to Charles Stetson, ended in divorce and a novella, *The Yellow Wallpaper.* The book tells the story of a depressed new mother (Gilman herself) who is told by both her doctor and husband to abandon her intellectual life and avoid any writing or stimulating conversation. The woman sinks deeper into depression and madness as she is left alone in the yellow-wallpapered nursery. Gilman's feeling

Photo 8.2 Charlotte Perkins Gilman

of hopelessness against the tyranny of male-dominated institutions is heard in the repeated refrain of "but what is one to do?" Between her first and second marriages, a period of about 12 years, Gilman had a number of passionate affairs with women. However, her second marriage—to George Gilman, a cousin—proved to be at least a somewhat successful and important relationship for Charlotte: some of her best work was done during the courtship and after the marriage.

In her lifetime, Gilman wrote over 2,000 works, including short stories, poems, novels, political pieces, and major sociological writings. She also founded *The Forerunner,* a monthly journal on women's rights and related issues. Among her best-known works are *Concerning Children, The Home, The Man-Made World,* and *Herland* (a feminist utopian novel). However, Gilman's most important theoretical work is *Women and Economy.* In Gilman's lifetime, the book went through nine editions and was translated into seven different languages.

Gilman was also politically active. She often spoke at political rallies and was a firm supporter of the Women's Club Movement, which was an important part of the progress toward women's rights in the United States. It initially began as a place for women to share culture and friendship. Before the development of these clubs, most women's associations were either auxiliaries of men's groups or allied with a church. In contrast, many of the clubs of the Women's Club Movement became politically involved, but others were simply public and educational service organizations. One such club became the Parent-Teacher Association (PTA).

Sick with cancer, Charlotte Gilman died by her own hand on August 17, 1935, in Pasadena, California.

Gilman's Perspective: Evolution With a Twist

Critical-functional evolution: We've seen that Harriet Martineau's perspective is somewhat informed by the idea of evolution. She sees society moving from organization by physical force to order through the moral force of education. Gilman, on the other hand, bases her entire perspective on a fully developed theory of evolution that in many ways is like that of Herbert Spencer. Gilman generally understands the history of human society as moving from simple to complex systems, with various elements being chosen through the mechanism of survival of the fittest. Gilman doesn't simply use the evolutionary perspective to explain societies up to this point. As we will see, she uses evolution to explain what is going on currently in society and where we are headed.

Gilman also takes a functional point of view. This isn't surprising, as evolutionist and functionalist thinking tend to go hand in hand. While Gilman's evolutionary point of view is complete, her functionalism is limited. The three clear functionalists in this book, Spencer, Durkheim, and Parsons, all propose requisite functions for society. In the case of Spencer and Parsons, the requisites are for any system, while Durkheim is only interested in social systems. Gilman, on the other hand, doesn't propose any requisite system needs. Her functionalism is limited to arguing that function determines structure and that structure within a system will tend to work to the overall benefit of the system. These two functionalist points are extremely important in her work on gender and yield interesting theoretical ideas.

We have seen this perspective several times in this book, with its clearest expression in Herbert Spencer. Like Spencer, Gilman argues that humans began as brute animals pursuing individual gain. Survival of the fittest at that point was based on individual strength and competition. Men would hunt for food and fight one another for sexual rights to women. It was a day-by-day existence with no surplus or communal cooperation. Wealth was, of course, unknown, as it requires surplus and social organization. Competition among men was individual and often resulted in death. Women, as is the case with many other life forms, would choose the best fit of the men for mating. There was no family unit beyond the provision of basic necessities for the young, mostly provided by women. Life was short and brutal.

> The proposition is that Society is the whole and we are the parts: that the degree of organic development known as human life is never found in isolated individuals, and that it progresses to higher development in proportion to the evolution of the social relation. (Gilman, as cited in Lengermann & Niebrugge-Brantley, 1998, p. 142)

As competition between species continued, humans gradually developed a social organization beyond its natural base. Very much like an organism, society began to form out of small cells of social organization that joined with other small cells. Gilman (1899/1975) says that "society is the fourth power of the cell" (p. 101). By that she means something similar to Spencer's notion of compounding: once societies started to grow and structurally differentiate, the process continued in a multiplicative manner. The reason for this is simple: complex societies, because of the level of social cooperation and division of labor, have a greater chance of survival then do individuals or simple societies. As societies became more complex, and thus more social, structures and individuals had to rely on each other for survival. Just like the different organs in your body must depend upon other organs and processes, so every social unit (individuals and structures) must depend upon the others.

An interesting thing happened for humans at this point. All animals are influenced by their environment. If the environment turns cold, then the organism will either develop a way of regulating body temperature (polar bears are well-suited for the artic), move to a different environment, or die out. Humans at one time were equally affected by the environment. But after we organized socially, our organization and culture allowed us to live in a controlled manner with regard to the environment: humans can now exist in almost any environment, on or off the planet. We thus distanced ourselves from the natural environment, but at the same time we created a new environment in which to live. Gilman argues that human beings have become more influenced by the social environment of culture, structure, and relations than by the natural setting—we have a new evolutionary environment.

This evolutionary approach with its organismic analogy has intuitive appeal. It certainly makes sense that differentiated structures and people with a high division of labor would be functionally dependent. Spencer named this as one of the main reasons that societies tend to integrate. However, Gilman has something specific that she wants us to see. To make it clear, let me state the principles of social evolution as a maxim: to say society is to say cooperation and dependency. The bottom line of this postulate is that, while at the beginning of our evolutionary path it was functional for individuals to struggle and compete for their own gain, it is no longer functional for them to do so now. As Gilman (1899/1975) says, "It is no longer of advantage to the individual to struggle for his own gain at the expense of others: his gain now requires the co-ordinate efforts of these others; yet he continues so to struggle" (p. 104). Gilman's point is in that last bit of the quote: we continue to struggle for individual gain even though it is no longer functionally advantageous. From an evolutionary standpoint, Gilman argues, there is something wrong. It is here that Gilman's critical viewpoint is most noticeable.

Gilman's critical perspective has clear Marxian overtones. Like Marx, she argues that the economy is the most important social institution. All species are defined by their relationship to the natural environment. Stated in functionalist terms, structures in an organism are formed as the organism finds specific ways to survive (structure follows function); and the same is true in society. The economy is the structure that is formed in response to our particular way of surviving, and it is thus our most defining feature. We are therefore most human in our economic relations. Marx argued that we are alienated from our nature as we are removed from direct participation in the economy, as with capitalism. Gilman also argues that alienation occurs, but the source of that alienation is different. Where Marx saw class, Gilman sees gender. Gilman, in fact, is quite in favor of capitalism—in no other system has our ability to survive been so clearly manifest. Even so, there are alienating influences that originate in our gendered relationship to the economy, most specifically the limitation of woman's workforce participation.

Gilman's perspective is thus similar to but different than Martineau's. Obviously, they are both concerned with gender, and we will see that more clearly as we go through Gilman's theoretical ideas. And they both have a sense of forward progress, from brute force to more peaceful, complex forms of society. Yet Gilman's evolutionary theory is far more developed than are Martineau's ideas, and that's true across the board. Gilman gives us a fully developed theory, whereas Martineau gives us more of a general perspective charged with pregnant ideas. Another way in which they are different is in their scope. In looking at society at one point in time, Martineau gives us a much fuller vision; but in terms of long-term social factors, it's Gilman who excels. Martineau's point of view is more in keeping with the Enlightenment. She sees herself as a cultured citizen who believes in the philosophical progress of humanity. Happiness is the natural law and education is our path to find it. Gilman, on the other hand, is a functional evolutionist who sees progress meted out over long periods of time by natural selection. Change for Gilman comes through massive movements of social structure over centuries.

Their most pronounced similarity is the idea of workforce participation for women. This is central to both women and is a key ingredient in contemporary feminist theory. Gilman, however, is much more detailed in her explication of this factor. Martineau argues that excluding women from the workforce reduces their social, political, and personal power, a concept that became a major tenet of contemporary gender theory. Martineau also gives us a more subtle cultural take. She says that where livelihood is an issue in marriage (where women must marry to survive), marriage can never be free from the taint of economy. Thus, marriage is debased from the ideal of two people in love to an economic relationship. Gilman focuses on this latter idea and details it in all its ramifications.

As we will see in the following pages, Gilman understands the exclusion of women from the economy in evolutionary terms. This perspective yields amazing results. We will see in the next section about *self- and race-preservation* that removing women from economic participation threw off the balance between the individual and society. This disequilibrium from a functionalist's view means that the evolutionary path of humanity is not functional. Gilman explores the main point of this dysfunction in her use of *gynaecocentric theory.* Gynaecocentric theory posits that the female is the basic social model rather than the male (which had been assumed for most of our scientific and philosophic history). Gilman argues that men and women have natural, intrinsic energies that are different from one another. Based on this postulate of different energies, Gilman contends that moving women out of the workforce actually ended up producing evolutionary advantages, which we'll cover in a moment. Thus, in the long run, it turned out to be functional to remove women from the workforce. However, it is certainly not the case that Gilman thinks everything is right and wonderful for women. In the section on *sexuo-economic effects,* we'll see that denying women access to the natural economic environment has had horrible consequences.

Dynamics of Social Evolution

Self- and race-preservation: I've diagramed Gilman's basic model of social evolution in Figure 8.1. As you can see, she, like Marx, argues that the basic driving force in humanity is economic production: creative production is the way that we as a species survive. In all species, economic needs push the natural laws of selection that result in a functional balance between self-preservation and race-preservation. Natural selection in **self-preservation** develops those characteristics in the individual that are needed to succeed in the struggle for self-survival. In the evolutionary model, individuals within a species fight for food, sex, and so on. Natural selection equips the individual for that fight. **Race-preservation**, on the other hand, develops those characteristics that enable the species as a whole to succeed in the struggle for the existence. Gilman is using the term *race* in the same way I'm using *species*; she isn't using it to make racial distinctions among ourselves. The most important point here is that the relationship between self- and race-preservation is balanced: individuals are selfish enough to fight for their own survival and selfless enough to fight for the good of the whole species.

Gilman's point here is interesting because most evolutionists assume that these two factors are one and the same. In other words, self-preservation is structured in such a way as to function as species-preservation. This assumption is what allows social philosophers like Adam Smith to take for granted that severe individualism in systems like capitalism is good. Gilman insightfully perceives that these two functions can be at odds with one another. She thus creates an empirical research question where many people don't see any problem or make a value judgment, such as the inherent goodness of capitalistic free markets.

The third box of Figure 8.1 is where human evolution becomes specific. Because of our overwhelming dependency on social organization for survival, we develop laws and customs that in turn support the equilibrium between self- and race-preservation, as noted by the feedback arrow. With the idea of "accumulation of precedent," Gilman has in mind art, religion, habit, and institutions other than the economy that tend to reinforce and legitimate our laws and customs. As you can see, there is a self-reinforcing cycle among the race/self-preservation proportion; social laws and customs; and the institutions, culture, and habits of a society. This social loop becomes humankind's environment. Of course, we still relate to and are affected by the natural world, but the social environment becomes much more important for the human evolutionary progress (or lack thereof). This

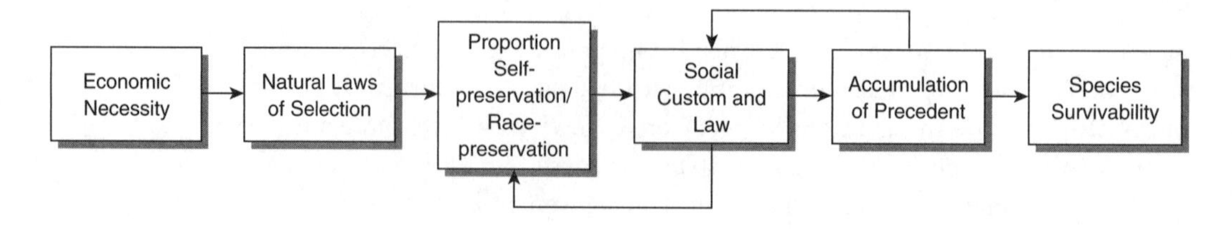

Figure 8.1 Gilman's Basic Model of Social Evolution

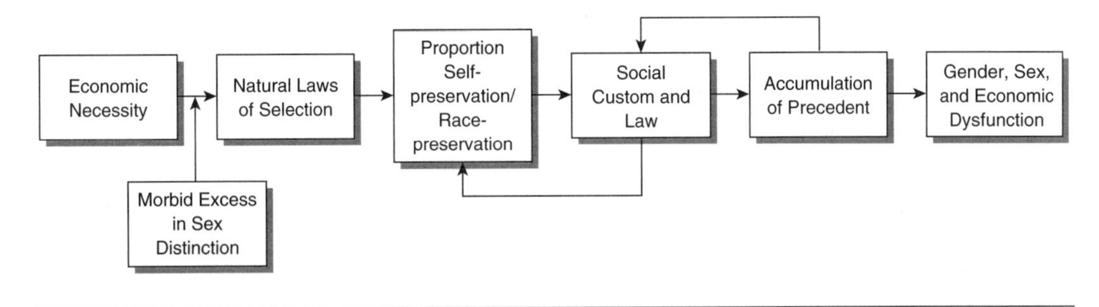

Figure 8.2 Gilman's Model of Sexuo-Economic Evolution

entire process works together to increase the likelihood of the survival of the human race.

That's the model of how things are supposed to work. Figure 8.2 depicts how things actually work under gender inequality. As you can see, the model is basically the same. The same central dynamics are present: economic necessity pushes the laws of selection that in turn produce the proportion of self- to race-preservation characteristics, which then create laws and customs that reinforce those characteristics and produce social behaviors, culture, and institutions that strengthen and justify the laws and customs. This time, however, the social loop doesn't result in greater species survivability, but, rather, in gender, sex, and economic dysfunctions.

I want to talk about those dysfunctions, but first we need to look at how Gilman sees the overall movement of evolution. Her conclusions are extremely interesting and perhaps startling, but in the overall way in which she makes her argument, she is years ahead of the functional arguments of her time. In fact, we can safely say that the kind of argument that she presents wasn't generally found in sociology until after Robert K. Merton's work around 1950. In general, Gilman argues that dysfunctions in some subsystems can have functional consequences for the whole.

Gynaecocentric theory: The dynamic difference between Figure 8.1 and Figure 8.2 is found in what Gilman calls a "morbid excess in sex distinction." Much like Spencer, Gilman argues that human sexuality and the kinship structure were evolutionarily selected. In the pre-dawn history of humankind, it became advantageous for us to have two sexes that joined together in monogamy. From an evolutionary perspective, the methods of reproducing the species are endless. They run from the extremes of hermaphroditism (both sexes contained in a single organism) to multiple-partnered egg hatcheries. Because human beings need high levels of culture and social organization to survive, the two-sex model was naturally selected along with monogamy. Thus, having two sexes which are distinct from one another is natural for us. In this model, the physical distinctions between the sexes produce attraction and competition for mates, just as in other species. Eventually, however, because of the need for social organization among humans, the brutal competition for mates was mitigated and the social bond extended past mother-child through monogamy.

This process, however, became tainted by patriarchy. Gilman presents a unique and fascinating argument as to why patriarchy came about. According to Gilman, there are distinct, natural male and female energies. The basic masculine characteristics are "desire, combat, self-expression; all legitimate and right in proper use" (1911/2001, p. 41). Female energy, on the other hand, is more conservative and is characterized by maternal instincts—the love and care of little ones. In this, Gilman sees us as no different than most other species—these energies are true of many male and female creatures. Yet the natural level of these energies posed a problem for human evolution. To survive as humans, we required a higher degree of social organization than simple male and female energy could provide. The male energy in particular was at odds with this need. As long as "the male savage was still a mere hunter and fighter, expressing masculine energy . . . along its essential line, expanding, scattering" (Gilman, 1899/1975, p. 126), social organization of any complexity was impossible. Men were aggressive and individualistic and thus had to be "maternalized" for social organization to grow.

> From the odalisque with the most bracelets to the débutante with the most bouquets, the relation still holds good,—women's economic profit comes through the power of sex-attraction. (Gilman, 1899/1975, p. 63)

In this argument, Gilman is using and expanding on Lester F. Ward's **gynaecocentric theory**: the female is the general race type while male is a sex type. In fact, Gilman dedicated her book *The Man-Made World* to "Lester F. Ward, Sociologist and Humanitarian, one of the world's great men . . . and to whom all women are especially bound in honor and gratitude for his Gynaecocentric Theory of Life." The word *gynaecocentric* means women-centered. Though Herbert Spencer didn't phrase it in quite this way, the theory is quite similar to his argument concerning monogamy. Family is the basis of social organization. All species, including humans, are related to one another biologically through descent. In our case, we used those biological relations to build social connections. The basic kin relationship is found between mother and child. This is the one relationship that has always been clear: it's obvious from whom the child comes. In order to build social relations on that basic tie, fathers had to be included through some mechanism. Of all the possible forms of marriage, monogamy provides the clearest connection between father and child, and thus the strongest, most basic social relationship.

What Ward and Gilman add is the idea of intrinsic male and female energy. Male energy is at odds with family and social ties. The "giant force of masculine energy" had to be modified in order to add and extend the social tie between father and child. This was accomplished through the subjugation of women (I told you it was interesting). When men created patriarchy and dominated women, they also made women and children dependent upon men. In natural arrangements, both men and women produce economically; but as men dominated women, they also removed them from the world of economic production. Women could no longer provide for themselves, let alone their children. Men, then, had to take on this responsibility, and as they did, they took on "the instincts and habits of the female, to his immense improvement" (Gilman, 1899/1975, p. 128). Therefore, men became more social as a result of patriarchy.

This functional need was a two-way affair. Not only did men need to take on some of the women's traits, the natural environment could be more thoroughly tamed by men rather than women. By taking women out of the productive sphere, and therefore being forced to be more productive himself, "male energy . . . brought our industries to their present development" (Gilman, 1899/1975, pp. 132–133). Both men and human survival in general benefited from patriarchal arrangements.

Gilman draws two conclusions from gynaecocentric theory. First, women should not resent the past domination of men. Great harm to individuals and society has come through this system, but at the same time necessary and great good has come as well,

> [I]n the extension of female function through the male; in the blending of facilities which have resulted in the possibility of our civilization; in the superior fighting power developed in the male, and its effects in race-conquest, military and commercial; in the increased productivity developed by his assumption of maternal function; and by the sex-relation becoming mainly proportioned to his power to pay for it. (Gilman, 1899/1975, pp. 136–137)

> It is what she is born for, what she is trained for, what she is exhibited for. It is, moreover, her means of honorable livelihood and advancement. *But*—she must not even look as if she wanted it! (Gilman, 1899/1975, p. 87)

The second thing Gilman draws from this theory is that the women's movements, and all the changes that they bring, are part of the evolutionary path for humanity. While women's economic dependence upon men was functional for a time, its usefulness is over. In fact, Gilman tells us, the women's and labor movements are misnamed; they ought more accurately to be called human movements. The reason that we've become aware of the atrocities associated with male dominance at this time in our history is that the dysfunctions are now greater than the functions. The behaviors and problems of patriarchy have always existed, but the benefits outweighed the costs and it therefore remained functional. Now, because of the level of social and economic progress, it is time to cast off this archaic form.

Sexuo-economic effects: However, it should be clear that Gilman's purpose is not to justify the subordination of women. Gynaecocentric theory gives her a way of understanding the effects, including the women's movements, in functional, evolutionary terms. It is, I think, to her credit that she maintains theoretical integrity in the face of the obvious suffering of women, and her own personal anguish as a woman. That said, a great deal of Gilman's work is occupied with exposing and explaining the current state of women.

Gilman characterizes the overall system and its effects as sexuo-economic. In **sexuo-economic** relations, two structures overlap, one personal and the other public. The social structure is the economy and the private sphere is our sexual relationships. The basic issue here is that women are dependent upon men economically; as a result, two structures that ought to be somewhat separate intertwine. Historically, as women became more and more dependent upon men for their sustenance, they were removed further and further from economic participation. As we've noted, the economy is the most basic of all our institutions: it defines our humanness. A species is defined by its mode of survival; survival modes, then, determine the most basic features of a species. In our case, that basic feature is the economy. The economy is where we as a species compete for evolutionary survival.

It is where we as a species come in contact with the environment and where the natural laws of selection are at work. Thus, when workforce participation is denied to women, they are denied interaction with the natural environment of the species. That natural environment, with all of its laws and effects, is replaced with another one. Man (meaning males) becomes woman's economic environment, and, just as any organism responds to its environment, woman changed, modified, and adapted to her new environment (form and function).

Gilman talks about several effects from living in this new environment, but the most important are those that contribute to **morbid excess in sex distinction**. As we've seen, all animals that have two sexes also have sex distinctions. These distinctions are called secondary sex characteristics: those traits in animals that are used to attract mates. In humans, according to Gilman, those distinctions have become accentuated so much that they are morbid or gruesome. In societies where they are denied equal workforce participation, the primary drive in women is to attract men. Rather than improving her skills in economic pursuits, she must improve her skills at attracting a man. In fact, expertise in attraction skills *constitutes* her basic economic skills (which is how prostitution can exist). Thus, the secondary sex characteristics of women become exaggerated. She devotes herself to cosmetics, clothes, primping, subtle body-language techniques, and so on. Most of these are artificial, yet there are also natural, physical effects. Because women are pulled out from interacting with the natural environment, their bodies change: they become smaller, softer, more feeble, and clumsy.

> As the priestess of the temple of consumption, as the limitless demander of things to use up, her economic influence is reactionary and injurious. . . . Woman, in her false economic position, reacts injuriously upon industry, upon art, upon science, discovery, and progress. (Gilman, 1899/1975, pp. 120–121)

The increase in sex distinctions leads to greater emphasis on sex for both men and women. The importance given to sex far exceeds the natural function of procreation and comes to represent a threat to both self- and race-preservation. The natural ordering of the sexes becomes perverted as well. In most animals, it is the male that is flamboyant and attractive; this order is reversed in humans. To orient women toward this inversion, the sexual socialization of girls begins early. In our time, young girls are taught that women should be beautiful and sexy, and they are given toys that demonstrate the accentuated sex distinctions (like Barbie), toy versions of makeup kits used to create the illusion of flamboyancy in women, and games that emphasize the role of women in dating and marriage. Another result of this kind of gender structure and socialization is that women develop an overwhelming passion for attachment to men at any cost; Gilman points to women who stay in abusive relationships as evidence. In the end, because of the excess in sex distinction brought on by the economic dependence of women, marriage becomes mercenary.

Obviously, the economic dependency of women will have economic results as well. Women are affected most profoundly as they become "non-productive consumers." According to Gilman, in the natural order of things, productivity and consumption go hand in hand, and production comes first. Like Marx, Gilman sees economic production as the natural expression of human energy, "not sex-energy at all but race-energy." It's part of what we do as a species and there is a natural balance to it when everybody contributes and everybody consumes. The balance shifts, however, as women are denied workforce participation. The consumption process is severed from the production process and the whole system is therefore subject to unnatural kinds of pressures. Non-economic women are focused on non-economic needs. In the duties of her roles of wife, homemaker, and mother, she creates a market that focuses on "devotion to individuals and their personal needs" and "sensuous decoration and personal ornament" (Gilman, 1899/1975, pp. 119–120). The economy, of course, responds to this market demand. Contemporary examples of this kind of "feminized" market might include such things as fashion items (like jewelry, chic clothing, alluring scents, sophisticated makeup, and so forth), body image products (such as exercise tapes, health clubs, day spas, dietary products, and the like), fashionable baby care products (like heirloom baby carriages, designer nursery décor, and special infant fashions by The Gap, Gymboree, Baby B'Gosh, Tommy, Nike, Old Navy, ad infinitum), and decorative products for the home(such as specialized paints, wallpaper, trendy furnishings, and so on). Gilman argues that the creative efforts of men, which should be directed to durable commodities for the common good, are subject to the creation and maintenance of a "false market." Women are thus alienated from the commodities they purchase because they don't participate in the production process nor are their needs naturally produced; and men are alienated because they are producing goods and services driven by a false market.

As Figure 8.2 shows, the morbid excess in sex distinction disables the laws of natural selection and puts self- and race-preservation out of balance; we've explored some of these shifts above. As a result, the customs and laws of society respond, as do the social institutions and practices (accumulation of precedent). Institutions and practices that properly belong to humankind become gendered. For example, the governing of society is a "race-function," but in societies that limit women's workforce participation, it is seen as the duty and prerogative of men. Decorating is also distinctly human; it is a function of our species, but it is perceived as the domain of women. Religion, an obvious human function, is dominated by men and has been used to justify the unequal treatment of women, as have law, government, science, and so forth.

Yet Gilman does tell us that things are changing. She sees the women's movements as part of our evolutionary path of progress. Implied in this conclusion is the idea that equality is a luxury. Under primitive circumstances, only the strong survive. The weak, the feeble, and the old are left behind or killed. As social relationships come into existence, cooperation is possible; and as people cooperate, a surplus develops. Yes, for the most part the powerful control the surplus and turn it into wealth. Yet, at the same time, it is then feasible to support some who had once been cut off. These people are integrated into society and society continues to grow. It becomes more complex and technically better able to control the environment and produce surplus. In turn, other groups are brought into the fold, and increasing numbers of disenfranchised groups are able to live and prosper. It is Gilman's position that just as technical progress is our heritage, so is ever-expanding equality. Evolution of the species not only involves economic advancement, but increasing compassion as well; ethical and technical evolutions are inexorably linked.

At the present evolutionary moment, we are only hindered by our own blindness. We can afford the luxury of complete equality, but we fail to see the problem. Gilman says that this is due to a desire for consistency. We don't notice what we are used to. Even evil can become comfortable for us and we will miss it when it's gone. This is one of the reasons why slavery wasn't seen as evil for so many years of our history. We also have a tendency to think in individual terms rather than general terms. It's easier for us to attribute reasons for what we see to individuals rather than broad social factors. This, of course, in C. Wright Mills' terms, is the problem of the sociological imagination. It is difficult to achieve.

> Being used to them, we do not notice them, or, forced to notice them, we attribute the pain we feel to the evil behavior of some individual, and never think of it as being the result of a condition common to us all. (Gilman, 1899/1975, p. 84)

Summary

- Gilman is a critical evolutionist. All species fight for survival and create organic structures to help them survive in their environment. In this case, the function produces the structure. For humans, the primary evolutionary structure is the economy. It is through the economy that we as a species survive.

- When women were removed from the economy through male domination, several things happened. First, women were taken out of the natural environment of economic production and given a false environment—men, and indirectly the home and family, became women's environment. Gilman is a social evolutionist and argues that, as every species will, women changed in response to change in their environment. This change in environment made the distinctions between the sexes more pronounced. Rather than being equipped to economically produce, women became equipped to pursue a husband (their survival depended upon it). Their bodies became smaller and softer from disconnection with the natural world, and women augmented these changes through artificial means (such as makeup, clothing, etc.). Human beings now have a morbid excess in sex distinction, and as a result, the natural order has been reversed. In most animals, it is the male that is marked for attraction; in humans, it is the female. In order to produce this artificial order, girls have to be socialized from birth to want to be attractive and to value having a husband over all else. Further, in societies where women's workforce participation is limited, women become unnatural consumers. The natural relationship is production followed by consumption. Women are simply consumers without production. As such, their perceived needs are radically changed and they produce a market focused on beauty, decoration, and private relationships. This market makes men produce dysfunctional goods, and they are thus alienated from the work of their hands; women, on the other hand, purchase goods that are not a natural part of their essential nature, thus alienating them from the commodity. Further, to support and legitimate this economy, the other institutions, laws, and customs in society have become gendered, and those things that are properly human (like government, religion, and education) are now seen as the domain of one sex.

- This evolutionary path has also had functional consequences. The natural energy of men is aggressive, individualistic, and dominating. The male is, then, ill-suited for social life. However, in making women and children dependent upon him, he also obligated himself to provide for them (to be socially involved and committed). Thus, through gender oppression, the natural energy of men has been modified: they now have maternal feelings and higher social abilities. Further, large societies require highly developed systems that can control the environment and extract resources. Gilman argues that men and women together could not have achieved the needed level of economic and technological development. In Gilman's evolutionary scheme men are naturally more aggressive and objectifying. Thus, having men solely responsible for the economic sphere meant that society produced more and achieved higher levels of technological control. Thus, pulling women out of the workforce was functional so that the economy could develop fully.

- Humans have, however, reached the point where the suppression of women is more dysfunctional than functional. The women's movements are sure signs that evolution is pushing us toward higher levels of equality.

W.E.B. Du Bois (American, 1868–1963)

William Edward Burghardt Du Bois was born in Great Barrington, Massachusetts, on February 23, 1868. His mother was Dutch African and his father French Huguenot-African. Du Bois' father, who committed bigamy in marrying Du Bois' mother, left the family when the boy was just two years old, but Du Bois grew up without experiencing the depth of suffering blacks felt at that time. Du Bois' education in the abject realities of blackness in America came during his college years at Fisk. He taught two summers in rural Tennessee and saw firsthand the legacy of slavery in the South. As a result, Du Bois became an ardent advocate of social change through protest. His belief in confrontation and dissent put Du Bois at odds with Booker T. Washington, a prominent black leader who favored accommodation. This disagreement with Washington influenced the early part of Du Bois' political life; he founded the Niagara Movement in large part to counter Washington's arguments.

Through most of his life, Du Bois generally favored integration, but toward the end he became discouraged at the lack of progress and increasingly turned toward Black Nationalism: He encouraged blacks to work together to create their own culture, art, and literature, and to create their own group economy of black producers and consumers. The cultural stand was directed at creating black pride and identity; the formation of a black economic community was the weapon to fight discrimination and black poverty. Du Bois was also a principal force in the Pan-African movement, which was founded on the belief that all black people share a common descent and should therefore work collectively around the globe for equality. As I mentioned, in the latter part of his life, Du Bois became disheartened at the lack

Photo 8.3 W.E.B. Du Bois

of change regarding the color line in the United States. In the end, he renounced his citizenship, joined the Communist Party, and moved to Ghana, Africa.

More than any other single person, Du Bois was responsible for black consciousness in America and probably the world during the twentieth century. His book, *The Souls of Black Folk,* defined the problem of the color line. He was a founding member of the NAACP and its chief spokesperson during its most formative years. Du Bois also produced the first scientific studies of the black condition in America, and he was the chief energy behind the Pan-African movement.

I have listed the major points in Du Bois' life in the timeline below. I encourage you to not skip over this information; read it and let it sink in. The timeline is but an outline of a life that spanned almost a century and helped to change the face of society.

- **February 23, 1868:** Born in Great Barrington, Massachusetts
- **1888:** Received his B.A. from Fisk University, Nashville (turned down by Harvard)—taught two summers in rural Tennessee, which introduced him to the deep poverty of the Southern black person
- **1888–1895:** Completed second B.A. and Ph.D. at Harvard, during which time he spent two years in Berlin, attending lectures by Max Weber and becoming friends with both Marianne and Max Weber

- **1894–1910:** Taught at Wilberforce, University of Pennsylvania, and Atlanta University (where he was a professor of economics and history for 14 years); published 16 research monographs
- **1899:** Published *The Philadelphia Negro: A Social Study* (the first ethnography of a black community in the United States)
- **1900:** Led the first Pan-African conference, in London
- **1903:** Published *The Souls of Black Folk,* which first set out the problem of the color line and Du Bois' opposition to Booker T. Washington's strategy
- **1905:** Founded the Niagara Movement, precursor to NAACP
- **1909:** Helped found the National Association for the Advancement of Colored People (NAACP); left academia and became editor of the association's journal, *The Crisis*
- **1910–1934:** Worked at the NAACP, edited *The Crisis,* became the principal leader of blacks in America; advocated Black Nationalism
- **1919–1927:** Engineered four Pan-African congresses
- **1934–1944:** Left the NAACP in a disagreement over assimilation versus separation; returned to Atlanta University
- **1935:** Published *Black Reconstruction: An Essay Toward a History of the Part Which Black Folk Played in the Attempt to Reconstruct Democracy in America, 1860–1880* (Marxist interpretation of the Reconstruction)
- **1940–1944:** Chaired the Department of Sociology at Atlanta University; editor of social science journal, *Phylon*
- **1940:** Published *Dusk of Dawn: An Essay Toward an Autobiography of a Race Concept* (viewed his career and life as representative of black–white relations)
- **1944–1948:** Returned to the NAACP; ended in disagreement; continued discouragement and movement toward political left
- **1951:** Indicted as unregistered agent for a foreign power; acquitted
- **1961:** Joined the Communist Party; moved to Ghana, continued work on his long-term project, *Encyclopedia Africana*
- **1962:** Renounced American citizenship
- **August 27, 1963:** Died in Ghana at age 95

As Charles Lemert (2000) says, "He died, as icons often do, at precisely the right moment—on the eve of the 1963 March on Washington. From the death of Frederick Douglass in 1895 to Martin Luther King Jr's coronation the day after Du Bois' death, no other person of African descent was as conspicuous in the worldwide work of contesting the color line." (p. 347)

> We push below this mudsill the derelicts and half-men, whom we hate and despise, and seek to build above it—Democracy! On such foundations is reared a Theory of Exclusiveness, a feeling that the world progresses by a process of excluding from the benefits of culture the majority of men, so that a gifted minority may blossom. (Du Bois, 1920/1996b, p. 543)

Du Bois' Perspective: The Experience of Oppression and Critical Knowledge

The centered subject: There are a couple of things that strike the thinking reader as she or he spends time with Du Bois' texts. The first is that Du Bois centers the subject in his writing. One of the things most of us are taught in college English is that we should write from a de-centered point of view. We are supposed to avoid using "I" and "me" and should always write from an objective perspective. Yet Du Bois begins one of his most famous works, *The Souls of Black Folk,* with "Between me and the other world." In another place he wonders, "Who and what is this I . . . ?" Du Bois isn't being self-centered nor is he unaware of the rules of composition. Du Bois is being quite deliberate in his use of personal pronouns and the centered subject. I believe one of the things he is telling us is that race is not something that can be understood through the cold, disassociated stance of the researcher. Race and all marginal positions must be experienced to be understood. Du Bois uses his life as the canvas upon which to paint the struggles of the black race in America and in the world.

The other thing that impresses the reader is that much of Du Bois' writing is a multimedia presentation. Du Bois moves back and forth among intellectual argumentation, song, prayer, poetry, irony, parable, data, riddles, analogy, and declaration. He weaves a tapestry for the reader, one that touches every part of the reader's being. He wants us to be able to understand the objective state of blackness as well as experience its soul. In this he reminds me of a colleague of mine. My colleague is a large black man who has done some amazing work with Los Angeles gang members. One time I asked him to present a guest lecture to one of my classes. You know what a lecture looks like, right? Imagine my surprise when he asked us all to close our eyes and lay our heads down on the desks. His lecture consisted of a dramatic reading of a poem—replete with gunshots and cries—that left the class stunned and some in tears.

It might well be coincidence that Du Bois, who is one of the few writers I've seen use such varied venues in writing, and the only professor I've seen break the mold of lecture so soundly, are both African American. Yet it does give us pause and it provides me with a transition into one of Du Bois' important points. Though Du Bois undoubtedly sees the color line as something that is socially created, he also acknowledges and honors the unique characteristics of and contributions from the "souls of black folks." He maintains that American music and folklore, faith and reverence, light-hearted humility, literature and poetry, speech and styles of interaction, practices of sport, and our political fabric are all strongly influenced by African-American culture. Yet it isn't a simple listing of accomplishments or influences alone that Du Bois has in mind; it is the soul of a particular kind of lived experience: "that men may listen to the striving *in the souls of* [italics added] black folk" (Du Bois, 1903/1996a, p. 107).

Standpoint of the oppressed: There are two things I think this subjective stance implies: the first is my own comment, the other is something I think Du Bois has in mind. First, any secondary reading of Du Bois, such as the book you have in your hands, falls short of the mark. This is generally true of any of the thinkers in this book—you would be much richer reading Durkheim than reading what somebody says about Durkheim—but it is particularly true of Du Bois. Part of what you can acquire from reading Du Bois is an experience, and that experience is a piece of what Du Bois wants to communicate.

The second implication of Du Bois' multidimensional, subjective approach is theoretical. Du Bois (1903/1996a) says that "the Negro is a sort of seventh son, born with a veil, and gifted with second-sight in this American World,—a world which yields him no true self-consciousness, but only lets him see himself through the revelation of the other world" (p. 102). Du Bois employs spiritual language here. The veil of which he speaks is the birth caul. In some births, the inner fetal membrane tissue doesn't rupture and it covers the head at delivery. This caul appears in about 1 in 1000 births. Due to its rarity, some traditional cultures consider such a birth spiritually significant and the caul is kept for good luck. The same is true of the seventh son reference. The seventh son is considered to have special powers and references to such are to be found in many folk and blues songs as well as in the Bible. The "second-sight" is a reference to clairvoyant or prophetic vision.

Thus, Du Bois is saying that because of their experiential position, African Americans are gifted with special insight, a prophetic vision, into the American world. They see themselves not simply as they are in themselves; they also see their position from the perspective of the other world. In other words, blacks and other oppressed groups have a particular point of view of society that allows them to see certain truths about the social system that escape others. This idea of critical consciousness goes back to Marx. Marxian philosophy argues that only those on the outside of a system can understand its true workings; it is difficult to critically and reflexively understand a system if you accept its legitimation. In other words, capitalists and those who benefit from capitalism by definition believe in capitalism. It is difficult for a capitalist to understand the oppressive workings of capitalism because in doing so the person would be condemning her- or himself.

Contemporary feminists argue that it is the same with patriarchy: men have a vested interest in the patriarchal system and will thus have a tendency to believe the ideology and have difficulty critiquing their own position. Dorothy Smith, a contemporary feminist, refers to this as **standpoint theory**. In general, this refers to a theory that is produced from the point of view of an oppressed subject; "an inquiry into a totality of social relations beginning from a site outside and prior to textual discourses. Women's standpoint [is seen] here as specifically subversive of the standpoint of a knowledge of ourselves and our society vested in relations of ruling" (Smith, 1987, p. 212).

Du Bois is thus arguing that African Americans have, by virtue of their position, a critical awareness of the American social system. This awareness can be a cultural resource that facilitates structural change. As Du Bois (1903/1996a) puts it, "This, then, is the end of his striving: to be a co-worker in the kingdom of culture, to escape both death and isolation, to husband and use his best powers and his latent genius" (p. 102).

Generally, Du Bois' perspective is more cultural and critical than either Martineau's or Gilman's; and in many ways his is much more in keeping with contemporary theories of difference. One of the things that oppression in modernity has done is deny the voice of the other. In the Durkheim chapter, we saw that one of the characteristics of modern society is the grand narrative. Nation-states provide all-encompassing stories about history and national identity. The purpose of such narratives is to offer a kind of Durkheimian rallying point for social solidarity. This sort of solidarity is necessary for nations to carry out large-scale programs, especially such things as colonization and war. The problem with such a narrative is that it hides inequities. For example, the grand narrative of equality in the United States was actually a story about white Anglo-Saxon Protestant males. Hidden in the national narrative and history was (and still is) the subjugation of Native Americans, African Americans, women, Mexican Americans, homosexuals, and so forth.

Contemporary theories of difference, then, focus on the subjective experience of the disenfranchised in contrast to this grand narrative. One of the things that is important in the fight for equality is allowing multiple voices to be heard; thus, we have recently moved from the cultural picture of the "melting pot" to that of the "salad bowl" in the United States. Du Bois' perspective is quite in keeping with this emphasis. In his own work, he uses the subjective mode to express the experience of the oppressed. He becomes a representative figure through which we might understand the plight of black people in America. Both Martineau and Gilman did use multiple approaches in expressing their ideas, and some of those used the first person, but not to the degree of Du Bois. Part of what this multiple-voice approach entails is valuing the outsider's point of view. And, interestingly, Du Bois is much more in tune with this feminist idea of standpoint theory than either Martineau or Gilman, neither of whom privilege outsider knowledge. In that sense, Du Bois' work contains a more critical edge, again in keeping with much of contemporary analysis.

Du Bois was a prolific writer, as were Martineau and Gilman, so I have chosen to focus on particular points of his discourse about race. In addition, Du Bois was at least as much a political figure as he was an intellectual one, so I've had to be even more selective with his writings than with Martineau's or Gilman's. In my opinion, I think that Du Bois' lasting contribution to social theory is his understanding of cultural oppression. In the section of this chapter on *cultural oppression,* we will see that it is just as necessary as structural oppression in the suppression of a social group. Du Bois' understanding of this process is quite good. He argues that cultural oppression involves exclusion from history, specific kinds of symbolic representations, and the use of stereotypes and their cultural logic of default assumptions. This cultural work results in a kind of double consciousness wherein the disenfranchised see themselves from two contradictory points of view. However, Du Bois isn't only interested in cultural oppression; he also gives us a race-based theory of world capitalism (see the section on *the dark nations and world capitalism*). There we will see that it isn't only the elite capitalists that benefit from the exploitation of blacks and other people of color; the middle class benefits as well.

Cultural Oppression

Exclusion from history: If we can get a sense of the subjectivity that Du Bois is trying to convey, we might also get a sense of what horrid weight comes with cultural oppression. Undergirding every oppressive structure is cultural exclusion. While the relative importance of structure and culture in social change can be argued, it is generally the case that structural oppression is legitimated and facilitated by specific cultural moves—historical cultural exclusion in particular. In this case, African Americans have been systematically excluded from American history, and they have been deprived of their own African history.

History plays an important part in legitimating our social structures; this is known as **history as ideology**. No one living has a personal memory of why we created the institutions that we have. So, for example, why does the government function the way it does in the United States? No one personally knows; instead, we have an historical account or story of how and why it came about. Because we weren't there, this history takes on objective qualities and feels like a fact, and this facticity legitimates our institutions and social arrangements unquestionably. But, Du Bois tells us, the current history is written from a politicized point of view: because women and people of color were not seen as having the same status and rights as white men, our history did not see them. We have been blind to their contributions and place in society. The fact that we now have Black and Women's History Months underscores this historical blindness. Du Bois calls this kind of ideological history "lies agreed upon."

Du Bois, however, holds out the possibility of a scientific history. This kind of history would be guided by ethical standards in research and interpretation, and the record of human action would be written with accuracy and faithfulness of detail. Du Bois envisions this history acting as a guidepost and measuring rod for national conduct. Du Bois presents this formulation of history as a choice. We can either use history "for our pleasure and amusement, for inflating our national ego," or we can use it as a moral guide and handbook for future generations (Du Bois, 1935/1996c, p. 440).

It is important for us to note here that Du Bois is foreshadowing the contemporary emphasis on culture in studies of inequality. Marx argued that it is class and class alone that matters. Weber noted that cultural groups—i.e., status positions—add a complexity to issues of stratification and inequality. But it was not until the work of the Frankfurt School (see Marx) during the 1930s that a critical view of culture itself became important, and it was not until the work of postmodernists and the Birmingham School in the 1970s and 1980s that representation became a focus of attention. Yet Du Bois is explicating the role of culture and representation in oppression in his 1903 book, *The Souls of Black Folk.*

Representation: Representation is a term that has become extremely important in contemporary cultural analysis. Stuart Hall, for example, argues that images and objects by themselves don't mean anything. We see this idea in Mead's theory as well. The meaning has to be constructed, and we use representational systems of concepts and ideas. Representation, then, is the symbolic practice through which meaning is given to the world around us. It involves the production and consumption of cultural items and is a major site of conflict, negotiation, and potential oppression.

Let me give you an illustration from Du Bois. Cultural domination through representation implies that the predominantly white media do not truly represent people of color. As Du Bois (1920/1996d) says, "The whites obviously seldom picture brown and yellow folk, but for five hundred centuries they have exhausted every ingenuity of trick, of ridicule and caricature on black folk" (pp. 59–60). The effect of such representation is cultural and psychological: the disenfranchised read the representations and may become ashamed of their own image. Du Bois gives an example from his own work at *The Crisis* (the official publication of the NAACP). *The Crisis* put a picture of a black person on the cover of their magazine. When the readers saw the representation, they perceived it (or consumed it) as "the caricature that white folks intend when they make a black face." Du Bois queried some of his office staff about the reaction. They said the problem wasn't that the person was black; the problem was that the person was *too* black. To this Du Bois replied, "Nonsense! Do white people complain because their pictures are too white?" (Du Bois, p. 60).

While Du Bois never phrased it quite this way, Roland Barthes (1964/1967), a contemporary semiologist (someone who studies signs), explains that cultural signs, symbols, and images can have both denotative and connotative functions. Denotative functions are the direct meanings. They are the kind of thing you can look up in an ordinary dictionary. Cultural signs and images can also have secondary, or connotative, meanings. These meanings get attached to the original word and create other, wider fields of meaning. At times these wider fields of meaning can act like myths, creating hidden meanings behind the apparent. Thus, systems of connotation can link ideological messages to more primary, denotative meanings. In cultural oppression, then, the dominant group represents those who are subjugated in such a way that negative connotative meanings and myths are produced. This kind of complex layering of ideological meanings is why members of a disenfranchised group can simultaneously be proud and ashamed of their heritage. Case in point, the black office colleagues to whom Du Bois refers can be proud of being black but at the same time feel that an image is *too* black.

Stereotypes and slippery slopes: In addition to history and misrepresentation, the cultural representation of oppression consists of being defined as a problem: "Between me and the other world this is ever an unasked question. . . . How does it feel to be a problem?" (Du Bois, 1903/1996a, p. 101). Representations of the group thus focus on its shortcomings, and these images come to dominate the general culture as stereotypes: "While sociologists gleefully count his bastards and his prostitutes, the very soul of the toiling, sweating black man is darkened by the shadow of a vast despair. Men call the shadow prejudice, and learnedly explain it as the natural defense of culture against barbarism, learning against ignorance, purity against crime, the 'higher' against the 'lower' races. To which the Negro cries Amen!" (Du Bois, p. 105).

I want to point out that last bit of the quote from Du Bois. He says that the black person *agrees* with this cultural justification of oppression. Here we can see one of the insidious ways in which cultural legitimation works. It presents us with an apparent truth that once we agree to it can reflexively destroy us. Here's how this bit of cultural logic works: The learned person says that discrimination and prejudice are necessary. Why? They are needed to demarcate the boundaries between civilized and uncivilized, learning and ignorance, morality and sin, right and wrong. We agree that we should be prejudiced against sin and evil, and against uncivilized and barbarous behavior, and we do so in a very concrete manner. For example, we are prejudiced against allowing a criminal in our home. We thus agree that prejudice is a good thing. Once we agree with the general thesis, it can then be more easily turned specifically against us.

In this movement from general to specific, Du Bois hints at another piece of cultural logic that is used in oppression. Douglas R. Hofstadter (1985) calls this the "slippery slope of sexism." Hofstadter argues that there can be a relationship between the general and specific use of a term. So, some of the connotations of each will rub off on the other. Hofstadter's specific concern is gender. You are aware of examples of this process in gender, such as in the statements "all men are created equal" and "there was a four-man crew on board." Is the use of the masculine pronoun meant in its specific or its general meaning? Are we referring to men specifically or to man*kind*? When such slippery slopes of language occur, it is easy for society to obliterate or oppress a cultural identity, which is why feminists talk about the invisible woman in history.

Du Bois has a similar slope in mind, but obviously one that entails race. In this section from *The Souls of Black Folk* from which we have been quoting, Du Bois says that the Negro stands "helpless, dismayed, and well-nigh speechless" before the "nameless prejudice" that becomes expressed in "the all-pervading desire to inculcate disdain for everything black" (Du Bois, 1903/1996a, p. 103). In *Darkwater* (1920/1996b) Du Bois refers to this slippery slope as a "theory of human culture" (p. 505). This work of culture has "worked itself through warp and woof of our daily thought." We use the term *white* to analogously refer to everything that is good, pure, and decent. The term *black* is likewise reserved for things despicable, ignorant, and fearful. There is thus a moral, default assumption in back of these terms that automatically, though not necessarily or even deliberately, includes the cultural identities of white and black.

In our cultural language, we also perceive these two categories as mutually exclusive. For example, we will use the phrase "this issue isn't black or white" to refer to something that is undecided, that can't fit in simple, clear, and mutually exclusive categories. The area in between is a gray, no-person's land. It is culturally logical, then, to perceive unchangeable differences between the black and white races, which is the cultural logic behind the "one drop rule" (one drop of black blood makes someone black). Again, keep in mind that this movement between the specific and general is unthinkingly applied. People don't have to intentionally use these terms as ways to racially discriminate. The cultural default is simply there, waiting to swallow up the identities and individuals that lay in its path.

Double consciousness: Du Bois attunes us to yet another insidious cultural mechanism of oppression: the internalization of the **double consciousness**. With this idea, Du Bois is drawing on his knowledge of early pragmatic theories of self. He knew William James and undoubtedly came in contact with the work of Mead and

Cooley. We can think of Mead's theory of role-taking and Cooley's looking-glass self (see below) and see how the double consciousness is formed. According to Du Bois, African Americans have another subjective awareness that comes from their particular group status (being black and all that that entails) and they have an awareness based on their general status group (being American). What this means is that people in disenfranchised groups see themselves from the position of two perspectives.

Let's think about this issue using Cooley's notion of the looking-glass self. In Cooley's theory, the sense of one's self is derived from the perceptions of others. There are three phases in Cooley's scheme. First, we imagine what we look like to other people; we then imagine their judgment of that appearance; and we then react emotionally with either pride or shame to that judgment. An important point for us to see is that Cooley didn't say that we actually perceive how others see and judge us; rather, we imagine their perceptions and judgments. However, this imagination is not based on pure speculation; it is based on social concepts of ways to look (cultural images), ways to behave (scripts), and ways we anticipate others will behave, based on their social category (expectations). In this way, Du Bois argues that African Americans internalized the cultural images produced through the ruling, white culture. They also internalize their own group-specific culture. Blacks thus see themselves from at least two different and at times contradictory perspectives.

In several places, Du Bois relates experiences through which his own double consciousness was formed. In this particular one, we can see Cooley's looking glass at work:

> In a wee wooden schoolhouse, something put it into the boys' and girls' heads to buy gorgeous visiting-cards—ten cents a package—and exchange. The exchange was merry, till one girl, a tall newcomer, refused my card,—refused it peremptorily, with a glance. Then it dawned upon me with a certain suddenness that I was different from the others; or like, mayhap, in heart and life and longing, but shut out from their world by a vast veil. (Du Bois, 1903/1996a, p. 101)

We can thus see in Du Bois' work a general theory of cultural oppression. That is, every group that is oppressed structurally (economically and politically) will be oppressed culturally. The basic method is as he outlines it: define and label the group in general as a problem, emphasize and stereotype the group's shortcomings and define them as intrinsic to the group, employ cultural mechanisms (like misrepresentation and default assumptions) so the negative attributes are taken for granted, and systematically exclude the group from the grand narrative histories of the larger collective.

The Dark Nations and World Capitalism

Exporting exploitation: Du Bois also sees the social system working in a very Marxian way. Like Marx, Du Bois argues that the economic system is vitally important to society and understanding history. Du Bois (1920/1996b) says that "history is economic history; living is earning a living" (p. 500). He also understands the expansion to global capitalism in Marxian, world systems terms. Capitalism is inherently expansive. Once limited to a few national borders, it has now spread to a worldwide economic system. One of the things this implies is that exploitation can be exported.

> Thus the world market most wildly and desperately sought today is the market where labor is cheapest and most helpless and profit is most abundant. This labor is kept cheap and helpless because the white world despises "darkies." (Du Bois, 1920/1996b, p. 507)

Capitalism requires a ready workforce that can be exploited. Goods must be made for less than their market value, and one of the prime ways to decrease costs and increase profits is by cutting labor costs. In national capitalism, this level of exploitation is limited to the confines of the state itself. Consequently, as the national economy expands and the living wage increases, the level of profit goes down. In other words, the average real wage of workers within a country increases over time—the average worker in the United States makes more today than she or he did 100 years ago. This proportional increase in income would lower profit margins. However, with global capitalism, exploitation can be exported. It costs less for a textile company to make clothing in China than in the United States, yet the market value of the garment remains the same. Profits thus go up.

Du Bois (1920/1996b) argues that this expansion of exploitation is due to increasing levels of "education, political power, and increased knowledge of the technique and meaning of the industrial process" (p. 504). There are, according to Marx and Du Bois, certain inevitable "trickledown" effects of capitalism. As industrial technologies become more and more sophisticated, better educated workers are needed. Increasing levels of education attune workers not only to the methods of production, but also to the ideologies behind the system (for example, most of you reading this text will work in the economy, not in education, yet here you are reading critical theory). Industrialization also brings workers closer together in factories and neighborhoods where they can communicate and become politically active.

The need for color: As a result, the economic system inexorably pushes against its national boundaries and seeks a labor force ready for exploitation. This notion of expanding systems and exporting exploitation is part of Immanuel Wallerstein's world systems theory. He argues that core nations exploit periphery nations so places like the United States can have high wages and relatively cheap goods. Wallerstein sees nations, but Du Bois tells us that the reality of those nations is color. "There is a chance for exploitation on an immense scale for inordinate profit, not simply to the very rich, but to the middle class and to the laborers. This chance lies in the exploitation of darker peoples" (Du Bois, 1920/1996b, pp. 504–505). These are "dark lands" ripe for exploitation "with only one test of success,—dividends!" (Du Bois, p. 505). In other words, middle class wages in advanced industrialized nations are based on there being racial groups for exploitation. Before we move on in our consideration of Du Bois' theory, I want to note once again his foresight. Wallerstein began publishing his work in 1974; Du Bois published his work on global racial exploitation in 1920.

> I see the [Souls of White Folk] undressed and from the back and side. I see the working of their entrails. I know their thoughts and they know that I know. (Du Bois, 1920/1996b, p. 497)

Du Bois thus sees race as a tool of capitalism. While Du Bois perceives distinct differences between blacks and whites that can be characterized in spiritual terms (the souls of black and white folk), he also argues that the color line is socially constructed and politically meaningful. The really interesting feature of Du Bois' perspective here is that he understands that both black and white social identities are constructed. Again, Du Bois beat contemporary social theory to the punch: we didn't begin to seriously think of white as a construct until the 1970s, and it didn't become an important piece in our theorizing until the late 1980s.

Du Bois argues that the idea of "personal whiteness" is a very modern thing, coming into being only in the nineteenth and twentieth centuries. Humans have apparently always made distinctions, but not along racial lines. Prior to modernity, people created group boundaries of exclusion by marking civilized and uncivilized cultures, religion, and territorial identity. As William Roy (2001) notes, these boundaries lack the essential features of race—they were not seen as biologically rooted or immutable and people could thus change. Race, on the other hand, is perceived as immutable and is thus a much more powerful way of oppressing people.

As capitalism grew in power and its need for cheap labor increased, indentured and captive slavery moved to chattel slavery. Slavery has existed for much of human history, but it was used primarily as a tool for controlling and punishing a conquered people or criminal behavior, or as a method of paying off debt or getting ahead. The latter is referred to as indentured servitude. People would contract themselves into slavery, typically for seven years, in return for a specific service, like passage to America, or to pay off debt. In most of these forms of slavery, there were obligations that the master had to the slave, but not so with chattel slavery. Under capitalism, people could be defined as property—the word *chattel* itself means property—and there are no obligations of owner to property. In this move to chattel slavery, black became not simply *a race* but *the race* of distinction. The existence of race, then, immutably determined who could be owned and who was free, who had rights and who did not.

> Till the devil's strength be shorn,
> This some dim, darker David, a-hoeing of his corn,
> And married maiden, mother of God,
> Bid the black Christ be born!
> Then shall our burden be manhood,
> Be it yellow or black or white
> (Du Bois, 1920/1996b, p. 510)

Summary

- Du Bois' perspective is that of a black man. His subjective experience of race is central in his work. He often uses himself as a representational character and he is vastly interested in drawing the reader into the experience of race. Du Bois also believes that being a member of a disenfranchised group, specifically African

American, gives one a privileged point of view. People who benefit from the system cannot truly see the system. Most of the effects of active social oppression are simply taken for granted by the majority. They cannot see it.

- All structural oppression must be accompanied by cultural suppression. There are several mechanisms of cultural suppression: denial of history, controlling representation, and the use of stereotypes and default assumptions. As a result of cultural suppression, minorities have a double consciousness. They are aware of themselves from their group's point of view, which is positive, and they are conscious of their identity from the oppressor's position, which is negative. They are aware of themselves as black (in the case of African Americans), and they are aware of themselves as American—two potentially opposing viewpoints.

- Capitalism is based on exploitation: owners pay workers less than the value of their work. Therefore, capitalism must always have a group to exploit. In global capitalism, where the capitalist economy overreaches the boundaries of the state, it is the same: there must be a group to exploit. Global capitalism finds such a group in the "dark nations." Capitalists thus export their exploitation, and both capitalists and white workers in the core nations benefit, primarily because goods produced on the backs of sweatshop labor are cheaper.

Thinking About Modernity and Postmodernity

Modernity and identity: Martineau, Gilman, and Du Bois are clearly modernists. Du Bois believes in the hope of a scientific approach to history, Martineau champions the cause of positivism and the hope of the Enlightenment, and Gilman is a social evolutionist—a theory born and bred for modernity. Yet, they are modern in a more fundamental way, as well. The notions of difference, identity, and social movements are themselves modern.

Human identity and self are built upon difference. According to psychologists, one of the fundamental things an infant must learn is that she or he is a separate and distinct entity, different from the rest of the world and from significant others. This type of self is discovered through the gradual unfolding of the person's psychological makeup, which is the inner core of the individual. In this way of thinking about things, the individual is always identical to the inner self and is thus essentially different from others. Most of us, to one degree or another, have bought into this way of thinking about the self, unless we have been thoroughly brainwashed by sociology (the fervent hope of all your sociology professors). The self we are talking about here is the one with which you are born and the one that constitutes you all through your life. Your self is inside you, and the self inside the other person is different by definition, and that is what makes us essentially different.

On the other hand, sociology understands identity and self in relation to society and culture. A person doesn't *discover* the self through one's lifespan; rather, she or he *constructs* the self in interaction with society and culture. This is an entirely different perspective. Here, the process of identity involves identifying one's self with an external element of culture, such as gender or race, not an internal characteristic. In other words, people are different not because they are *born* different, but because they relate to different social and cultural elements. There are two important points that we can glean from the sociological scheme: the construction of identity and self is carried out with reference to social structures and culture; and the construction of identity and self is always to some degree exclusionary. Let's expand a bit on the former point first.

According to sociologists, identity isn't a given; it isn't something we are born with and it isn't defined by the boundaries of our body. For the psychologist, self and identity are embodied, and it is a simple matter to see where the self ends and the alter (the other) begins. Infants have to discover this, but it is nonetheless a given. Sociologists, however, argue that identities based on race, gender, religion, politics, music, work, and so forth are social constructions with which the individual identifies and through which the individual understands and gives meaning to her or his subjective experiences—in other words, one's self. That being the case, the boundaries upon which identities and selves are created are themselves cultural and social and not defined intrinsically by one's body. For psychologists, difference isn't a big deal because it is an intrinsic and natural part of our psychological make up that is contained within a physical body; for sociologists, difference is important precisely because it is a function of culture. We can't take difference for granted as a psychologist can; we have to understand how it is created.

Part of what we have in the works of the Gilman and Du Bois are theories about how these differences are produced. Du Bois gives us a cultural understanding of how difference is produced and maintained by the ruling class (through history, representation, stereotypes, cultural logic, and so on). Gilman's theory has cultural components, but it is basically an evolutionary theory of difference: difference is produced by men and women interacting with different environments. From this perspective, culture is produced by social structures and functions only to reinforce culture, although Gilman does seem to indicate that culture can have independent effects. Martineau doesn't give us a strong theory of difference, nevertheless her ideas of courage and chastity fit in nicely with Du Bois' insights into cultural logic and Gilman's separation of spheres.

In terms of creating difference, race and gender are among the strongest social distinctions we make. But as we will see when we get to Mead and Schutz, race and gender are social constructions. There are obvious biological features, but as Mead and Schutz will tell us, it's the meaning that matters. Even Gilman's theory with its genetic features (male/female energy, women's bodily changes) recognizes that the meaning we place on those characteristics is vitally important. The meaning is our reality, not the genetic or biological attributes. What makes race and gender our strongest social distinctions are the *legitimations* we can attach to them. We make them *appear* natural and unquestionable.

The postmodern twist: The idea of postmodernism gives the construction of difference an interesting twist. Postmodernists buy into the sociological idea of the relationship between the self and culture. Identities and selves are clearly created in relation to culture. Accordingly, as long as the culture of a society is fairly cohesive and stable, then the individual's self and identities will be, too. In fact, modernity requires such an identity and self. We have to see ourselves as part of some larger, abstract scheme of things, some imagined community called the nation. We have to have such an identity to participate in national projects, like going to war and paying taxes. If you want me to lay down my life for something, you'd better make sure that I think we are all on the same page about everything. I have to emotionally and psychically relate to the whole of society and its meaning.

More than that, modernity requires a stable, core identity in order to fulfill its goal of equality. As I've pointed out, included in the project of modernity are the ideas of democracy and equal treatment before the law. Ironically, these ideas initially pushed society to make distinctions between those who could vote and receive equal rights and those who could not. That's why the U.S. Constitution wasn't meant for everybody. It's paradoxical, but there are some theoretical reasons for it. The modern movements for democracy and equality were social movements, and social movements require people to have a strong sense of identity.

One of the fundamental ways in which identity and difference are constructed is through exclusion. In psychology, this can by and large be taken for granted: I am by definition excluded from you because I am in my own body. For sociologists, exclusion is a cultural and social practice; it's something we *do,* not something we *are.* This fundamental point may sound elementary in the extreme, but it's important for us to understand it. In order for me to be me, I can't be you; in order for me to be male, I can't be female; in order for me to be white, I can't be black; in order for me to be an American, I can't be communist, ad infinitum. Cultural identity is defined over and against something else. Identity and self are based on exclusionary practices. The stronger the practices of exclusion are, the stronger will be the identity; and the stronger my identities are, the stronger will be my sense of self. Further, the greater the exclusionary practices, the more real will be my experience of identities and self.

Hence, the early social movements for equality and democracy had clear practices of exclusion. It gave the people the needed level of identity to make the sacrifices necessary to fight. The distinctions between royalty and commoner were torn down, and the differences between the recipients of "equal" rights were created. Let's note that no matter how limited they were in who should be included, they took the idea of equality further than it had ever been taken before. And let's also note that no matter how upset we might be at the limitations of these early social movements, the people in the United States are *still* making distinctions about who can have civil rights and who cannot (the debate over gay marriage is a case in point).

The twist that postmodernism gives to all this is found in the ideas of cultural fragmentation and de-centered selves and identities. We've talked about all this before, particularly in our chapters on Durkheim and Weber, so I don't want to belabor the point. But, the entire idea of gender or race is modernist. It collapses all the individualities into an all-encompassing identity. However, a woman isn't simply female; she also has many identities that crosscut that particular cultural interest and may shift her perceptions of self and other in one direction or another. For us to claim one or more of these identities—to claim to be female or black (or male and white)—is really for us to put ourselves under the umbrella of a grand narrative, which includes very strong exclusionary tactics.

More to the point, the construction of centered identities is becoming increasingly difficult in postmodernity. Remember the connection between identity and culture? As we've seen, postmodernists argue that culture in postindustrial societies is fragmented. If the culture is fragmented, then so are our identities. The idea of postmodernism, then, makes the issues of gender and race very complex. Racial and gender identities may thus be increasingly difficult to maintain. The culture has become more multifaceted, and so have our identities. This means that we have greater freedom of choice (culture and structure are not as determinative in postmodernity), which we all think we enjoy, but freedom of choice implies that the distinctions between gender and racial identities aren't as clear or as real as they once were (remember that claiming and maintaining a racial identity is a cultural practice). Thus, social movements around race and gender are difficult to create nowadays. If our identities are fragmented, then emancipatory politics with its social movements is difficult to pull off. We looked at this idea specifically in our Weber chapter. The end result here is that the culture of race and gender becomes more vital than the structure.

Further, this multiplicity takes Du Bois' idea of double consciousness to a new level. For example, of her mixed racial heritage, Gloria Anzaldúa (as cited in O'Brien & Kollock, 2001) says,

> Cradled in one culture, sandwiched between two cultures, straddling all three cultures and their value systems, *la mestiza* undergoes a struggle of flesh, a struggle of borders, an inner war. . . . These numerous possibilities leave *la mestiza* floundering in uncharted seas. In perceiving conflicting information and points of view, she is subjected to a swamping of her psychological borders. (p. 567)

As I said, gender and race become more complex under postmodern considerations: the cultural boundaries aren't as clear, social movements become increasingly difficult to create, and life politics rather than emancipatory politics becomes the standard. I've heard it said, "Just when feminism made women's subjective experience important, postmodernism de-centered the subject." Yet, postmodernism also contains a value about identities. It may be a utopian vision, but it may also contain something we can hope for. Modern identities are strong and clear. We can talk about women's rights under modernity. But in doing so, we create a grand narrative that simultaneously excludes other identities, such as sexuality, religion, race, ethnicity, tastes, and so on. In postmodernity, the severity of structural and cultural differentiation may make identities difficult to maintain, but we should value complex identities, multiple voices, and local rather than grand identities.

Building Your Theory Toolbox

Conversations With Martineau, Gilman, and Du Bois

There are many interesting details of Harriet Martineau's life on the Web. One such site, http://cdl.library.cornell.edu/cgi-bin/moa/moa-cgi?notisid=ABK4014–0053–102, contains an article that was published in *Harper's* after Martineau's death. Using that site, the recommended sites below, and your search engine, answer the following questions:

- Why did Martineau write *Illustrations of Political Economy*? Why did the Society for the Diffusion of Useful Knowledge refuse to publish the work? Herbert Spencer read these works as a child. Who was an influence on both Spencer and Martineau?
- What is Unitarianism? Why did Martineau become an atheist?
- At one time, Martineau became incapacitated due to illness. She was offered, but refused, state aid. Why did she refuse state aid?
- Find the homepage for the National Organization for Women (NOW). What political agendas are prominent? How are these different than or the same as the issues Martineau and Gilman addressed?

One particularly interesting site for Charlotte Perkins Gilman is http://www.scribblingwomen.org/cgwallpaperfeature.htm. There you can listen to the play version of her book *The Yellow Wallpaper*. Using your search engine and recommended sites, answer the following questions:

- Who were Henry Ward Beecher and Harriet Beecher Stowe? How do you think being related to the Beecher family affected Gilman?
- Why did Gilman refuse to call herself a feminist? What did she prefer call herself?
- Why did Gilman's father abandon the family? How do you think that influenced Gilman and her work?
- How did Gilman earn her living after completing her training at the Rhode Island School of Design?
- Who is Edward Bellamy and how did he influence Gilman's work?
- A number of Gilman's poems can be found on the Web. Read and report on one of her poems. What was the poem about? What is the current state of the topic of the poem?
- Describe the Women's Club Movement: What were their concerns? What effects did they have? To what were most of these clubs organizationally related? How were black women treated by these clubs?

Of particular interest with W. E. B. Du Bois is his political life. Use your search engine and the recommended sites to find information on the following:

- Who was Booker T. Washington? What did he accomplish? What was his relationship to Du Bois? What were their points of agreement and/or disagreement? What was the Tuskegee Machine? What was the Niagara Movement? How was it formed? What did it eventually lead to?
- What are Black Nationalism and Pan-Africanism? What was Du Bois' relationship to these movements? Use your search engine to find out what is currently going on with Black Nationalism and Pan-Africanism.
- When Du Bois was editor of *The Crisis,* who made up the board of directors for the NAACP? Find *The Crisis* online (careful, typing in "the crisis" will pull up a variety of sites. Make sure you get the right one). What articles are in the current issue? How does the journal reference Du Bois' theory and thought? Find the Web site for the NAACP. What kinds of issues are currently being addressed?
- How did the U.S. military treat blacks during WWI? How did Du Bois respond? What resulted from Du Bois' work in this area?
- Who was Marcus Garvey? What was Du Bois' relationship with Garvey?
- How many organizations can you find on the Web that are in some way related to Du Bois? What different kinds of work are associated with his name?

Passionate Curiosity

Engaging the World (using the theory)

- Martineau is very much a daughter of the Enlightenment. She is particularly in tune with the Enlightenment's idea of progress and human rights. What did Martineau say was the basic human right? How did she measure a society's progress toward that right? If we used her methods of measurement, how would our society fare today? Are there other groups or situations that we could use as measuring rods today in addition to the ones that Martineau used (race, gender, and marriage)? If so, what benefits would be gained by using them?

- Go home and watch TV. Intentionally watch programs that focus on African Americans and pay particular attention to commercials that feature people of color. Using Du Bois' understanding of representation and Barthes' ideas of denotation and connotation, analyze the images that you've seen. What are some of the underlying connotations of the representations of blacks, Chicanos, Asians, and other minorities? How do you think this influences the consciousness of members of these groups?

- Evaluate the idea of standpoint or critical knowledge. If we accept the idea that knowledge is a function of a group's social, historical, and cultural position, then is this idea of critical knowledge correct? If so, what are the implications for the way in which we carry on the study of society? If you disagree with the idea of standpoint knowledge, from what position is true or correct knowledge formed?

• According to Du Bois, one of the ways cultural oppression works is by excluding the voices and contributions of a specific group in the history of a society. The Anti-Defamation League has a group exercise called "name five." The challenge is to name five prominent individuals in each category: Americans; male Americans; female Americans; African Americans; Hispanic Americans; Asian or Pacific Islander Americans; Native Americans; Jewish Americans; Catholic Americans; pagan Americans; self-identified gay, lesbian, or bisexual Americans; Americans with disabilities, Americans over the age of 65. For which categories can you name five prominent people? For which can't you name the five? What does this imply about the way we have constructed history in this country? In addition to trying this activity on yourself or a friend, go to the Biography channel's Web page (http://www.biography.com/ae). There you will find a searchable database of over 25,000 people whose lives are deemed important. Try each of the categories that we mentioned. What did you find?

• Martineau was very clear about the correlation between education and social progress. *The Chronicle of Higher Education* has a searchable Web site (http://chronicle.com). Go to the site and search for articles concerning minority enrollment in higher education. You might also search for articles that give an account of minority teachers and professors. Using this site, get a sense of what is going on with minorities in higher education. Are the percentages of enrollment going up or down?

Weaving the Threads (synthesizing theory)

• Gilman gives us a very specific theory of gender oppression. Compare her theory first with what Marx said about gender oppression and then with what Simmel said. How are these three thinkers similar and different on the issue of gender? Which do you think is more accurate? Why? How do the different theories create different ideas about how to bring about gender equality?

• Thinking about how Martineau, Gilman, and Du Bois talk about inequality, do you think there are some common features in the way in which both gender and race are oppressed? If so, what are those common features? What would they indicate about a general theory of domination? How would those common features fit in with what we've already seen about domination and inequality, especially from Weber and Marx?

• Compare and contrast Martineau and Weber on the issue of religion. In some ways, of course, they are two different approaches: Weber's is a theory of religious change and Martineau's a religious typology. But in what ways do they overlap? Do you think Martineau takes us one step further in the evolution of religion than did Weber? If so, what factors do you think are at work? How does Martineau's typology compare to what Durkheim expected in terms of the future of religion?

Further Explorations—Books

More than with any of our other theorists, you should read Martineau, Gilman, and Du Bois, rather than read *about* them. For Du Bois, I recommend that you pick up *The*

Oxford W.E.B. Du Bois Reader, edited by Eric Sundquist. It contains two of Du Bois' books in their entirety (*The Souls of Black Folk* and *Darkwater*) as well as a myriad of other important writings. For Martineau, I recommend that you start with her books *Society in America* and *How to Observe Morals and Manners.* For Gilman, start with *The Yellow Wallpaper,* and then move on to *Women and Economics, The Home, Human Work,* and *The Man-Made World, or Our Androcentric Culture.*

Each of these thinkers also published autobiographies (in fact, Du Bois wrote two, at different points of his life): For Du Bois, *The Autobiography of W.E.B. Du Bois: A Soliloquy on Viewing My Life from the Last Decade of Its First Century;* and *Dusk of Dawn: An Essay Toward an Autobiography of a Race Concept;* for Martineau, *Harriet Martineau's Autobiography;* and for Gilman: *The Living of Charlotte Perkins Gilman: An Autobiography.*

In terms of introductions to other neglected theorists, I recommend Howard Brotz's reader, *Negro Social and Political Thought, 1850–1920;* and *The Women Founders: Sociology and Social Theory, 1830–1930,* by Patricia Madoo Lengermann and Jill Niebrugge-Brantley.

Further Explorations—Web Links

For Du Bois:

http://members.tripod.com/~DuBois/ (Site maintained by Jennifer Wager; the Du Bois virtual university)

http://www.duboislc.org/index.html (Site maintained by The Du Bois Learning Center)

For Martineau:

http://cepa.newschool.edu/het/profiles/martineau.htm (Site maintained by the department of economics, New School)

http://www.webster.edu/~woolflm/martineau.html (Site maintained by Women's Contributions to the Study of Mind and Society)

For Gilman:

http://www.cortland.edu/gilman/ (Site maintained by The Charlotte Perkins Gilman Society)

http://www.womenwriters.net/domesticgoddess/gilman1.html (Site maintained by the online journal *Domestic Goddesses,* edited by Kim Wells)

CHAPTER 9

The Problem of Meaning and Reality—Alfred Schutz (Austrian, 1899–1959)

It is somewhat odd for me to include Alfred Schutz in a classical theory textbook. He certainly lived later than most of the people we consider, and his work didn't become well-known in the English-speaking world until the 1960s. However, the primary criterion for including someone in this book is her or his influence on contemporary theory, and there is an important segment of contemporary theory that we couldn't understand without Schutz.

In general, we could probably term this segment of theory "phenomenological sociology," but it is more informative to talk about two areas in particular: the social construction of reality and ethnomethodology. The social construction of reality is most clearly associated with Peter Berger and Thomas Luckmann. Luckmann, you may have noticed, published with Schutz and continued publishing his work after the theorist died. Berger and Luckmann are particularly interested in how reality is objectified. As we have seen, meaning isn't synonymous with the objective world and it is something that is attributed or created. Because meaning is produced, it can change; and because it can change, it is also precarious. So, how do people overcome the precarious nature of their reality?

Berger and Luckmann argue that intrinsic within the reality-producing process is objectification, and this objectification makes the created reality seem as if it carries the same ontological status as brute, material reality. This objectification process extends to the individual subjective experience through the use of language. Further, because humans are instinctually deprived, they are driven to internalize this objectively real culture. Thus, a subjective and truly precarious reality is transformed into an *inter*subjective one. Yet despite these processes that push for objectification, reality is still precarious for humans and must be maintained in consciousness. It must be continually replicated and guarded. People must engage in a willful suspension of doubt and must continue in routine, thus giving reality a taken-for-granted character. In particular, people have to engage in talk. Our talk with both anonymous and significant others provides a background of taken-for-grantedness—talk presents a world that is spoken of as if everything were real and therefore unquestionable. Talk, like ritual, also keeps before the consciousness the objectivity of social reality, thus making it an object of intentional consideration.

Berger and Luckmann's theorizing, and the idea of constructed reality, have produced an array of studies that deconstruct certain elements of what we take for granted. There are studies, for example, that take apart patriarchy and show how gender inequality is produced through sets of ideological assumptions. There are also studies that trace the historical construction of race or sexuality, demonstrating that race and sex could have been created differently. The most theoretically grounded of these studies find their base in Schutzian ideas.

The other approach, ethnomethodology, is principally concerned with the "methods" people use in everyday interactions that allow them to produce a sense of intersubjectivity and reality. There is a fine but important distinction between the social construction of reality and ethnomethodology: researchers in the constructivist vein are generally interested in how we construct an objective reality out of symbolic elements, and ethnomethodologists are concerned with social order as an achievement that is locally produced, *in situ,* right here/right now/in just this manner. Ethnomethodologists are concerned with the commonsense methods and reasonings whereby people in face-to-face situations produce a social order (event), which includes a sense of shared reality (facticity). An ethnomethodologist may in fact claim that the notion of an objective, cultural reality is itself reification.

Ethnomethodological research thus tends to focus on those patterned kinds of behaviors that allow us to leave unquestioned our cultural reality and to accept it in a taken-for-granted manner. One of the insights that comes from this perspective is the "et cetera principle." With the et cetera principle, people often refer to things using a shorthand method with the assumption that "I could go on," but that assumption is never questioned or proven. For example, when you are talking to someone, have you ever said "you know"? You know, like when you say something like, "You know how people lie even to themselves? Well, the other day . . ." "You know" functions as an et cetera device. It allows us to create a sense of shared, objective reality, even though if it were questioned, we would end in infinite regression (like when the two-year-old asks, "Why?"). Some important sociological thinkers in this perspective include Harold Garfinkel, Harvey Sacks, Don H. Zimmerman, Melvin Pollner, Hugh Mehan, and Houston Wood. Ethnomethodology has in turn influenced such contemporary thinkers as Anthony Giddens.

Alfred Schutz is one of those rare thinkers that move us to consider the wellspring out of which other concerns and ideas come. For example, many sociologists investigate race, and their concern is how race creates stratification—the unequal distribution of scarce resources. Schutz's ideas move us to a more basic level: how does "race" become meaningful and real to people? As Mary Douglas says, "Schutz's focus was the 'primordial foundation . . . that makes all understanding possible' (in Grathoff, 1989, p. 212), namely the lifeworld (*Lebenswelt*) or the world of everyday life. He aimed to show what makes the everyday world possible at all and how it in turn makes understanding possible, including scientific understanding" (p. 367).

Schutz in Review

- Alfred Schutz was born on April 13, 1899, in Vienna, Austria. He was raised an only child in an upper-middle-class Jewish family. As a child, Schutz began playing the violin, which became a lifelong interest.
- Schutz studied law and business at the University of Vienna, from where he received his law degree. While in Vienna, Schutz attended lectures given by Max Weber. Weber sparked Schutz's interest in social science and the problem of meaning. Schutz felt that Weber left the problem of meaning unanswered, and he combined ideas about consciousness and time from Henri Bergson and Edmund Husserl to create his own phenomenological theory of meaning.
- In 1927, Schutz became the chief financial officer for Reitler and Compan a Vienna banking firm. He continued to work full time for the bank, first in Vienna and then in New York, until the 1950s, when he took a full-time position at the New School. A major portion of Schutz's research and writing was actually done before he became a professor, which means that much of his academic work was done "part time," while working full time at the bank. During the years prior to his time at the New School, he typically worked 15 to 16 hours every day, including weekends. His wife, Ilse, assisted by transcribing Schutz's working notes and letters from his taped dictations.
- His major work, *The Phenomenology of the Social World,* was published i 1932. The fourth volume of his collected works was published in 1996, and a fifth volume is planned.
- On July 14, 1939, Schutz immigrated with his family to the United States
- In 1943, Schutz began teaching sociology and philosophy courses at th New School for Social Research; he served as chair of the Philosophy Department from 1952 to 1956.
- Alfred Schutz died May 20, 1959, in New York City

The Perspective: Social Phenomenology—Seeing Reality From a Human Viewpoint

> There will be, however, different opinions about whether this behavior should be studied in the same manner in which the natural scientist studies his object.... [W]e take the position that the social sciences have to deal with human conduct and its commonsense interpretation in the social reality, involving the analysis of the whole system of projects and motives, of relevances and constructs.... Such an analysis refers by necessity to the subjective point of view (Schutz, 1962, p. 34)

The problem of meaning: The social world is not synonymous with the natural world. For example, I am holding an exceedingly flat, rectangular object in my hand. But that 2½- by 6-inch object isn't all we see when I hold the $50 bill out to you (What do I have to do to get that $50? What does he want? How could I use that money?). Here's another illustration: There are two people standing in front of us. But we see more than those two discrete mammalian objects as they speak words that somehow create a binding relationship called marriage between them (Where does "marriage" exist? Does it exist in the words, or in some middle physical space between the two people?).

We could come up with myriad examples (such as the meaning of the flag, or a tree, or water, or sex) because our world is made up of meanings, not simple, physical objects. The social world is a meaningful world, and meaning is always something other than the thing-in-itself. Through meaning, we turn paper into money and dirt into farmland and biological sex into making love or into rape. The important thing for us to note is that meaning is attributed, or laid on top of, objects and experiences.

There is little doubt that the social and natural worlds are different. What *is* in doubt is the relationship between them. Some people, such as natural scientists, assume that we can create a language that provides a window into the natural universe. The relationship between the universe and this language would be immediate. Others, like philosophers, argue incessantly about the relationship. Some philosophers posit that there is a necessary relationship between the social and physical worlds—that every element of the social world is built up from and based upon the physical. Other philosophers say that the human world is absolutely unique, with no connection to the outside world at all.

The existence of meaning unquestionably creates a problem—is the human world real in the same way the physical world is? Yet, while this may be a dilemma over which philosophers argue, there are vast numbers of people for whom this problem of meaning doesn't exist at all. In fact, most of us readily accept and live in our world without question. We accept our world as given and never see or experience the problem of meaning. And that is the world in which Alfred Schutz is interested.

> I am afraid that I do not exactly know what reality is, and my only comfort in this unpleasant situation is that I share my ignorance with the greatest philosophers of all time. (Schutz, 1964, p. 88)

Schutz, of course, recognizes this problem of meaning, but he isn't concerned with it directly. In some ways, perhaps, he recognizes that the question is unanswerable, because it is impossible to ever know for certain whether we are thinking outside of our own culture. Regardless of this more philosophical issue, Schutz isn't concerned with issues of ontology because "the person on the street" isn't concerned with them. Schutz is interested in the natural attitude that people have about their lifeworld and how they experience that world. This emphasis on conscious experience and the naturalness of experience comes from phenomenology.

The phenomenological method: **Phenomenology** originated with Edmund Husserl (1859–1938), a German philosopher with whom Schutz became friends. Husserl formed the phenomenological approach in response to both empiricism and rationalism. *Empiricism* is the philosophy that all knowledge comes from and is tested by sense data gathered from the physical world. *Rationalism*, on the other hand, posits that reason and logic are the tools through which true knowledge is attained. Reason can lay hold of truths that exist beyond the grasp of sense perception.

Phenomenology cuts a middle road between the two by focusing on human consciousness. Consciousness is more than sense data. Husserl argued that the only things that can exist for humans exist in consciousness. That is, only those things of which we are intentionally aware can exist for us. This isn't quite as esoteric as it sounds. Think of it this way: the world of the frog is very different than the human world. Part of the difference is based on sense perceptions (like the frog can only see moving objects), but, more importantly, much of the disparity is founded on the brain and the process of awareness.

However, while consciousness for Husserl is more than sense data, it is less than reason. In reason, humans use cultural tools such as logic and categories to understand and make sense of the world. Husserl saw those cultural tools as obstructions to investigation. He wanted to set those aside and investigate pure consciousness through epoché, or transcendental phenomenological reduction.

Epoché is a Greek word meaning to stop or cease, or to suspend judgment. It also refers to a position in space or set point in time. As a methodological device, Husserl was asking us to suspend our belief in the reality of the human world and to direct our investigative view to our consciousness of the world alone. Husserl felt that we could get to pure consciousness through bracketing. In **bracketing** we set aside or disconnect all of the accumulated experiences and reflections concerning the world of objects, all the cultural tools through which the lifeworld is recognized, organized, and understood as meaningful. Phenomenology, then, emphasizes immediate experience (or the phenomenon of pure consciousness), apart from all assumptions, language, or theories.

The reason Husserl wanted to bracket and set aside all cultural tools of understanding is because he argued that reality and truth can be found only in pure consciousness. Unlike thinkers who located social order, meaning, and reality external to people, Husserl believed that the basis of all human existence is in the individual consciousness. Husserl thus wanted to make possible "a descriptive account of the essential structures of the directly given." In epoché, a researcher suspends belief in the reality of this world and directs attention exclusively to her or his own consciousness of the stream of experience so that its reality can come forth.

> As we proceed to our study of the social world, we abandon the strictly phenomenological method. We shall start out by simply accepting the existence of the social world as it is always accepted in the attitude of the natural standpoint, whether in everyday life or sociological observation. (Schutz, 1967, p. 97)

Let's use an illustration to understand what Husserl is getting at. Let's say you are walking across campus. There are innumerable objects and things happening around you, but in a real way these don't exist for you because you don't notice them. Suddenly you notice a woman. How will that woman appear to you? In other words, how will you see and understand her? You might notice her clothes, hair, figure, or walk and compare it to the cultural understanding you have about women (gleaned from media, parents, peers, and prior experiences).

What Husserl is asking you to do is to bracket all that cultural baggage. He is arguing that there is a pure experience waiting for you underneath all of your preconceived ideas. He wants you to transcend the culture and reduce the objective phenomenon to its subjective essentialness. In other words, see the woman without any preconceived ideas or categories of what constitutes a woman. It's important to note that the very fact you notice "clothes, hair, figure, or walk" with reference to "female" is itself a cultural act. If you could put all of this aside, you would be able to experience the woman as directly given and be able to create an account of her essential structure.

Schutz took up the general emphasis of phenomenology. He argues that everyday life, apart from scientific or philosophical theories, is the most important focus of analysis. However, there is also a difference between Husserl's phenomenology and Schutz's. Husserl wanted to reduce everything to pure consciousness and direct experience. But Schutz isn't convinced that the lifeworld as given is that of pure consciousness (and it's extremely difficult if not impossible to get to, as you probably observed from our exercise).

The natural attitude: The **lifeworld** for Schutz is exactly that world that Husserl wanted to bracket. It is the world as it is experienced immediately by the person. It is a world determined by culture and filled with meaning. Schutz takes the human world as it presents itself to us as his subject of analysis. It is the commonsense world of everyday reality. Obviously, this theoretical focus can exist and is exciting because of the problem of meaning. This issue of the natural attitude toward the lifeworld wouldn't be remarkable if the relationship between the natural and social worlds wasn't a puzzle. So we aren't getting away from the question, we're just approaching it from a different position or issue. And this is a question that can be answered. Rather than asking about reality per se, Schutz is asking about *how people experience their lifeworld as real.*

People experience their world as real primarily by taking a natural attitude toward it. The **natural attitude** is one where the world is taken for granted and doubt is suspended. Schutz ironically notes that in the natural attitude, we have a special kind of epoch. Remember that in Husserl's epoch, doubt about reality is suspended as a methodological device. In Schutz's concept of the natural attitude, rather than bracketing our belief in culture, we *bracket any doubt* in the reality of culture. Doubt is suspended even in the face of the different worlds in which humans exist. We know that other people believe different things about the universe. These beliefs are so different that they actually constitute an entirely different world than ours (the Yanomamö, for example, believe that there are four different layers to the universe).

Yet rather than seeing these different realities as reason to doubt our world, we willfully suspend doubt and pragmatically move through life. We wouldn't be able to live socially or individually if we didn't. And here we can begin to see a bit of what Schutz is after. The natural attitude is a *device* through which our world appears to us as real. It can't appear to us as such otherwise, and every culture throughout time has been made to appear real by just such an attitude.

To think like Schutz, then, is to understand the problem of meaning. The human world is a meaningful world, which implies a problematic relationship between the world (human) and the universe (physical). But to think like Schutz also means to be concerned with how the world is taken, perceived, or experienced as real, rather than entertaining direct questions of ontology (how things exist).

Creating meaning: But to see through Schutz's perspective is not only to understand the problem of meaning. Schutz is also concerned with meaning itself. Max Weber was also concerned with meaningful action. Weber based quite a bit of his understanding about society on his typology of action. According to Weber, there are four types of action: instrumental rational, value rational, traditional, and affective. In Weber's eyes, both traditional and affective actions lie very close to the borderline of what we can call meaningful. The rationale for this line of reasoning is that these actions are either performed out of habitual (traditional) or emotional (affective) response. They don't involve much in the way of thinking and planning.

Schutz criticizes Weber's understanding of meaning attribution. Of particular concern for Schutz (1967) is the way in which Weber defined meaningful behavior: "when Weber talks about meaningful behavior, he is thinking about rational behavior and, what is more, 'behavior oriented to a system of discrete individual ends' *(zweckrational).* This kind of behavior he thinks of as the archetype of action" (pp. 18–19). So, for Weber, the meaning of an action is determined beforehand. With Weber, action is meaningful if it is motivated by explicit goals. And motivations and goals occur before the action.

On the other hand, Schutz makes the point that meaning is always associated with individual consciousness. It's kind of obvious when you think about it. If we aren't conscious of something, then we aren't aware of it; and if we aren't aware of it, it can't have any meaning for us. Pretty simple, but it is also very revealing in terms of how meaning is produced. We can only be conscious or aware of something that already exists. If meaning is linked to consciousness, and we can become conscious only of things that already exist, then meaning attribution is always after the fact. We can think about future events, but when we do, we actually think about them as already past. For example, when I think about completing this book, I think about it as already completed.

Thus, to think like Schutz is to realize that meaning isn't determined beforehand. Meaning is always produced after the fact through a backward glance. Meaning occurs through the conscious act of an individual picking out from the stream of experience a particular object as the focus of attention.

I want you to notice something about the above statement (be conscious of it: turn your attention to the past statement). Schutz understands meaning attribution as an individual, subjective process. Thus Schutz's understanding of meaning attribution is not only different from Weber's, it is also different from Mead's. Mead argued that meaning is produced in social interaction. It emerges from the three phases of the speech act: giving a cue, responding to the cue, and responding to the response. Thus for Mead, meaning is social and flexible. But for Schutz, meaning is subjective and is much less likely to change. One of the things this emphasis implies is that Schutz is also concerned with the problem of intersubjectivity. When we talk of subjectivity, we are referring to something that exists in the consciousness of one person. Intersubjectivity, then, is the linking or interrelating of two consciousnesses or subjectivities.

Intersubjectivity refers to the sense that we experience shared worlds, that the emotions and thoughts that I experience are understood and shared by another.

Schutz explains a number of ways in which intersubjectivity is produced, but his basic argument is that it is produced through language, or what he refers to as "stocks of knowledge." Language is social and by its very nature has the same meaning for the sender and the receiver—it is thus reciprocal: it means the same to you and to me. And it is through this feature of language that we can experience the other person. Language, because it is social, creates a sense of shared inner worlds.

Before concluding this introduction to Schutz, it would serve us well to talk a bit about how he characterizes the scientific viewpoint. Recall that science is based upon certain assumptions: the universe is empirical, operates according to law-like principles, and humans can discover those principles and use them to control the environment of their lives. Collapsing these together, we can say that science is based upon an objectifying point of view. That is, scientists look at phenomena as if they are not part of what they are observing. That position is more intuitively acceptable if the scientist is viewing an atom than if she or he is looking at society. How the researcher can stand outside of society while the individual is a social being is the sticky problem that every social scientist must face.

Schutz is cautious about the scientific perspective in "social science." It isn't so much that he denies scientific knowledge about the social world; rather, he argues that all knowledge of society must begin with the subjective actor. Schutz characterizes subjective meaning attribution as first-order constructs and scientific meaning as second-order constructs. According to Schutz, all knowledge about the world is constructed. In labeling common knowledge as first-order, he is recognizing that everyday knowledge is more immediate to the human experience in the world. Second-order is more abstract and subject to other considerations, like the laws of logic. These second-order constructs are basically the same as Weber's ideal types. In seeking to understand the social world, the scientist must construct concepts with an eye to the subjective interpretation of the actor. In other words, our ideas about what is going on in any situation must match how someone in that situation would interpret the same thing. If we ignore this tenet of subjective relevance, we run the risk that "the subjective world of social reality will . . . be replaced by a fictional non-existing world constructed by the scientific observer" (Schutz, 1964, p. 8).

Taken together, then, Schutz's perspective focuses on meaning, and there are three main concerns that result from this focus. The first issue surrounding meaning is that the social world and the physical universe are not the same. Therefore, the reality of the social world is problematic. However, Schutz isn't concerned with objective ontology of the world, but rather how people experience a meaningful world as real. Schutz is interested in the lifeworld of everyday existence and how it appears real. The second issue that comes from focusing on meaning is its production. Schutz argues that meaning is produced as individuals become conscious of certain elements in the flow of experience through a backward glance. Meaning is thus created after the fact and exists in the individual consciousness. The third issue emerges from the second: if meaning is subjective, how can people have a sense of intersubjectivity? To think like Schutz, then, is to be concerned with how a sense of meaning, reality, and intersubjectivity are produced. And to think like Schutz is to have the position that *sociology must begin with the subjective experience of the individual.*

The rest of this chapter is divided into three main sections. In *Being Conscious in the Lifeworld,* we will see that human beings have a particular way of being aware of the world around them. And it is through this special awareness, called intentionality, that meaning and reality are formed. Yet, even though we as individuals must become intentionally aware of the world for it to exist, the lifeworld of which we are aware is socially fashioned. We consider this important point in *Ordering the Lifeworld—Creating Human Reality.* And because our reality and meaningful world are social, it means that we are intrinsically connected with others. But what we will see in *Connecting With Others* is that the way in which we are aware of and connect with other people varies by anonymity. All through our discussion, I would like you to be aware of five important subthemes: consciousness, objectivity, intersubjectivity, ordering, and continuity. These ideas weave themselves throughout Schutz's works and our discussion as well.

I also want to advise you about something in advance. We are going to be building our concepts layer by layer throughout the chapter. In other words, I will introduce a concept in one section and then return to it again in a later section. So you will see concepts repeated and expanded. There are many good reasons for doing this, but the most important is that this approach to Schutz will help us understand his concepts and theory more fully. So, remember to pay more attention, not less, when you see an idea repeated—we are building our understanding step by step.

Being Conscious in the Lifeworld

> [T]he biographical articulation [of the course of the day] is *superimposed* over the rhythm of the day. If I reflectively attend to past periods of life, summarily to survey them and examine their meaning, then I obtain monothetically within my grasp, in such *post hoc* interpretations of the greatest span, huge stretches of polythetically built up courses of days (how has it "come about" that I am a drunkard?). (Schutz & Luckmann, 1973, p. 57)

Lived experience and intentionality: A common experience for most of us is "telling someone about our day." We come home from work or school and a loved one asks, "How was your day?" At that point, we will tell them a story about our experiences. We even tell *ourselves* stories about our experiences in a day, a week, or a lifetime. However, what we experience and how we experience it are quite different than our telling of those lived experiences.

We live life as a stream of undifferentiated experiences, in what Schutz terms "internal time-consciousness." For Schutz, lived experience is like a river that flows through a valley: it has neither beginning nor end and every drop of water is connected to every other drop of water by a never-ending flow. That's how our day exists in its natural state. One experience is connected to every other experience by our continuing movements through time.

Telling such a tale about our day in its natural state would be exceedingly boring (and it would take an entire day to tell it) because it would lack meaning: "I moved my left arm forward at the same time I moved my right leg and foot. As the arc of my left arm and right leg reached its highest point, I began to move them back and my right arm and left leg forward." Yuck. This kind of pure experience is called **durée** (pure duration or unmarked time). And pure experience is without meaning. To tell the story of our day, we must bring it into our consciousness. For Schutz, meaning is a constant construction that can only be understood by analyzing this tension between lived experience and consciousness (what happened versus what we're going to make of it). To understand and explain the tension between consciousness and lived experience, Schutz employs Husserl's concept of intentionality.

From a phenomenological point of view, reality and meaning come to exist for humans through **intentionality**: turning our attention to an already lapsed experience. The concept of intentionality comes from the work of Franz Brentano and it refers to a kind of mental exertion that is directed toward an object. Intentionality is the foundation of human consciousness and meaning, according to Schutz. One of the keys to understanding intentionality is that it is always aimed at an object, experience, or emotion that has already occurred. Meaning construction, then, always takes place after the fact.

For example, we become conscious of being awake in the morning only after we are already awake. Or let's use another example: You're walking home from campus late at night. As you pass a dark alley, you think you hear a noise and see movement. You catch your breath and your heart races. When do you become aware that you are afraid? You become conscious of fear *after* you experience "fear" (that particular set of reactions and sensations). As Schutz (1962) says, "there is no such thing as thought, fear, fantasy, remembrance as such; every thought is a thought *of*, every fear is fear *of*, every remembrance is remembrance *of* the object that is thought, feared, remembered" (p. 103). Human consciousness, then, has an intentional characteristic about it. We not only feel pain; we can think about that feeling and intend ourselves toward it.

The intentional object, then, is that toward which the person is directing her or his attention and intended actions. But this attention is just the beginning point for a series of related intentional objects and actions. Each intentional object appears to us in an "intentional horizon of later phases of activity," each of which we see to its completion or to a point where we say to ourselves "and so on."

For example, outside my home it's currently raining hard with high winds. I perceive rain and wind as threats to my house, because there is another object that has been the subject of my intentionality: the large branch that is hanging by a thread from one of the neighbors' trees. The wind and rain become real and meaningful to me as I see them in a matrix of other intentional objects that include past (last year our house was hit hard by falling tree limbs) and future objects of intention (I need to call the tree service; if I don't, I'll have to call the insurance company; how much is my deductible?).

The notion of intentionality for Schutz also includes the idea of "noema." The *noema* is like an intermediate set of mental properties that would make a target object suitable for intentionality, even if the exact object does not exist. In other words, we can take a thought or belief as an intentional object. We don't necessarily need a physical object. What this means is that it is possible for Santa Claus or the Goddess to be an object of intentionality and thus become a real object for

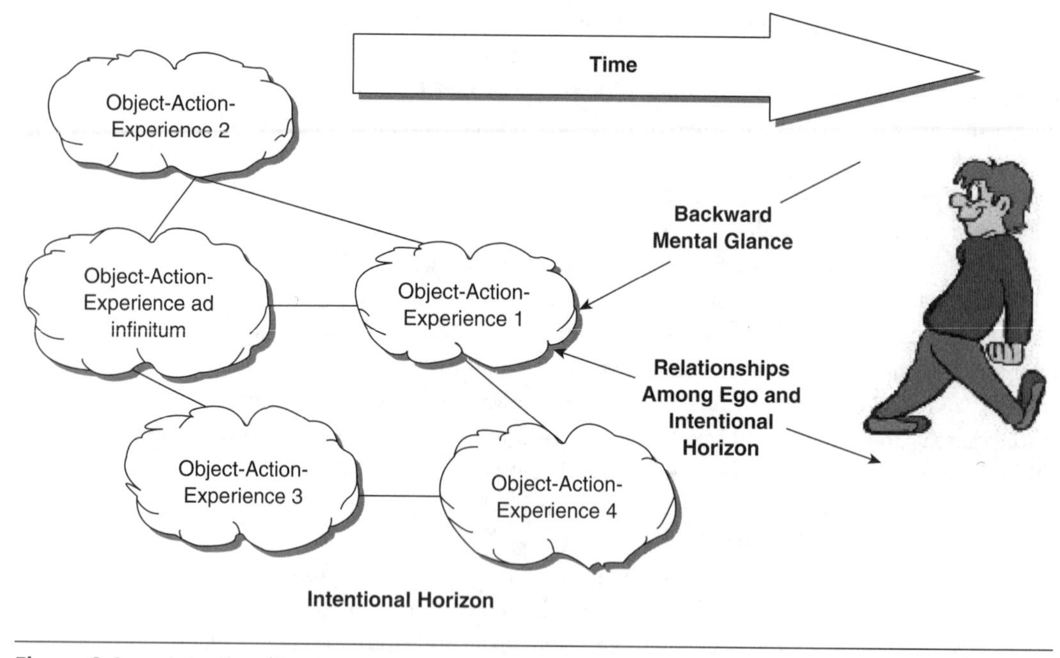

Figure 9.1 Intentionality

people. This conceptual movement, of course, implies that entire symbolic worlds can exist for humans, worlds that have no basis in brute or physical reality.

The concept of intentionality also implies a relationship between ego and object; or the I-Pole and the Object-Pole of intentionality. In other words, consciousness isn't simply attention, it also contains relationships. It is thus awareness with purpose. We relate our self to an object through possible actions and goals. This relationship also implies that all meaning is subjective—it exists internally to the individual.

I have illustrated these three characteristics of intentionality in Figure 9.1. You will see that time is moving forward, but meaning attribution moves back through time. Notice also that when an object becomes the focus of our intentional gaze, we see not only it but also all other objects and possible actions associated with it. This horizon forms the meaning context. And, finally, we can see that intentionality creates pragmatic relationships among the intended objects, actions, goals, and so forth, and the individual ego.

However, this idea of subjectivity does not lead us to solipsism, an extreme version of idealism where the self can know nothing except its own thoughts and states. Nor does this notion of intentionality leave us with only internal processes, which would mean that Schutz would be interesting just to philosophers and psychologists. No, Schutz's ideas of meaning, consciousness, and intentionality lead us quite naturally to sociological concerns and problems.

Here is the important bridge that Schutz forms from phenomenology to sociology. Intentional objects, objects of which we are aware and which are meaningful, are cultural objects, according to Schutz. He maintains that the ego becomes meaningfully conscious through an already existing lifeworld of culture. And our subjective intentionality confirms this to us as we orient ourselves pragmatically to this world. In other words, in the natural attitude, I not only take the world of culture as a given, I organize my behaviors as if culture is real. And when I do, I get the results I expect, thus confirming my belief in that cultural world. For example, when I act as if the $50 bill is real, the reality of the culture is confirmed. Reality, then, is reflexively constructed.

All cultural systems are built up through **reflexive construction**. That is, they are produced from their own elements. There are countless ways in which this is true, but let me give you just one example. The evidence for the reality of any system is always reflexively provided. Consider the incident of two automobiles colliding. What is it? Well, you might say that it is an accident. But what does that simple statement assume about the universe as a whole? Accidents can only happen if there is no such thing as fate or spiritual forces behind the universe. In other words, accidents can only happen in a natural world. But let's suppose that the driver of one car is a born again, evangelical Christian. What does the event then become? It's "God's will." God is either testing or chastening the individual.

Notice how the incident *becomes* what it is *assumed* to be. The incident becomes either an accident or an act of God, depending on the assumptions of the people experiencing or viewing it. Once the definition is applied, the meaning of the incident is taken as proof that that is how the universe works. The person who calls it an accident will say to herself and others "See? S*** happens." On the other hand, the Christian will see proof of God's hand in their lives. Either way, the collision is used to legitimate an already existing reality system or its definition, as proof for that system is provided by that self-same structure. I'm not just singling out religion here. How does science prove itself true? Science proves itself by using the scientific method. An odd sort of proof system, isn't it? All elements of our culture become real reflexively.

Ordering the Lifeworld—Creating Human Reality

Reality as social order: In the middle of the Southwest desert, there is a place called Four Corners. It is the only place in the continental United States where four state borders come together. Arizona, New Mexico, Colorado, and Utah all touch there. There are no rivers, mountains, or other natural boundaries there. People act really funny when they visit the place. They spread out on all fours with a hand or a foot in each state and have their picture taken. They look like giant spiders with smiles.

Let me give you another illustration before we get to our subject matter. I work in Greensboro, North Carolina. The population of the area is somewhere over 100,000 people. As with many Southern cities, we have a noticeably large population of African Americans. It seems pretty easy to tell who is African American and who isn't. Yet if we took every adult living in the area and lined them up from the lightest skin tone to the darkest, what do you think we would see? Chances are we would see a gradual gradation of color with no clear breaks among groups.

What do these two illustrations have in common? In both cases, we see the use of symbolic boundaries to order the human world. As Robert M. Pirsig says in *Zen and the Art of Motorcycle Maintenance,*

> [T]here is a knife moving here. A very deadly one: an intellectual scalpel so swift and so sharp you sometimes don't see it moving. You get the illusion that all those parts are just there and are being named as they exist. But they can be named quite differently and organized quite differently depending on how the knife moves. (p. 66)

Of course, what we are in the presence of in my two illustrations is what Eviatar Zerubavel (1991) calls a "social scalpel."

We live in an ordered world. But the order isn't a natural one; it is a cultural order, which is why people in different cultures live in different worlds. Sometimes we will use natural borders, like the Rio Grande between the United States and Mexico. But they don't in and of themselves order our world. The world is ordered symbolically through what Schutz refers to as stocks of knowledge at hand.

Ordering through stocks of knowledge: **Stocks of knowledge** are the sum of known rules, recipes for action, ideas, typifications, interpretive schemes, and so on that is generally available to all of us. In short, it is every piece of cultural knowledge that we have at our disposal for knowing about and negotiating our way through the world. It's like the toolkit you might have in your garage, or the sewing kit you have in the closet. In the kit, you have what you need to work on a car or sew a shirt.

> Now, to the natural man all his past experience are present as *ordered,* as knowledge or as awareness of what to expect, just as the whole external world is present to him as ordered. Ordinarily, and unless he is forced to solve a special kind of problem, he does not ask questions about how this ordered world was constituted. (Schutz, 1967, p. 81)

These stocks of knowledge are fashioned through experience. The largest part of our stocks of knowledge is formed through what has already been experienced, and most of what we have experienced has been experienced by others. We learn our basic stock of knowledge through families and peers. This knowledge, then, is collective and is held to be available to everyone who belongs to us. And because it is there for everyone, it is seen as objective and is taken for granted. By the way, the phrase "everyone who belongs to us" acts as a kind of technical term for Schutz. He employs it to emphasize three ideas that we see repeated throughout his work: his concern for the everyday lifeworld; the fact that culture is objective because it is not dependent upon individuals but is available *to all;* and that stocks of knowledge are group-specific—they are available *to us as a group.*

The most important element found in one's stock of knowledge is the typification. **Typifications** function like average expectations. They are built up from past experiences, create a general type upon which future anticipations can be based, and they ignore individual and unique characteristics. Typifications are applied to situations, objects, others, and even to our self.

Here's how it generally works. Survival is a group problem for human beings; we survive because we are social. In attempting to solve problems of survival, we work collectively. As we come up with a solution (like how to plant food or socialize babies) that works for the group, we habitually behave in the same way in order to solve the same problem. That way we don't have to come up with new solutions all the time.

As we collectively act in this habitualized manner, we begin to see the behaviors of the individuals as *types* of behaviors, rather than *individual* behaviors. And we begin to see each other as social types, not individuals, filling positions and acting out scripts. These typifications come to be held reciprocally so that we can survive more easily as a group. Through our mutually holding and using these typifications, they come to be seen as objective, living outside of the individual. And these reciprocally held typifications produce the relations that we have with other people. How do you relate to the person standing in front of your class? Or, even closer to home, how do you relate to your significant other? We relate through the typifications of gender, spouse, professor, and so on. We even understand our self as a type of male or husband or musician.

People rarely examine their stocks of knowledge. They are there, and as long as they work, people leave them alone. In other words, as long as recipes for action generate satisfactory results and as long as the beliefs give satisfactory explanations, the stock of knowledge will remain intact. When a problem does arise and the relevant stock of knowledge can't explain it or doesn't provide a script for action, the stock is then examined. But it is examined pragmatically: knowledge is only examined enough to suffice.

In other words, most people are not interested in constructing a coherent, consistent, or clear body of knowledge. The typical person on the street is only interested in what works. Even in the face of a problem, we don't begin to doubt the reality of our culture. In fact, according to Schutz, we aren't all that efficient or rational when a problem does arise. When faced with a problem, we don't question the world as it exists; we simply search for a solution that suffices. We are only interested in the pragmatics of the moment.

In the natural attitude, the self is experienced as the working self, working at and completing projects. So, for example, my knowledge of the human body is sketchy and incoherent at best. But that really doesn't matter to me. What is important is that I know what to do to get a problem fixed: go to the doctor. The same is true about my knowledge about computers or botany. What matters for us is having sufficient knowledge to make our way pragmatically through this life.

> Through every "problem's solution" something "new" becomes consequently something "old." (Schutz & Luckmann, 1973, p. 231)

New or problematic situations are thus built upon already existing stocks of knowledge. Schutz gives the example of a dog. We have a typified image of a dog—it is four-footed, wags its tail, and barks. When we encounter something new or not contained in the type, we don't reject the type but simply modify it. We add a new element to our typification of dog: dog is four-footed, wags its tail, barks, and bites.

The ways in which we modify our typifications and stocks of knowledge help to create a sense of continuity in the lifeworld. Our meaning context for actions remains basically the same. We either subjectively experience those portions of the objective stocks (that which we have learned from others) as they pragmatically fit, or we modify them as little as possible. We thus synthesize various elements of the stocks of knowledge to form contexts of meaning for our actions. For example, when you began college, you had a typification of teacher and classroom. But as you brought those into the university setting, they were modified because the teachers are now professors and the expectations of the classroom are different. And you continue to modify and synthesize them as you encounter further differences.

Each synthesis can be added to other syntheses, and the meaning-context grows larger with every new experience. Schutz (1967) argues that "this constitution is carried out, layer by layer, at lower levels of consciousness no longer penetrated by the ray of attention" (p. 77). These different meaning-contexts form our stock of knowledge, are always taken for granted in terms of their existence, and provide an ever-expanding context for experience.

Social distribution of stocks of knowledge: Stocks of knowledge are socially distributed. There is a general stock of knowledge that constitutes the culture of any given society at any given time. But the actual stock of knowledge at hand differs according to the particular place an individual occupies in the social structure. For example, my stock of knowledge is different than the mechanic who works on my car. The general way these differing stocks are transmitted and known is through the vocabulary and syntax of everyday language. The way we talk about things in the everyday world of our social groups contains a language of named things and events. Each name contains typifications that refer to things that the social group deems relevant.

Let's take a commonplace example. I hang out with a group of people who drink beer. There are some beers I really like and others I don't. And there is a stock of knowledge we use to convey our likes and dislikes. So when I take a drink of beer and say, "that has a sweet malty nose," I can share that experience with others in my group, and they would understand what I mean, even though they might not have experienced that beer yet.

What would happen if I gave you a taste of this beer? Would you tell me about the "sweet malty nose"? Perhaps not. If you wouldn't, why wouldn't you? Is there something wrong with your taste buds? Well, presumably, there isn't anything wrong your taste buds. The reason you wouldn't tell me about or even taste the malt is that we interact with different kinds of groups around beer. In my group, we've accessed a stock of knowledge, a language, to convey these tastes and you probably haven't. So we can use that language to share our personal beer experiences with the group, thus creating a group-specific world and intersubjectivity. Notice something important here: Because we have language to talk about a sweet malty nose, it exists for my group and it exists for me; but this sweet malty nose does not exist for you. You haven't divided the beer world up like that. It isn't really the beer that contains the experience; the experience is produced in my group through our language.

This is a pedestrian example and is of little consequence. But there are groups and divided stocks of knowledge that are important, like race, class, and gender. Each of these groups and subgroups has within it a different stock of knowledge. And these different stocks create different worlds for each of these groups to inhabit.

The work of Pierre Bourdieu (1984) may help us see how stocks of knowledge based on class create different lifeworlds. Bourdieu argues that one of the things class position affects is our level of cultural capital. Cultural capital refers to the informal social skills, habits, linguistic styles, and tastes that an individual demonstrates, and is thus similar to Schutz's stocks of knowledge. Bourdieu contends that cultural capital is distributed differently through education and what he calls the "distance from necessity," both of which are ultimately functions of class position.

Culture, in particular language, is standardized through the education system. Schooling is used to impose restrictions on popular modes of speech and to propagate a standard language. So all through your education experience, teachers have been changing the way you talk and write, forcing you into a standard mold. This intent of education is nowhere seen more clearly than in the debates surrounding Ebonics, or African American Vernacular English (AAVE).

Beyond this general function of education, higher levels of education are associated with a particular kind of language and thinking. Bourdieu argues that schooling provides different linguistic and conceptual tools to individuals. The amount of scholasticism (simple knowing and recognition of facts) and classicism (critical and aesthetic knowledge based on the classics) required varies by the level of schooling. At the lower levels of education, the degree of scholasticism required is high; at the higher levels, classicism is high.

In other words, in your primary and secondary education, you were basically taught facts and your job was to memorize them. But as soon as you hit college, things began to change. You are no longer asked to accept things at face value or as simple issues. You begin to see things in terms of abstract relations. For example, as a school child, you might hear a piece of music and characterize it as boring, adult, or elevator music. If you have a bit of knowledge, you might say that it is classical. But as the result of taking a music appreciation course at the university, you can now understand that piece as baroque and see it in terms of its relation to the evolution of Western music. Thus, individuals with higher education tend to be disposed to see additional levels of meaning in objects, to understand them in complex sets of relations, and to classify and experience them abstractly.

In addition to education, cultural capital is differentially distributed through *distance from necessity*. This concept gets at how close we live to only meeting the basic needs of life (food, shelter, and clothing). Distance from necessity enables people to experience a world that is free from urgency and to practice activities that constitute an end in themselves. This ability to conceive of form rather than function—aesthetics—is dependent upon "a generalized capacity to neutralize ordinary urgencies and to bracket off practical ends, a durable inclination and aptitude for practice without a practical function" (Bourdieu, 1984, p. 54).

"Practice without practical function" is a provocative way of capturing these class differences. People in the upper classes can spend their time doing things that have no practical outcomes. They are done for the pure enjoyment of doing them. I know a man who fishes. But he only fishes to have fun. He spends hours upon hours tying his own flies and reading books and seeking the prime fishing spots. Yet he always sets the fish loose and never eats them. Because of his distance from necessity, my friend is more concerned with the aesthetic experience of fishing. Thus, the upper classes prefer abstract art; lower classes tend to like representational art. Upper-class members prefer music that is complex and provocative, while the lower classes generally prefer music that one can dance or sing to. Distance from economic necessity implies that all natural and physical desires can be sublimated and dematerialized. The working class, because it is immersed in physical reality and economic necessity, interacts in more physical ways than the distanced elite.

Taken together, we can see how the level of education and distance from necessity create distinctly different stocks of knowledge based on class position. It isn't simply that the upper classes use fancier words; class-based stocks of knowledge determine how the world appears to us. The greater the level of education and distance from necessity, the greater will be the tendency to see the world as contingent, abstract, and complex. Conversely, the lower the level of education and the closer the distance from necessity, the more likely will be the propensity to see the world as certain, concrete, and straightforward.

Individual ordering through relevance structures: Beyond the differences based on status or class, our particular social group's stock of knowledge and lifeworld is further ordered through an individual's structures of relevance. **Relevance structures** are made up of those contexts or domains that the individual finds important through her or his multiple interests and involvements. There are three different kinds of relevance structures, all of which are interrelated: thematic, interpretational, and motivational.

Thematic relevance simply refers to something that becomes the object of our intentionality. Remember that it is intentionality that creates meaning and a sense of reality. Different parts of our stock of knowledge may become thematically important to us—and thus more meaningful and real—through imposition by unfamiliarity (like your first date) or social pressure (like the time you were first offered a cigarette), or they may become relevant because we voluntarily take them up, or because we are presented with a hypothetical case (as when you consider the possibility of time travel).

Different elements of the stock of knowledge are also routinely important for interpretational activities. As our life course brings different objects into our world, those elements of our stock of knowledge needed for attributing meaning become relevant. This generally happens automatically. But when a match cannot be made between an already constructed meaning-context and an observed phenomenon, the individual becomes "motivationally conscious" and begins the "process of explication." The process continues until the problem is sufficiently solved. And, as we've seen, this solution gets appended onto the already existing typifications, recipes, and so on.

We also make certain elements of the available stock of knowledge relevant due to our individual motivations. There are two kinds of motivational relevances, "in-order-to" and "because-of." In-order-to motivations are explicitly linked to personal projects. You have the personal project of getting a college degree. In order to complete that project, there are different arenas in the available stock of knowledge that you were motivated to make relevant, such as curriculum requirements and financial aid. Thus, at my university, you would access knowledge about GEC (General Education Core) and CAR (College Additional Requirements). If you attended this university, these elements in the available stocks of knowledge would

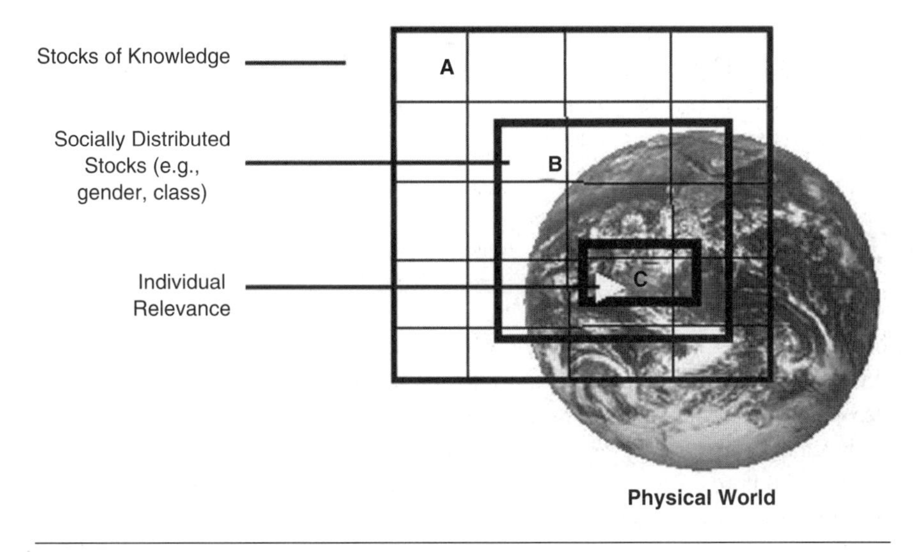

Figure 9.2 Ordering the World

become the subject of your intentionality and would be meaningful and real, and they would order your lifeworld.

Because-of motivations are linked to attitudes. According to Schutz and Luckmann (1973), "an attitude consists in the readiness to steer toward particular goals and to set in motion certain modes of conduct," and they contain "typical expectations concerning hypothetical events which appear more or less 'vital'" (p. 217). Most of you have an attitude toward college. You view education as the primary path for "getting ahead" or "getting a good job." As a result, when in high school, you viewed this then hypothetical event of attending college as vital and you had an inclination to steer toward it and take the necessary classes and steps to make it happen.

Attitudes, according to Schutz and Luckmann (1973), come about due to sedimented life experiences that may be "inaccessible to the immediate grasp of reflective consciousness" (p. 229). Schutz uses this picturesque word, "sedimented," to refer to historical or biographical events and experiences that without thought or action simply get laid down on top of one another, layer after layer after layer. In the example of your attitude toward college, your life has been filled with various conversations and media images that made college a "natural" part of your lifeworld. Most of these experiences you probably can't even remember. They simply form a taken-for-granted background of conversation.

The lifeworld, then, is a humanly ordered world. I've pictured the successive ordering stages in Figure 9.2. Each step represents additional arrangements. The lifeworld is ordered through stocks of knowledge that are available to "all of us." The ordered world takes the place of the natural one for human beings; and because stocks of knowledge are by and large left unexamined and taken for granted, they form the basis of human reality (A). The lifeworld is additionally ordered, as these stocks are distributed differently based on social position and experience (B). And, finally, the lifeworld is further ordered based on individual structures of relevance (C).

> No longer can man confront reality immediately; he cannot see it, as it were, face to face. Physical reality seems to recede in proportion as man's symbolic activity advances. Instead of dealing with the things themselves man is in a sense constantly conversing with himself. (Cassirer, 1944, p. 42)

Connecting With Others

The problem of solipsism: I have a question for you. Have you ever felt alone in your head? Sometimes it seems like no one truly understands our experiences. Let's make this question more concrete. Have you ever looked at or tasted something and wondered how it really tastes or looks to someone else? Next time you're in class, look at the chalkboard. Maybe it's green like the ones at my school. You see green and I see green. But how do we know that it is the same green? Can we ever truly know how the other experiences or is conscious of the world?

If you've ever thought these kinds of things, you are in good company. It is either directly or indirectly a big concern in philosophy. It's called the problem of solipsism and it began with Descartes. I'm sure you've heard his great pronouncement, "I think, therefore I am." Descartes came to this conclusion as the result of his quest to have only true beliefs.

Descartes wanted to believe only those things that are true. His method of determining truth is called "hyperbolic doubt." He considered every possible field of knowledge and asked whether it has ever in the past failed or could be imagined to fail under possible conditions. He began by considering our senses and pointed out that the possibility of optical illusions makes it so that they cannot be trusted absolutely. He systematically moved through different ways of knowing (for example, have you ever thought that this life is just a dream?) and came to the conclusion that the only thing of which we can be certain is our own existence. Everything else is dependent upon that; it alone cannot be questioned. Because I think and every other experience and fact is dependent upon that one, I must therefore be the one existence I cannot doubt.

Solipsism, then, is the idea that existence is for me *my* existence and *my* consciousness. Everything that I experience—all the objects, events, other people, and so on—is constituted *by me* as part of *my consciousness.* Beyond Descartes, the basic premise of solipsism is that our ideas, motivations, perceptions, and the like are all abstractions from our own inner experience. In other words, as the individual encounters the world, she or he forms a map of the world that represents the relationship between the object (anything of which we become aware) and the organism (her or his individual self). The individual then makes generalizations about these relationships and creates a language to represent them.

Solipsism thus begins with the ego. The individual "I" encounters the world and is driven to make sense of it. An important implication of this idea is that we understand the mental states of others from analogy: If this is how I experience the world, then by comparison it must be similar to the way others experience the world. To go back to our chalkboard example, if you are a conscious being, then your experience of it must be analogous to mine.

Yet the problem of solipsism begins with a false premise. It assumes that the individual experiences the world primitively or independently. It assumes that human experience works from the inside out. And that, according to Schutz, is backwards. The primary data that the person has to master isn't the physical world but the world of language. And language isn't an individual creation; it is a social entity. A solitary individual doesn't need language. I didn't come up with language so that I could understand the world; we developed language so we could communicate and work collectively. All our concepts and ideas, whether about our inner state or the external environment, are derived from the same public language system.

> It is the fundamental ontological category of human existence in the world and therefore of all philosophical anthropology. As long as man is born of woman, intersubjectivity and the we-relationship will be the foundation for all other categories of human existence. (Schutz, 1966, p. 82)

Further, the self of our experience is not ultimately individualistic; it is social. It is social because we become conscious of our self and our experiences through language. As we've seen, human consciousness is consciousness *of* something. It is not a primitive experience; it is the experience of an entity that is linguistically constituted and situated. Thus, when we become conscious of the chalkboard, it isn't the simple sense data of which we are aware. When we ask, I wonder if you see the same green as I do, the question itself is based on the way in which we are aware of the board as "green"—*as an element of language.*

Let's use a different example to drive this point home. When you are in pain, when you first identify and are conscious of pain, you don't recognize it as pain that you have distinct from everyone else. You experience pain as something that is common to everyone that "belongs to us." The statement you make, "I am in pain," is the same statement that is available to me.

Subjectivity is basically objective and intersubjective: I've said all of the above to emphasize a point here: our subjective awareness is fundamentally intersubjective. The way in which we are conscious of the lifeworld intrinsically connects us to others. The lifeworld is linguistically constructed and is experienced through a matrix of relations in which other and self are essentially bound. This is not to say that you don't have personal and private experiences; you do. But the process we are involved in is making personal an already existing lifeworld built from public stocks of knowledge.

Remember that in our discussion of the I-Pole and Object-Pole of intentionality, we focused on how this characteristic of intentionality implies that meaning is subjective. The I-Pole and Object-Pole of intentionality also means that the lifeworld is objective and intersubjective. It is a world of relationships between and among the ego and other entities, both objects and other people. Thus, we add others to the intentional horizon in Figure 9.1. These other entities are objective because we encounter the lifeworld as already given and meaningful. We are born into it and thus we are always conscious of its historicity. The lifeworld is built up through the sedimentation of human experience. I receive these experiences through traditions and habits, and modify them somewhat through my own meaningful experiences.

This lifeworld is *there* for everyone that belongs to us. It is there to be examined and is thus objective. It is also there for each of us and is therefore intersubjective. The lifeworld is intersubjective because we live in it as a person among people. The lifeworld exists for me as sets of relationships: my connections to meaningful objects and people. When I experience my world, I experience it through these relationships. My lifeworld is arranged into associates (friends), contemporaries (living people I don't know), predecessors (people who lived before me), and successors (people I assume will live after me). My subjective awareness, then, is of one living in the midst of multiple others.

> It is obvious that both idealizations, that of the interchangeability of the standpoints and that of the congruency of relevances—together constituting the *general thesis of reciprocal perspectives*—are typifying constructs of objects of thought which supersede the thought objects of my and my fellow-man's private experience. (Schutz, 1962, p. 12)

Schutz also notes that the lifeworld is experienced as intersubjective through a particular device called the **reciprocity of perspectives.** The reciprocity of perspectives is based on two idealizations or assumptions. First, we assume that our standpoints are interchangeable. We assume that the world that has meaning for me also has the same meaning for you. What is different is the distance from or relationship to the world. So, while there are idiosyncrasies, I believe that you would interpret the world as I do if you stood in my place and had my experiences. In other words, we believe and act as if there is a fundamental, objective world available to all of us (even though in truth there isn't). Any differences are due only to differences in unique standpoints; the differences would "of course" disappear if we stood in the other's place.

The second assumption we make is that whatever differences may exist in our positions in the lifeworld, they are irrelevant for the purposes at hand. People are pragmatically motivated, according to Schutz. Each situation is in some way guided by practical concerns. What matters to us, then, is the issue at hand, not any peculiarities that may exist in our relationship to the world. This means that we never question the validity or actuality of the general world that is culturally created.

Some of this might appear as painfully obvious. But keep in mind that these insights are as true for every other world as they are about yours. The reason these things might appear obvious is that we take our world for granted. This **taken-for-grantedness** makes our world appear real and obvious. We live naively in the lifeworld as it is presented to us, whether that world is twenty-first-century America where an African American is seen as a full citizen, or eighteenth-century America where the same person was seen as subhuman. And the same is true of the "American way of life," or the world of Communist China.

An example that hits closer to home concerns my visit to the doctor's office. My experiences and my doctor's experiences are generally quite different. But I assume that if she were to change positions with me, her experiences would be like mine. And I also assume that whatever differences might exist are not important for the purpose at hand: getting me better. I really don't care about the pharmacodynamics or the physiology of the human body. Nor do I care about the Lexus she drives or the upper-class neighbors she has. And she doesn't care about the social construction of reality or my religious experiences. What matters is the task at hand and all the rest is left unquestioned and intact. That's all pretty obvious, huh? It might appear so, except when you consider that the same exact processes happen when a native Brazilian goes to a shaman to be healed by spirits, or hekura. The medical world and the world of the hekura are both produced through the same cultural processes: intentionality, reciprocity of perspectives, and taken-for-grantedness.

Making others anonymous: Thus, when considering how the lifeworld is ordered and how we connect with others, we start from the premise that *we are fundamentally connected to others* through the way in which we exist and are conscious. Beyond that, Schutz explains that we experience others through various degrees of **anonymization**—the process through which individual and personal differences are ignored.

You remember that typifications are average expectations that are built up from past social experiences. Typifications, then, function like categories. They include and exclude, contain legitimate or normally expected behaviors, and they imply an intrinsic value hierarchy. When I understand you, for example, as a male (a social typification), I include you with others like you and simultaneously exclude others *not* like you, I expect certain kinds of behaviors from you as a male, and I will evaluate your performance as a male based on those expectations. But notice something: as I use the typification, you become increasingly anonymous. I don't understand your behaviors as Pete (the individual); I understand them as male (the type).

People are more or less anonymous to us, based on our social relationship to them. Schutz conceptualizes two basic fields of social relations. The first is the sphere in which we are closest to others. It is the field of "I-Thou-We." When people are near us in time and space and they become the object of our intentionality, we enter into a **Thou-orientation**. This is direct social observation, but it is one-sided. We can observe the actions of another, say a student sitting on a bench reading a book, without the other person being reciprocally aware. And we can infer the subjective motivations for the student's behavior, by referring to our own subjective stock of knowledge regarding in-order-to and because-motives. In other words, we can use our own memory of why we have sat on a bench and assume that it is similar enough for the other. Or we can use general objective stocks and attribute motive.

> But they, on their part, never turn up as real people, merely as anonymous entities defined exhaustively by their functions. Only as bearers of these functions do they have any relevance for my social behavior. (Schutz, 1967, p. 185)

However, when the typifications are reciprocal, we enter a **We-orientation**. The We-relationship is the least mediated of all. In the Thou-orientation, we observe the other but do not interact with the other. In our example, we typically understand the person's behavior as a student, but we do not observe these behaviors immediately, nor do we ask the student for clarification. In We-relations, we encounter the individual as most available and real. We can observe the minutia of the person's facial expressions, hear the tone of voice, ask for clarification, and so on. In fact, the other in a We-orientation "is in a certain sense presented to me as more 'alive' and more 'immediate' than I am to myself" (Schutz & Luckmann, 1973, p. 66). Of course, I know more about myself than I do about the other. My autobiographical story is available in its completeness to me. But to observe myself, I must stop and reflect. In We-relationships, the other is immediately available and I can unreflectively live in and merge with in the actual experience.

Because We-associations are defined by immediate, face-to-face interaction where both parties are mutually aware of each other, participants are presented with the possibility that they may have to modify their typifications of the other. In the Thou-orientation, the other is seen and experienced firsthand, but is understood completely through typifications. In the We-orientation, conversely, the relationship is constituted as a union, with both parties immediately available to each other.

Thus, in We-orientations we can individualize our types. But, Schutz tells us, even in face-to-face relations, individuals usually do not have to seriously reflect on and challenge their stock of knowledge. And when changes are made, the modifications are slight, they generally occur over long periods of time, and the other is still understood in terms of a type. For example, after 25 years of marriage, I will know my wife pretty well. But notice how I just referred to her—she's my *wife.* She is Jen, an individual person, but I understand her as a specific kind of wife and woman, both of which are typifications.

In the **They-orientation,** we find the most anonymous of all others. They-relations take place in the indirectly experienced, not the face-to-face, lifeworld. These associations are stratified and occur in a series of ever-increasing anonymizations. Because of this increasing distance from immediate experience, an individual's understanding of who the other is becomes increasingly typified, so much so that the others in the They-orientation are not regarded as free agents. They can do nothing to change the characterization we place on them. This locked-in typing cannot be overcome unless they become part of the directly experienced lifeworld. The way in which They-relations work is very important for understanding how stereotypes work and why there is no easy way for us to rid ourselves of them.

Thinking About Modernity and Postmodernity

The postmodern problem: Like Mead, Schutz does not have a specific theory of modernity, and thus there is little that is directly related to the idea of postmodernity. The reason for this is probably due to the level of abstraction at which these two men think. Each of them is interested in the foundations that lie behind social thinking, and each of them uses philosophical issues to get at these underpinnings.

As a side note, I'd like to point out the value of philosophical thinking for the sociologist. Not only did Schutz and Mead turn to philosophy for inspiration, so did Marx, Simmel, Weber, and many other classical thinkers. Philosophy is a good beginning place, but for the sociologist, we need to bring it home to where and how people live in the social world, which is exactly what Schutz does in forming his general theory of meaning.

So what we have with Schutz is an abstract and general explanation of how meaning is created. Presumably, this theoretical understanding would apply across time and space, whether in modernity or postmodernity. So while we may not be able to see direct connections to the idea of postmodernity, we can extend Schutz's thoughts and see a socially grounded way in which people mitigate the effects of living in what might be a postmodern society.

According to much postmodern theory, we are living in an age where meaning and reality are doubtful. Though there are many factors, this doubt is mostly generated by a shift that has inextricably linked culture with the economy. Thus, Perry Anderson (1998) notes that "culture has necessarily expanded to the point where it has become virtually coextensive with the economy itself . . . as every material object and immaterial service becomes [an] inseparably tractable sign and vendible commodity" (p. 55). We can see this shift most clearly through Marxian theories of postmodernity.

But capitalist economy is a poor producer of culture. In more traditional societies, culture was primarily produced in identifiable social groups and in such orthodox institutions as religion. Capitalism, on the other hand, rips indigenous culture out of its social embeddedness and reduces the sacred to commodities that are bought and sold. The result of producing or modifying culture through markets and advertising is that it becomes de-centered and virtually meaningless. For example, last week I saw a billboard announcing "The World's Greatest Concert." Yet the concert was to be held at a local fairground and featured unknown artists. In advertising, the real or direct references for words and ideas, such as "world's greatest," disappear. As Jean Baudrillard (1993) notes, "the systems of reference for production, signification, the affect, substance and history, all this equivalence to a 'real' content, loading the sign with the burden of 'utility,' with gravity—its form of representative equivalence—all this is over with" (p. 6).

And because of the link between culture and self—the meaning of the self is produced through a variety of cultural identities—the self in postmodernity becomes fragmented and lacks any core. Thus, if culture is fragmented and lacks a center, then so does the self: "as an older industrial order is churned up, traditional class formations have weakened, while segmented identities and localized groups, typically based on ethnic or sexual differences, multiply" (Anderson, 1998, p. 62). These segmented relationships "pull us in myriad directions, inviting us to play such a variety of roles that the very concept of an 'authentic self' with knowable characteristics recedes from view. The fully saturated self becomes no self at all" (Gergen, 1991, p. 7).

The micro-foundations of meaning: But Schutz's theory challenges this direct relationship between culture and the self. According to Schutz, meaning is produced subjectively. And while we use our stocks of knowledge to create this meaning, this use is driven by the individual's subjective awareness of the world. In other words, it is what people do with culture that is important, not the state of the structure of culture. Remember, individual need for cultural cohesion is far less than many suppose. According to Schutz, we are interested in pragmatically sufficing and maintaining the taken-for-grantedness of culture, rather than creating a culture that is not fragmented.

In addition, beginning with the individual, Schutz sees that there are three levels of coherence or unity produced by the actor. The first is based on durée, pure duration or unmarked time. This is the primal unity of living through essential or undifferentiated time. In durée, we live in a moment-by-moment stream of time, and "[h]owever diverse the lived experiences may be, they are bound together by the fact that they are *mine*" (Schutz, 1967, p. 75). While these experiences and their unity exist below the level of conscious awareness, Schutz intimates that they form a basic sense of unity in life and self.

The second level of unity begins with the reflective glance. This is the level of intentionality where meaning is bestowed upon our experiences and our self within those experiences. As we've seen, the basic motivation behind this kind of consciousness is pragmatic. Thus "the knowledge of the man who acts and thinks within the world of this daily life is not homogeneous; it is (1) incoherent, (2) only partially clear, and (3) not at all free from contradictions" (Schutz, 1944, p. 501). Now, it is true that "experts" and "well-informed" citizens will strive for coherency, but they will only be concerned with consistency in well-defined, extremely narrow areas.

The final level of unity that the subject produces is the synthesis of separate acts into meaningful wholes or contexts, which then become objects for intentionality. For example, when I tell you about my life, I take various experiences and link them together into a meaningful whole. In other words, I will tell you stories about myself that create meanings that appear to be connected but are not linked by duration and may have had different meanings in the intentionality of the time.

As a case in point, let me tell you about why I am a theorist. My mind has always thought in generalities and looked for hidden connections. It's kind of intuitive for me. I remember as a kid that my father used to sit me down after I had done something wrong and ask, "Why, Kenneth, why did you do that? There is a reason for everything." Even at a young age, I was inclined to look for connections between behaviors and circumstances. I also remember that I used to sit in the garage with my chemistry set and think about wild experiments to perform. I would look at my dog and think, "what would happen if I did this?" Of course, I never did, I'm too much of a humanitarian, but I wondered about the hypothetical possibilities. Another illustration is from graduate school. I remember giving a presentation in class. We were supposed to dissect a series of articles and compare and contrast the different concepts and operationalizations, except in my presentation, I ended up making the concepts more general. Rather than looking at the particulars, I saw the abstract relations. The professor looked at me and said, "You really *are* a theorist." He gave me a B for the presentation because I thought theoretically. This tendency to think theoretically even happens when I watch TV, especially the news. I don't ever really see or hear what they are saying; I'm observing the methods through which they are producing a ready set of meanings for the general population. It can be pretty frustrating sometimes, but thinking theoretically is just something that I have always done.

That really is my story (and I'm sticking to it). But notice what I did. I took different experiences that are obviously not linked to one another in pure duration and undoubtedly had different intentional meanings in the circumstance (I almost quit sociology after that remark from the professor), and I connected them together in a coherent self-narrative that produces a meaningful context for me to understand other experiences. I took discrepant happenings and put them together in a story that essentializes my way of thinking ("I've always been a theorist."). Thus, "[w]hat was polythetic and many-rayed has now become monothetic and one-rayed. We now have a *configuration of meaning* or meaning-context" (Schutz, 1967, p. 75). Through my storytelling, I took things that are contingent and unconnected and made them appear essential.

We all do this all the time. The motivation for doing it is again pragmatic. We want to create a configuration of meaning that allows us to act, interpret, and exist in a particular way, whether looking back and reinterpreting certain looks and feelings in order to create the meaning-context of "being in love," or, as I have done, to justify and legitimate a particular way of existing in the world.

The importance for us here is that these meaning-contexts are produced by the individual. As long as the individual has pragmatic motivations, she or he will be able to produce configurations of meaning that are sufficiently coherent and authentic even in the face of postmodern dynamics, at least, according to Schutzian theory.

Summary

- To think like Schutz is to focus on the problem of meaning. Meaning always implies a difference between objects and experiences on one hand, and the meaning attributed to them on the other. In order to understand how people cope with the problem of meaning, Schutz concentrates his work on the natural attitude in the lifeworld.

- Meaning is specifically created through intentionality. Intentionality implies that meaning is subjective and is created through a backward accounting. But intentionality involves more than simply paying attention to a thing. Intentional objects are perceived in fields of actions and relationships. Because intentionality places the individual in an action mode toward the object, meaningful objects end up being reflexively constructed. In addition to physical objects, experiences, and other people, symbolic worlds can be the object of intentionality.

- The world toward which we become intentional is an already-existing, ordered lifeworld. Stocks of knowledge are socially produced and impose an order on the universe and human experience within it. These stocks of knowledge are available to every one of us and are minimally modified as individuals encounter problems or differences that their stocks of knowledge do not cover. Pragmatically sufficing generates a taken-for-grantedness about and continuity for social meaning, thus making it appear real and objective. Stocks of knowledge are further ordered through social differentiation and individual structures of relevance.

- The production of meaning may imply the problem of solipsism—reduction of all meaning and reality to individual experience. However, posing this problem overlooks the fact that collective stocks of knowledge—in particular, language—are the primary data that individuals must master. That being the case, meaning and reality are intrinsically objective and intersubjective. In other words, meaning is always formed in, around, and through relations with others. We understand the meaning of self and other through stocks of knowledge and, specifically, typifications. This implies that varying degrees of anonymity are associated with meaning production.

• Quite a bit of postmodern theory focuses on the relationships among capitalism, mass media, and meaning. The argument is basically a structural one wherein problems created at the macro level produce almost automatic effects for people. Thus, because capitalism and mass media have trivialized and fragmented culture, the individual's experience of meaning, reality, and self are fragmented and flat, as well. However, Schutz's theory is a micro-level theory of meaning production. Because people are pragmatically motivated to create meaning, and because meaning is produced through intentionality (a reflexive glance that synthesizes discrete and diverse experiences into a whole narrative), then the postmodern concerns about culture at the macro level may be misplaced.

Building Your Theory Toolbox

Conversations With Alfred Schutz—Web Research

Information about Schutz's life is scarce on the Web, but by using your search engine and the recommended sites, you should be able to answer the following questions:

- Schutz taught at the New School for Social Research. What is the New School? When and why was it founded? What is the University in Exile?
- What circumstances and events brought Schutz to the United States in 1939? In other words, how did Schutz and his family escape Hitler's regime?
- How did Schutz characterize his relationship between philosophy and sociology? In what discipline was Schutz's degree? What were Schutz's complaints about the American academic system? How did having to work in a second language affect Schutz?
- What was the Mises Circle? Describe a typical evening for Schutz with the Mises Circle.
- Find a comprehensive bibliography of the works of Alfred Schutz on the Web. How many works were published before Schutz's death? How many works were published after his death?

Passionate Curiosity

Seeing the World (using the perspective)

• Let's think about race, class, gender, and heterosexism. People that approach such topics from a Marxian or Durkheimian perspective would see these things as structures of inequality, social facts that determine people's lives. On the other hand, researchers that approach such topics from a Meadian point of view would see race, class, gender, and heterosexism as social objects whose meaning emerges out of social interaction. Schutz presents us with yet another way of conceptualizing such things. How would Schutz see race, class, gender, and heterosexism? According to Schutz, where do these things "live" and how do they influence our lives? Compare and contrast these three approaches to race, class, gender, and heterosexism. When and why would you use one approach over another? What specific insights can be gained through each of these approaches?

- Using Schutz's perspective, how would you approach the study of religion? In other words, what would a Schutzian study of religion look like? What kinds of questions would be asked? Come up with a definition for religion using Schutz's perspective. (Note: my use of religion in this chapter doesn't begin to answer this question.) After you've come up with the Schutzian perspective on religion, how does that compare to Marx, Weber, and Durkheim?

Engaging the World (using the theory)

- We talk a lot about race, class, and gender. But they are not the only categories of distinction. How many social categories of distinction can you name? Now look at your list and mark the ones that appear to you as having clear distinctions. For example, you may think that the distinctions between heterosexuals and homosexuals are clear. Now, using Schutz's theory, explain how all of the distinctions are created and maintained. According to Schutz's theory, from where does the order we perceive come? How are these distinctions maintained or legitimated (think about reflexivity)?

- One of the things we considered under Du Bois' theory is the influence of media representations of minorities on the consciousness of the minorities themselves. Using Schutz's theory, how would media representations affect how whites perceive minority populations, particularly if whites do not have regular interactions with the group? What about the reverse? How are whites (or males, or heterosexuals) perceived by minorities? Is the process of perception the same or different?

- Humberto Maturana, currently a professor of physiology at the University of Chile, once said, "When one puts objectivity in parenthesis, all views, all verses in the multiverse are equally valid. Understanding this, you lose the passion for changing the other" (You can see the complete quote at http://www.oikos.org/maten.htm). Use Schutz's theory to explain Maturana's comment. What applications do you think this might have for the social world in which we live?

- Pick a group of which you are a member. It could be a general group, like male, or a more specific group, like hip-hop. Analyze the group's culture in terms of stocks of knowledge. Be certain to address the issues of typifications and social distribution. What are the effects of having such a stock of knowledge? In what ways do you think this stock is different from what we might call the common stock of knowledge? What implications do you think this might have? What do you think is happening to your stocks of knowledge as a result of your education? In what ways does your world present itself differently to you now than before your college education?

Weaving the Threads (synthesizing theory)

- One of our continued concerns in this book is the assumption we make about the social world—is it objective or subjective? Schutz very clearly argues that the work of sociology must focus on the subjective experience and world of the actor. Using Schutz's theory, explain why Schutz would characterize scientific investigation of society as "fictional." Obtain a copy of an introduction to sociology textbook. How much of that book would Schutz characterize as fiction?

Further Explorations—Books

Grathoff, R. (Ed.) (1979). *The theory of social action: The correspondence of Alfred Schutz and Talcott Parsons.* Bloomington: Indiana University Press. (Letters between Schutz and Parsons)

Natanson, M. (Ed.). (1970). *Phenomenology and social reality: Essays in memory of Alfred Schutz.* The Hague: Nijhoff. (Essays indicating different approaches in using Schutz's theory)

Psathas, G. (Ed.). (1973). *Phenomenological sociology: Issues and applications.* New York: Wiley. (Good source for learning about Schutz's influence as well as for understanding phenomenological sociology)

Rogers, M. (2000). Alfred Schutz. In G. Ritzer (Ed.), *The Blackwell Companion to major social theorists.* Malden, MA: Blackwell. (Good chapter-length introduction to Schutz)

Wagner, H. R. (1983). *Alfred Schutz: An intellectual biography.* Chicago: University of Chicago Press. (Biography of Schutz that emphasizes his intellectual roots)

Further Explorations—Web Links

http://plato.stanford.edu/entries/schutz/ (Site maintained by Stanford Encyclopedia of Philosophy)

http://www.iep.utm.edu/s/alfredschutz.htm (Site maintained by The Internet Encyclopedia of Philosophy)

Further Explorations—Web Links: Intellectual Influences on Schutz

Henri Bergson: http://en.wikipedia.org/wiki/Henri_Bergson (Site maintained by Wikipedia Internet Encyclopedia good brief introduction to Bergson as well as extensive hypertext)

Max Weber: http://www.faculty.rsu.edu/~felwell/Theorists/Weber/Whome.htm (Site maintained by Frank W. Elwell at Rogers State University; site specifically oriented toward undergraduates and contains good explanations of some of Weber's concepts)

Edmund Husserl: http://www.husserlpage.com/ (Site maintained by Bob Sandmeyer; extensive information and Web links for Husserl)

CHAPTER 10

The Social System—Talcott Parsons (American, 1902–1979)

> *The announcement of* Structure's *death turns out to have been premature, as was the announcement of Parsons.' In fact, it is possible that, with the exception of Habermas, no post-classical sociological theorist is more talked about today in Europe and the United States than Parsons himself.* (Alexander, 1988a, p. 96)

There are some sociologists who might be troubled to see Parsons included in a textbook such as this. They might find it curious for a few reasons, I think, but the most apparent issue is that Parsons does not fall into the normal definition of classical theory. Yet he isn't contemporary either. Parsons and people like him now fall into a middle ground, a kind of no-person's land, neither classic nor contemporary. The result is that Parsons often gets left out of the picture all together. That would be all right, except that for much of the twentieth century, Parsons was "the major theoretical figure in English-speaking sociology, if not in world sociology" (Marshall, 1998, p. 480). And, as Victor Lidz (2000) notes, "Talcott Parsons . . . was, and remains, the pre-eminent American sociologist" (p. 388).

Parsonian functionalism held sway for some 20 to 30 years until the 1960s, at which time social thinkers began to question Parsons' structural-functionalism. His scheme did not seem to adequately account for the almost ubiquitous social movements centered on race, gender, and the Vietnam War. Society was in upheaval and Parsons was accused of being politically conservative and his functionalism was seen as little more than a justification of the status quo. Many theorists turned to various forms of conflict theory to understand and explain the social movements. Yet, Parsons' work was so influential that in order to present a different theoretical framework, people felt compelled to first criticize Parsons (this practice can still be seen). A clear example of this approach is in C. Wright Mills's *The Sociological Imagination,* where an entire chapter is dedicated to a critique of Parsons' grand theory.

However, not everybody was critical of Parsons. Robert K. Merton was a student of Parsons and formed many of his ideas in conversation about Parsons' theory. Merton (1967) criticized a number of his teachers' apparent assumptions, including the functional unity of society postulate. The functional unity of society assumes that all standard social activities or cultural symbols are functional for the entire society or cultural system. In other words, everything that can be studied by a sociologist is functional for everyone in the society—if it is standard across society, then it must be functional for society. Merton contends that this is empirically not the case, and, further, it assumes a level of social integration that very few societies, especially modern ones, have achieved. According to Merton, some practices and symbols are actually **dysfunctional** for some groups. Functional analysis must therefore specify the units for which an item is functional and must allow for diverse consequences, both functional and dysfunctional. While Parsons does acknowledge that some things can be dysfunctional, he sees this as a threat to the system rather than a normal part of society. Thus, it is Merton who gives us the notion of dysfunction as part of functional analysis (the idea of the "dysfunctional family" is based on this part of Merton's work).

Merton also moved functional analysis forward by softening some of Parsons' ideas, or ideas that seemed to stem from Parsons' theorizing. Rather than only understanding society in terms of functions that are both necessary and determinative, Merton opens the door for us to see things as potentially dysfunctional or non-functional; he also helps us see that there can be functional alternatives. In addition, Merton gives us the conceptual tools to understand functions as either latent or manifest. You might recall from your introductory course that a manifest function is one that is explicit and acknowledged. On the other hand, a **latent function** is an observable consequence that is neither intended nor recognized by the social system but nonetheless leads to adaptation and balance. The university as a marriage market is just such a function. Universities allow some people in and keep others out, and different universities have different standards. Thus, when you attend a specific university, you are generally assured of meeting only those who would be "suitable" for you to marry, people who have the same socioeconomic status as you. Yet that function of higher education is nowhere officially acknowledged; it is a latent function.

Merton isn't the only sociologist who is using Parsons' functionalism. Since the 1980s, Parsons' theory has sparked the interest of other thinkers, not only in the United States but in the rest of the world as well. His work has now become the basis of "neo-functionalism" and is being used by such important contemporary theorists as Jeffrey C. Alexander and Niklas Luhmann. Yet, despite this continued interest and theory building, Alexander (1985) tells us that Parsons' work still evokes strong negative reactions, even in its neo-functionalist form. Of Alexander's work one critic said, "This is only one example of the revival of functionalist theorizing which has recently surfaced, a development of which I am fully aware even while I find it appalling" (as cited in Alexander, 1985, p. 7); and, another said that "'the library of critiques' of functionalism 'must be taken from the shelves again.'"

According to Alexander (1985), functionalism, or neo-functionalism, "indicates nothing so precise as a set of concepts, a method, a model, or an ideology. It indicates, rather, a tradition" (p. 9). What he means by that is that at present, functionalism is more of a perspective than a precise theory. It's a way of looking at or analyzing the social world. The interesting thing about this perspective is that, if you are going to look at the interrelationships among different social structures, you almost have to use it. For example, even though theories of globalization are not explicitly functional, they nonetheless are based on the perspective that Alexander is talking about.

The neo-functionalist approach provides us a general picture of the interrelation of social parts. It is a way of describing the parts of the social system as symbiotically linked to one another and interacting without direction from an outside force. This is our systems model, but here Alexander conceptualizes it as a sensitizing device rather than a mechanism that determines how society works. Functionalism, then, discourages us from looking at individual social phenomena as if they existed in a vacuum. Rather, as you will recall from our first chapter, sociological inquiry ought to always consider the wider context and the relationships that might exist between one's research topic and the other elements in society; using the functionalist perspective forces us to take just such a point of view. In this vein, neo-functionalism also maintains Parsons' distinctions among the personality, culture, and social systems. Parsonian theory sees you, culture, and society intertwined and influencing one another. The relations and interactions among and between these subsystems are vital for understanding society and social change—the tensions produced by their interpenetration are a source of change and control.

Thus, neo-functionalism is not simply concerned with equilibrium (this is a general critique of functionalism). Equilibrium is taken as a reference point for the system but not necessarily for the actors; thus, this perspective is as concerned with deviance and social control as it is with integration. Let's use Merton as an example again. Merton proposes a functional theory of deviance. He argues that society values certain goals and the means to achieve those goals above others. In the United States, for example, we value economic success and believe that education and hard work are the proper means to achieve that goal. However, in any social system, there are those who do not have equal access to the legitimate means to achieve success. People who are disenfranchised will respond with conformity (acceptance of both goals and means), innovation (acceptance of goals but use of innovative means, such as robbery), ritualism (abandon the goals and perform the means without any hope of success), retreatism (deny both the goals and the means and retreat through such means as alcoholism), or rebellion (the individual creates her or his own goals and means).

Finally, functional analysis continues to recognize differentiation as a major source of system change. We've seen this issue in different forms with Spencer and Durkheim. Ironically, differentiation is in many ways at the heart of the postmodern critique as well. Differentiation causes individuation and institutional strains. Postmodernists see social and structural differentiation as having reached a threshold point past which fragmentation rather than integration and cultural pluralism rather than cohesion dominate. Functionalists, on the other hand, will point to the fact that archetypically postmodern societies, such as the United States, continue to exist and operate and must therefore be functionally integrated. In the hands of someone like Parsons or Merton, the continued existence of a society under conditions of postmodernity becomes a frightfully interesting empirical question. On the one hand, there are factors that seem to indicate that society is fragmenting; yet, one the other hand, society remains intact. How can society still work? The functionalist will then look for those systemic processes and prosperities that maintain cohesion and integration, because the perspective gives us eyes to see such a question and motivation to search for an answer.

Thus, part of the reason I've put Parsons here is because he is without a doubt one of the most influential sociological theorists; and if not here, where? Like classical theorists, Parsons continues to inspire theorizing: many contemporary sociologists are using modifications of his theory or asking Parsonian questions. Parsons gives us a clear set of questions to ask and a powerful perspective from which to answer them.

If we are going to ask questions about the interrelationships among different social entities, such as how the state influences education, Parsons' approach is one of our best beginning points. True institutional or structural analysis will always be oriented toward the interconnections among and between the institutions. For example, if you want to study family, looking at the internal workings of kinship in your society is only a small part of understanding that institution. If you were to look at family in U.S. society, you might find that divorce rates and the number of "latchkey kids" have increased. So what? Well, you say, that's bad. How do you know? Unless you are simply going to make a moral argument, chances are you will have to make a functional argument about how divorce and latchkey kids are dysfunctional.

You can't simply say that divorce is breaking apart the family and therefore it is bad. For example, let's say we lived several hundred years ago. At that point in time, we might look around us and see that our form of government and economy is falling apart. We could look at any number of indicators (like the use of standardized money, free markets, centralized taxation, bureaucratization, and so on) and say, "See? All these things are bad because they are leading to the downfall of feudalism." But we would have been misguided. The downfall of feudalism was necessary for the advent of democracy and capitalism, and most of us seem pretty enamored with those. Thus, in order to understand anything like changes in government or changes in family, it is imperative that we place them in their context. In order to correctly understand how and why the kinship phenomena are occurring, you must place family firmly in its context, at least the interinstitutional relations that exist among family, economy, and government (you may want to add religion and education in there, too).

Parsons also needs to be in a book like this because in some ways he, more than anyone else, set the stage for contemporary theory. He gave us our beginning structure of theory in at least two ways. First, Parsons set the agenda. In order to do our work, sociologists (as well as all researchers and theorists) argue against someone else. When we first set out to do research or to theorize, we map out the field. We find out who said what and why they said it. Then, in order to make our own mark, we argue against either what they said or why they said it. Think about it this way: a publication that simply said, "the other guy is right," wouldn't be very interesting (and it wouldn't lead to tenure). This isn't a negative thing; almost all of us do it: remember that Marx argued against Hegel, and Weber against Marx. My point here is that few other people have spurred as much debate as has Parsons. He helped set our agenda by having a theory so powerful that in order to do theory in the sixties, seventies, and eighties, you almost had to argue against Parsons; and people are still arguing about Parsons today.

The second way in which Parsons set our agenda is that he told sociologists in the United States who to read. He isn't alone in this, but more than any other person, Parsons set our canon. We are debating the canon now, but previously it was pretty clear who you had to read. Because of the place Parsons' work held in the twentieth century, and because Parsons built his theories on the work of European thinkers like Weber and Durkheim, others had to read those folks, too. In this, Parsons is like the crossover point for sociological theory in the United States—Parsons told us who to read from the past and his work became a foundation for work in the future. As we will see, this issue of the canon in which Parsons is so entwined is one of the key issues in postmodern thinking.

Parsons in Review

- Born December 13, 1902, in Colorado Springs, Colorado, to Edward Smith Parsons Sr., a Protestant clergyman and college professor who also served as President of Marietta College, and Mary Augusta Ingersoll. Talcott Parsons had two brothers and two sisters.
- Parsons began his university studies at Amherst, planning on becoming a physician but later changing his major to economics. He received his B.A. in 1924.
- Beginning in 1924, Parsons studied political economy abroad, first at the London School of Economics. There he came in contact with the anthropologist Bronislaw Malinowski, who was teaching a modified Spencerian functionalism. Parsons also met his wife-to-be, Helen Walker, while in London. Parsons then studied at the University of Heidelberg, where Max Weber had attended and taught. At Heidelberg, Parsons studied with Karl Jaspers, who had been a personal friend of Weber's. The dissertation that Parsons completed for Heidelberg was called "Capitalism in Recent German Literature: Sombart and Weber."
- After teaching a short while at Amherst, Parsons obtained a lecturing position at Harvard. He was one of the first instructors (along with Carle Zimmerman and Pitirim Sorokin) in the new sociology department in 1931. Parsons became department chair in 1942 and began work on the Department of Social Relations—formed by combining sociology, anthropology, and psychology. It was Parsons' vision to create a general science of human behavior. The department was in existence from 1945 to 1972 and formed the basis of other interdisciplinary programs across the United States.
- After 10 years of work, Parsons' first book was published in 1937: *The Structure of Social Action.* More than any other single book, it introduced European thinkers to American sociologists and created the first list of "classical" theorists. Parsons was particularly responsible for bringing Max Weber to the attention of U.S. sociologists; Parsons translated several of Weber's works, including *The Protestant Ethic and the Spirit of Capitalism.*
- Parsons died on May 8, 1979, while touring Germany on the 50th anniversary of his graduation from Heidelberg.

The Perspective: Abstract Social Systems

> A society is a type of social system, in any universe of social systems, which attains the highest level of self-sufficiency as a system in relation to its environments. This definition refers to an abstracted system, of which the other, similarly abstracted subsystems of action are the primary environments. This view contrasts sharply with our common-sense notion of society as being composed of concrete human individuals. (Parsons, 1966, p. 9)

Society as a system: Notice Parsons' definition of society above. He explicitly tells us that society is not composed of individual human beings. This, as Parsons points out, is the common notion. Part of us wants to be able to point to concrete entities in order to understand society. Mead did exactly that when he told us that society is found in the observable symbolic interactions of people. However, most theorists ask us to move up at least a little from the concrete. Simmel, for example, asked us to move up one level of abstraction from Mead's social interaction. Simmel argued that society exists in abstract social forms. Parsons, on the other hand, asks us to move completely up the ladder of abstraction.

Parsons defines society in terms of its systemic properties. That is, society is a system in which each part is linked to every other part either directly or indirectly, and through these connections, each part contributes to the functioning of the whole. The definition of a system may sound a bit like the organismic analogy that we found in Spencer, and in some ways that's true; it is a more sophisticated organismic approach. However, there are different emphases that result when we think in terms of systems. One of the first things that systems theory attunes us to is the relationship between the system and its environment. All systems are set apart from their environment by a boundary, and all systems exist within an environment by virtue of their negotiation of the boundary and the environment. Further, if the theory takes a cybernetic approach, as does Parsons', then the boundary interchange is understood in terms of shared, controlled, and stored information.

Two problems for systems: There are two central problems at which systems theory looks. The first problem is external and concerns the boundary negotiation between the system and environment. All systems exist within and because of an environment. Let's take the idea of an ecosystem as an example. An ecosystem is constituted by two subsystems: a given set of living organisms and their physical environment, and the relationships between them. We can look at any element within either of those systems and see its survival connections to the rest. Thus, particular birds are dependent upon specific plants, and the plants are reciprocally dependent upon the birds (remember the birds and the bees?). In addition, both the birds and plants are dependent upon a particular kind of soil, measure of rainfall, and so on. Each subsystem draws what it needs from the other in order to maintain life. In a like manner, society must develop a system that can extract the needed raw materials, convert them to usable goods, and distribute them throughout the system. This is the boundary-maintenance function, and because society has two distinct boundaries—one with the natural environment and the second with other systems "in any universe of social systems"—Parsons actually divides this function into two parts: adaptation and goal attainment.

The other problem for systems is internal. This internal problem is sometimes phrased as the Hobbesian question of social order. Thomas Hobbes was an English philosopher of the Enlightenment who lived from 1588 to 1679. In Hobbes' view, human nature is governed by two forces: the fear of death and the desire for power. In their natural state (that is, apart from society), then, human beings cannot expect any goodwill from others, and they spend their lives in the antagonistic pursuit of power and stability. Hobbes said that such a life would be "solitary, poor, nasty, brutish and short." According to Hobbes, humans are intrinsically selfish and thus live unstable and unpleasant lives. The Hobbesian question of order begins with this kind of view of people and asks how we can alleviate human uncertainty and create harmony. This point of view begins with the individual and asks about social order. Durkheim started in this same position and argued that the collective consciousness is necessary for humans to live together. It provides us with meaning and order. If the collective consciousness becomes weak and enfeebled, both individuals and society become sick and die.

> [A] social system is a mode of organization of action elements relative to the persistence or ordered processes of change of the interactive patterns of a plurality of individual actors. (Parsons, 1951, p. 24)

However, the question of order is a bit different for people who think from a systems approach. Like Spencer, the problem for Parsons lies more in structural differentiation than in individual self-interests. This starting point is inherent in systems theory: a system by definition has multiple parts and these elements must be integrated in order for the system to work. It's kind of like my "surround sound" music system: there must be a certain degree of compatibility for the system to work. If you have one component that only works with digital signals and another that only works with analog signals, you won't get any music. Obviously, the more compatible are the different elements, the better the system will work. This isn't to say that Parsons doesn't think culture is important; he does. In fact, we will see that for Parsons, culture is the most important system. He understands it, though, in systems terms as well as individual terms.

Generalized media of exchange: Notice in our example that compatibility is determined by communication among and between the parts. This is another way that systems theory sensitizes us to certain things. In systems, the kind of information used and the way in which the information is shared are important. For example, I sometimes use a digital effects processor when I play electric guitar. When set up like that, the guitar, processor, and amp form a system. The thing that constitutes them as a system is the fact that they communicate with one another. *How* they communicate is extremely important. The owner's manual for the processor tells me not to use cables that contain resistors. The resistors will change the way the information moves and will influence the sound. The same is true for society. Society, or, as Parsons would put it, the subsystems within the social system, are integrated through "generalized media of exchange." What those media are and how they facilitate exchange is extremely important in a Parsonian view of society.

The generalized media facilitate exchanges among and between the subsystems. For example, polity (or government) provides the economy with capital and the economy provides productivity for polity. We usually think of capital as money, but it is really much more. The existence of capital is based on the existence of a government that standardizes, prints, distributes, controls, and guarantees currency; a polity that provides for a stable banking system oriented toward investment; a state that regulates both internal and external markets; a government that provides for national security in which capital can be accrued; and so on. The idea of productivity here is the same as in common usage: it is a measure of effectiveness in utilizing labor and equipment. The Gross National Product is such a measure. Thus, polity provides capital and economy provides productivity and both are facilitated by a particular generalized medium of exchange: money.

Equilibrium: Another issue that we can glean from this approach is that systems tend toward equilibrium. According to Merriam-Webster, *equilibrium* is a state of balance between or among opposing forces or processes resulting in the absence of change. We can make a distinction between smart and dumb systems (Collins, 1988, pp. 47–54). Dumb systems have little or no informational feedback and they are relatively closed. A mechanical clock is a good example of such a system. Once it is set and wound, there is no systemic mechanism through which the system can receive new information and make adjustments. Smart systems get some degree of information from the environment. A thermostat is a good example of a smart or open system. The thermostat has three important elements: a goal state (the temperature you set it at), an information mechanism (its ability to read the temperature in your home), and a control mechanism through which it turns the air conditioner or heater on and off. Open systems such as the thermostat tend toward equilibrium. The thermostat balances out the forces of hot and cold through its control mechanism to keep your house at a comfortable 73 degrees.

Parsons understands society as a kind of open system, at least in terms of its tendency toward equilibrium. Parsons argues that most systems will tend toward equilibrium because it is the natural state of life. If we look at the physical or animal world, we see a great deal of overall stability. In fact, that is one of the problems that face evolutionists: we don't see much change in our lifetime or even the lifespan of many, many generations. Evolutionary change occurs over very long periods of time. One of the reasons for this slowness is undoubtedly due to the fact that all things appear to exist within interrelated subsystems and systems. Sudden change in such a systemic environment would probably bring chaos instead of order. In the same way, sudden change in one societal subsystem will tend to bring chaos, unless the change is countered with equal changes in other subsystems. In other words, unless social changes are met with equilibrating pressures, they will lead to the demise of that society. The Roman Empire is an example, as is the Union of Soviet Socialist Republics. The issue of systemic equilibrium attunes us to different kinds of questions, processes, and mechanisms, all of which exist at the systems level, not the level of the individual.

Analytic theory: Parsons' work is actually a synthesis of three theorists. Parsons (1961) notes how he used each one: "for the conception of the social system and the bases of its integration, the work of Durkheim; for the comparative analysis of social structure and for the analysis of the borderline between social systems and culture, that of Max Weber; and for the articulation between social systems and personality, that of Freud" (p. 31). Thus, the work of Parsons, like that of Du Bois, is a product of theoretical synthesis; but unlike Du Bois, Parsons is a functionalist and a grand theorist. Like Spencer, Parsons feels that the observable world can be understood using very few abstract concepts; but unlike Spencer, Parsons is not convinced of positivistic theory. His theory, then, is quite abstract and is analytical rather than dynamic. That is, what he gives us are sets of concepts and definitions, much like Weber, and not much in the way of causal theories or prediction. All this, coupled with his systems approach, means that in Parsons' theory we end up with a number of abstract, categorical terms and schemes that are systematically embedded with one another.

Thus, to think like Parsons is to think in systemic terms. Society exists as a system composed of subsystems. These subsystems are seen in terms of their general function for the whole as well as their relationships to one another. The relationships between the subsystems are understood in terms of exchange and communication. Society is somewhat like a smart system in that the communication of information is the most important feature of intersocietal relations. The functions of the subsystems generally fall into two categories—those that negotiate external boundaries and those that provide for internal cooperation. Due to these functions, society, like all systems, tends to hold at a point of equilibrium. So most, but not all, of Parsonian theory is oriented toward explaining the forces and social control mechanisms that lead to social integration. Finally, to think like Parsons is to think in the most abstract terms possible. Parsons feels that all systems operate in basically the same way. Thus, all Parsonian terms will tend to be expressed at the highest possible level of abstraction and they will tend to be nested and understood together.

The rest of this chapter is divided into three main sections. In *The Making of the Social System,* we see that even though Parsons is a systems theorist, or perhaps because of it, he begins his theorizing at the lowest possible level, with individual behaviors. He looks at what he calls "voluntaristic action" and sees that human behavior is situated within an activity system that simultaneously creates and limits free will. Parsons understands **institutionalization** as the process of building up from these individual, voluntaristic actions different social systems that in turn set the stage for the unit act. In the section *System Functions and Control* we see that Parsons understands all social subsystems in terms of four requisite functions. He breaks the two primary needs concerning external and internal relations into two functions each. This scheme, Parsons argues, can be used to analyze any system, be it personality, social, or physical. Parsons then explains how all of the subsystems are related to one another through functional interdependency and generalized media of exchange, and how all of human existence is framed by a general system called the "cybernetic hierarchy of control." Finally, in *Social Change,* we see Parsons' theory of revolutionary change and reinstitutionalization.

The Making of the Social System

Voluntaristic action: If you were going to study society as a whole, where would you begin? You would probably start with what you know about society, or with your definition of society. Maybe you'd start with the idea of institutions or social structure. Like most of us, you'd probably start with a specific institutional problem, like institutionalized racism or class-based inequality, and study how it affects human behavior. You would, however, have to push beyond that level of analysis if you were going to study society as a whole. You would have to think more abstractly.

Perhaps you'd first come up with an abstract definition of society. That's what a number of our theorists did, like Georg Simmel. He defined society as existing in the social forms that guide interactions and then set out to define the different forms and look at how they influenced individuals as well as the interaction. Parsons gives us a clear definition of society as well. He defines it in system terms and not in terms of concrete individuals. Parsons' perspective is very abstract and we would probably expect him to begin his study of society at that level, but he doesn't. If fact, when he begins his theorizing, Parsons acts as if he doesn't know anything about society at all.

Let me put it to you this way: If you knew absolutely nothing about television and you were confronted with this box-like object that pictured moving people in it, how would you begin to understand it? You might begin with the picture itself, but the problem with that approach is that you would be looking at the effect rather than the object or process itself. That's kind of what we would be doing if we started thinking about the whole of society from the point of view of institutional effects. Another way to begin that makes sense is to start with one of the cords leading into the television, either the power cord or cable. You might unplug one and then the other to figure out which one is more basic and begin with it, building up from the smallest component to the largest. That's exactly where Parsons begins his understanding of society—at its most basic level.

> First a word should be said about the units of social systems. In the most elementary sense the unit is the act. This is of course true . . . of *any* system of action. (Parsons, 1951, p. 24)

Allow me a moment of reflection before we proceed with Parsons. I've taken you through this rather silly example of the television because *how* a theorist thinks is at least as important as *what* a theorist thinks. If we can understand our theorists' beginning points, assumptions, and particularly their questions, then we have begun to think about society. *Thinking about society is more important than knowing things about society.* Remember the quote at the beginning of Chapter 1? "Imagination is more important than knowledge." Parsons not only gives us a unique definition of society, he gives us an amazing place to start *theorizing* about society, especially given his definition. His perspective and beginning point can both inspire our thinking, if we will but stand a moment in his shoes, his questions, and his imagination.

Parsons credits Max Weber with his beginning point for theory. This area of theorizing is referred to as **action theory**. As we've seen, there are a number of polar opposites in social and sociological theories. In this case, we have the agency and structure continuum. **Action theory** references a group of theories that focus on human action rather than structure. Within this group are theories that center on meaning and interpretation, and theories that are concerned with the nature of human action.

Weber was concerned with both, but he explicitly developed a typology of social action. According to Weber, simple behavior is distinguished from social action by the subjective orientation of the actor. If, in the action, the person takes into consideration the meaning of the act for others, then it is social action. Weber argued that there are four distinct types of social action: traditional, affective, value-rational, and instrumental-rational. The focus of Weberian action theory is on the latter two types and is concerned with the degree of rationality in human behavior. Exchange and rational choice theories take up these issues in particular, and argue that action is best understood in terms of people making rational choices wherein they maximize their utilities (hence the name "utilitarianism"). In other words, people have clear preferences and try to get the most out of every encounter with other people by weighing costs and benefits. The question here becomes how rational can people be in exchanges.

George Herbert Mead also developed a theory of action. His is a theory that approaches the act from the perspective of the individual, yet Mead was still very much concerned with social action. He argued that human action, specifically social action, occurs in four phases: the impulse, perception, manipulation, and consumption. Mead saw symbolic meaning pertinent to all phases of the act, but particularly the perception phase. Humans perceive the meaning of the impulse in terms of situated social objects and other people. The focus of Meadian action theory, or symbolic interaction, is on the negotiation and attribution of meaning.

Parsons specifically names his approach "voluntaristic action theory." He isn't so much concerned about how meanings are negotiated in interaction, like Mead, but he wants to understand the context of human action. Like rational choice theories, voluntaristic action draws from utilitarianism in that it sees humans as making choices between means and ends. But it modifies utilitarianism by seeing these choices as being circumscribed by the physical and cultural environments.

> When a biologist or a behavioristic psychologist studies a human being it is as an organism, a spatially distinguishable separate unit in the world. The unit of reference which we are considering as the actor is not this organism but an "ego" or "self." The principal importance of this consideration is that the body of the actor forms, for him, just as much part of the situation of an action as does the "external environment." (Parsons, 1949, pp. 46–47)

An example of voluntaristic action is your behavior right now. In order to read this book, you had to enroll in class, pick up the syllabus, buy the book, schedule time to read, and actually sit down at the scheduled time and read it. All of this may be seen as voluntaristic action: you voluntarily acted, choosing among various ends and means—you could be drinking a beer and watching TV right now, but you selected this behavior. "But," you say, "I didn't volunteer to do this class work!" Yes, you did; nobody physically forced you. However, you volunteered under certain influences from the environment. The same with shaving or not shaving this morning, or the clothes you are wearing right now. They are all aspects of voluntaristic action; the questions are how much freedom do you have in making choices in action, and what goes into the decisions that you make in order to act.

Parsons' first theoretical work, *The Structure of Social Action,* explains this by giving us an analytical model of action. It's a framework or scheme through which we can view and understand human action. Parsons' scheme doesn't predict the kinds of actions in which people will engage; his thinking is more fundamental than that. He gives us an analytical model that we can take into any situation and begin to understand the myriad elements that go into human action.

The unit act: This analytical model in its completion is termed the **unit act** and is pictured in Figure 10.1. As you can see from the figure, Parsons argues that every act entails two essential ingredients: an agent or actor and a set of goals toward which the action is directed. We can see here Weber's notion of social action (in particular, instrumental rationality) and its influence on Parsons' theory. The initial state within which the actor chooses goals and directs her or his process of action has two important elements: the conditions of action and the means of action. The actor has little immediate agency or choice over the conditions under which action takes place. Parsons has in mind such things as the presence of social institutions or organizations, as well as elements that might be specific to the situation, like the social influence of particular people or physical constraints of the environment. For example, being at a fraternity party (physical setting) will influence your action, but so will the presence of your parents (social influence) at the same party.

About the means, on the other hand, the actor does potentially have choice. Some situations allow quite a bit of freedom, but others, like being in a jail, do not.

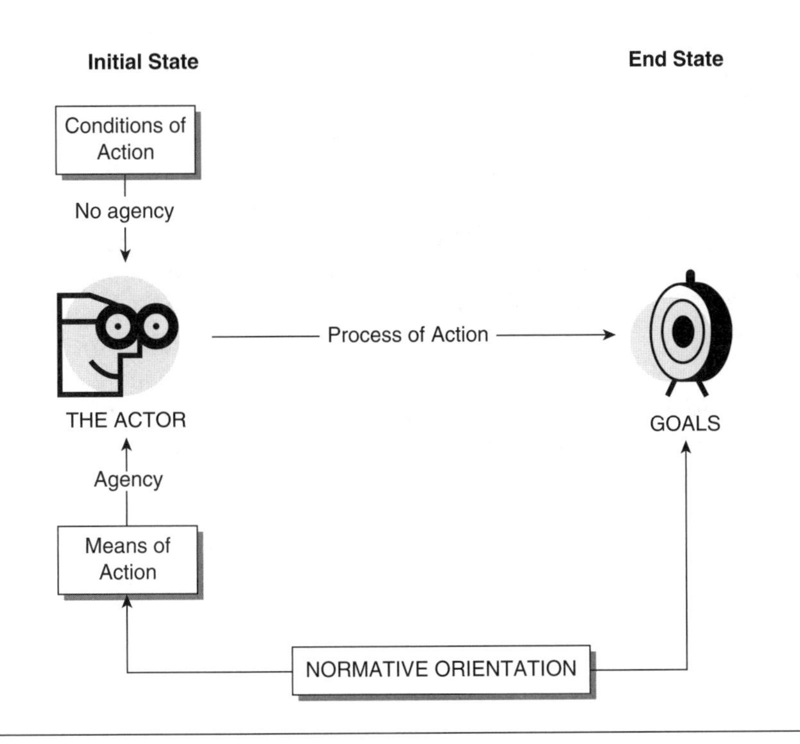

Figure 10.1 The Unit Act

However, notice that even in those situations where there is freedom of choice about the means and goals, the normative orientation of the action limits or defines the choices made. For Parsons, this is an extremely important point. Parsons says that the concept of human action demands that there be normative influence. In other words, Parsons is arguing that human action is distinctly *cultural* action. Remember that **norms** are behaviors that have sanctions attached to them, be they positive and/or negative. Such behaviors necessarily have social meaning attached to them, have a position on a value hierarchy, and are directly or indirectly related to some social group. Most of the rest of Parsons' work may be seen as an explication of the environment, particularly cultural, wherein social action takes place.

Notice also that Figure 10.1 does not contain any predicted relationships; none of the arrows have positive or negative signs. This kind of diagram is an analytical model. That is, its function is to call our attention to particular items in a given phenomenon. So, when a human being acts, we can understand that action by breaking it down into these different parts. For example, think about the reasons you decided (voluntaristic action) to come to college. What was your goal in such a decision? How were the methods you used influenced by cultural norms, values, and beliefs? Were there any conditions of action that were set for you by others, like your parents or peers?

Modes of orientation: Having understood that human action is circumscribed by various conditions, Parsons sets out to understand those circumstances. In explaining the conditions, Parsons in the end creates an abstract theory of the social system. Let's begin by thinking about the situation. Remember, Parsons is thinking abstractly, so we don't want to be too specific. So, what do you bring with you to a social interaction, whether you are meeting to practice for a play or to study for a test? Parson argues that you bring two things: motivations and values. He refers to these as **modes of orientation** because they orient or position us within the situation, whatever it might be. Motives, of course, refer to something within a person (such as a need, idea, or emotion) that stimulates her or him to action. Motives are the energy for action: a person without motivation is like a car without gas—nothing happens. Value refers to a thing's worth and it is always a position on a hierarchy—some things are valued more highly than others. Thus, in every situation we are motivated to do something and we have certain ideas about what will be valued in the encounter with regard to the action.

> Orientation to the situation is *structured* . . . with reference to its developmental patterns. The actor acquires an "investment" in certain possibilities of that development. It matters to him how it occurs, that some possibilities should be realized rather than others. (Parsons, 1951, p. 8)

According to Parsons (1990), all social action is understood in terms of some form of relation between means and ends: "This appears to be one of the ultimate facts of human life we cannot get behind or think away" (p. 320). The relationship between means and ends is formed though shared value systems. In this sense, cultural value systems function basically as a scale of priorities that contains the fundamental alternatives of selective orientations. This shared value system prioritizes means and ends; and, because it is shared, value hierarchies stabilize interactions across time and situations. Action and interaction would be disorganized without the presence of a value system that organizes and prioritizes goals and means. The evaluative aspect of culture is particularly important in this respect because it defines the patterns of role expectations and sanctions, and the standards of cognitive as well as appreciative judgments for any interaction. In other words, the values that we hold tell us the kinds of behaviors we can expect from others and how to judge those behaviors and other social objects.

Let me give you an example. I had coffee with a colleague the other day. We are in the planning stages of a book on sociological social psychology. There are a number of ways to look at social psychology from a sociological position, and in my department there are several people who hold these various views. I am not planning on writing a book with them, nor did I invite them to go for coffee. It's not that I don't like them; I do, but I have a particular perspective about social psychology based on valuing certain aspects of the literature more than others. My colleague shares those views. Because of that shared value system, we knew what to expect from each other and how to value and appraise what was said and done, both intellectually and aesthetically.

Parsons sees value systems as having multiple levels that correspond to various degrees of commitment. Action may be stabilized through a shared system of meanings and priorities, but for a society to be integrated, people need to be committed to paying the costs necessary to preserve the system. In any functional interaction, we can name and prioritize the things that are important, but this discursive or cognitive accounting isn't enough. We must also be committed to some things more than others in terms of willingness to sacrifice. People are compelled to sacrifice when the collective *means* something to them, that is, when they have a significant level of emotional investment in the group; the more meaningful is the collective, the more willing people are to make sacrifices. Thus, the value system of any group varies by degree of commitment, with orientations and preferences at the most basic level and ultimate meanings and values at the highest level.

Like Durkheim, Parsons recognizes the basic human need for "ultimate" meanings, yet his argument concerning the need for ultimate significance is more tied to group identification and Weber's concern with legitimacy. Parsons argues that interaction requires that individual actions have meanings that are definable with reference to a common set of normative conceptions. In other words, our behaviors become meaningful because they are related to a group and its expectations. Group identity is, of course, symbolic. Humans are not tied together by blood or instinct but by meaningful issues such as freedom and democracy. By their very nature, symbols require legitimacy, grounds for believing in the meanings and system. No symbolic or normative systems are ever self-legitimating, nor are they legitimated by appeal to simple utilitarian issues. For example, we rarely hear someone justify the institution of American education by saying that it is necessary to be such for the survival of the American system (unless you're in a sociology class). Legitimating stories always appeal to a higher source. Thus, Parsons (1966) argues that legitimation is always "meaningfully dependent" (p. 11) upon issues of ultimate meaning and therefore is always in some sense religious.

There are three general kinds of values we hold: cognitive, appreciative, and moral. In other words, in any situation, we will place importance on empirical, factual knowledge (cognitive); standards of beauty and art (appreciative); or ultimate standards of right and wrong (moral). There are also three kinds of motives: cognitive, cathectic, and evaluative. Cognitive motivation refers to a need for information. You might be motivated to meet with your advisor because you need information about which courses to take. Cathectic motivation is the need for emotional attachment. You might feel the need to call home some weekend in order to experience emotional attachment to your family. With evaluative motivation, we are prompted to act because we feel the need for assessment, such as talking with your boss halfway to your year-end evaluation to find out where you stand.

There are also three types of cultural patterns (as I mentioned in the introduction, Parsons tends to link and embed his concepts with one another). Culture acts as a resource for both our motivations and our values in action, so it shouldn't surprise us to find that Parsons' types of culture correspond to his types of motivations and values. Culture, then, contains a *belief system.* While we might think of beliefs in a religious sense, Parsons has in mind belief as cognitive significance. It's interesting that he would phrase cognitions in terms of belief. He's acknowledging that the things we hold in our head, things through which we see and know the world around us, are in fact beliefs about the way things are. Culture also contains *expressive symbols.* Thus, culture not only provides the things we know, it also patterns the way we feel. These feelings are captured, understood, and expressed through symbols like wedding rings to express love, or gang colors to symbolize aggression. Culture also contains systems of *value-orientation standards.* It is culture that tells us what to value and how to value it.

These different kinds of motives, values, and cultures combine to produce three distinct types of social action. These function much like Weber's ideal types in that they are ways of understanding action, and none of them usually appears in its pure form. I've pictured how these ideal types are formed in Figure 10.2. Each type of action—strategic, expressive, and moral—is formed by combining a motivation with a value. Each of the specific culture systems provides information and meaning for each of the action types as well as the corresponding needs and values. As you can see, the ideal type of instrumental action is composed of the need for information and evaluation by objective criteria. Expressive action is motivated by the need for emotional attachment and the desire to be evaluated by artistic standards. Moral action is motivated by the need for assessment by ultimate notions of right and wrong.

In any social encounter, then, we will be oriented toward it with varying degrees of motivation and values. There are three different kinds of motivations and three types of value systems. The motivations and values will combine to create three different types of action. Any social action will have varying degrees of each type, depending on the specific combination of motives and values, and can be understood in terms of being closer or further from any of the ideal types of action. However, the importance of this scheme of social action is not simply its potential for measurement. Parsons goes beyond Weber in proposing a typology of social action and uses it to form a broader theory of institutionalization.

Institutionalization: What Parsons is doing is building from the ground floor to the top of the system, from the actor in the unit act to society. He starts with one small

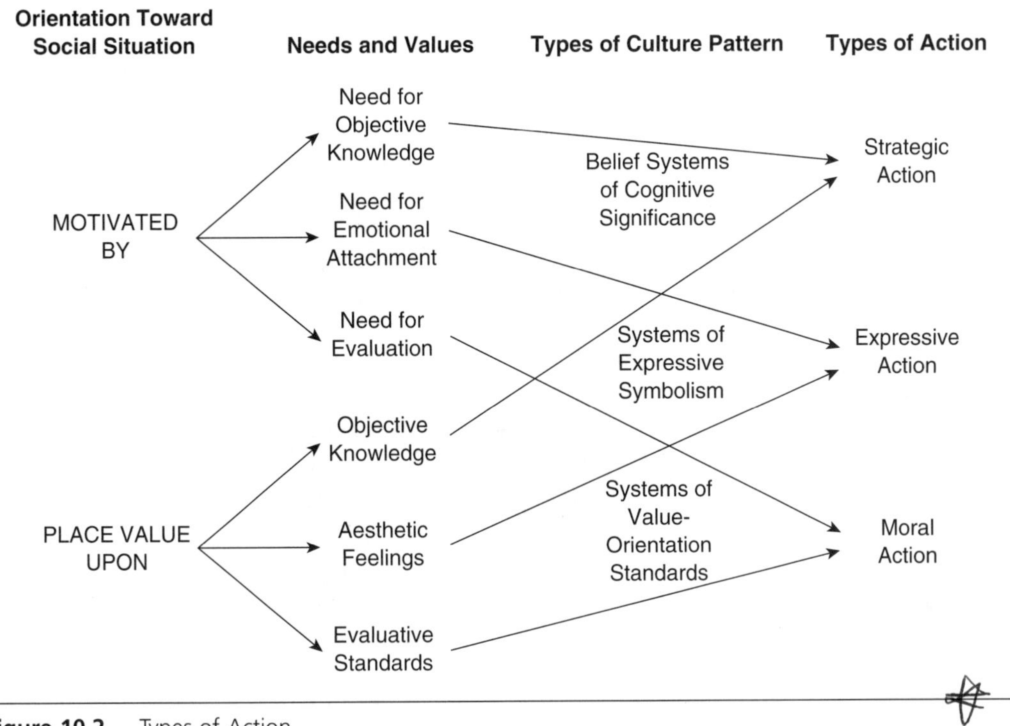

Figure 10.2 Types of Action

action and then argues that actors have discernable orientations toward their behavior. As a result, actions tend to fall into specific types. In brief, we have come this far: voluntaristic action → unit act → modes of orientation → types of action. From this point, Parsons argues that people tend to interact with others who share similar orientations and actors. So, if I want to engage in strategic, instrumental action, with whom will I be most likely to interact? For instance, if I'm interested in buying a guitar that has been advertised in the paper (strategic), then I'm not interested in interacting with someone who wants to talk about the evils of rock music (moral) or the beauties of a Shakespearian sonnet (expressive). Obviously, I will seek out others who want the same kind of thing out of the interaction. As we interact over time with people who are likewise oriented, we produce patterns of interaction and a corresponding system of status positions, roles, and norms. **Status positions** tell us where we fit in the social hierarchy of esteem or honor; **roles** are sets of expected behaviors that generally correspond to a given status position (for example, a professor is expected to teach); and **norms** are expected behaviors that have positive and/or negative sanctions attached to them. Together, these form a **social system**—an organization of interrelated parts that function together for the good of the whole. Society is composed of various social systems like these.

For this, Parsons gives us a theory of institutionalization. The notion of institutionalization is very important in sociology. Generally speaking, *institutionalization* is the way through which we create institutions. For functionalists such as Parsons,

institutions are enduring sets of roles, norms, status positions, and value patterns that are recognized as collectively meeting some societal need. In this context, then, **institutionalization** refers to the process through which behaviors, cognitions, and emotions become part of the taken-for-granted way of doing things in a society ("the way things are").

I've diagrammed Parsons' notion of institutionalization in Figure 10.3. Notice that we move from voluntaristic action within a unit act and modes of orientation to social systems. In this way, Parsons gives us an aggregation theory of macro-level social structures. One of the classic problems in sociological theory is the link between the micro and macro levels of society. In other words, how are the levels of face-to-face interaction and large-scale institutions related? How do we get from one to the other? Most sociologists simply ignore the question and focus on one level or another for analysis. Here Parsons gives us the link through the process of institutionalization. Large-scale institutions are built up over time as individuals with particular motivations and values interact with like-minded people, thus creating patterns of interaction with corresponding roles, norms, and status positions.

> In so far as ego's gratifications become dependent on the reactions of alter, a conditional standard comes to be set up of what conditions will and what will not call forth the "gratifying" reactions, and the relation between these conditions and the reactions becomes as such part of the meaning system of ego's orientation to the situation. (Parsons, 1951, p. 11)

There's a follow-up question to the micro–macro problem: once created, how do institutions relate back to the interaction? For Durkheim, the collective consciousness becomes an entity that can act independently upon the individual and the interaction. It does this through moral force. People feel the presence and pressure of something greater than themselves and conform. For Parsons, it's a bit different, even though he does acknowledge moral force. According to Parsons, all social arrangements, whether micro or macro, are subject to system pressures. Thus, institutions influence interactions not so much because of their independent moral force, but rather, because interactions function better when they are systematically embedded in known and accepted ways of doing things. In addition, rather than being dependent upon individual people or interactions, or having its own whimsical nature as Durkheim would have it, society is subject to self-regulating pressures (see the *Equilibrium* section under Parsons' perspective) because it is a system.

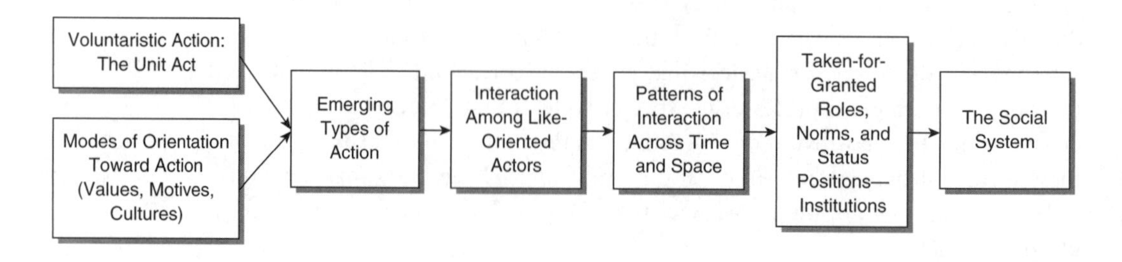

Figure 10.3 The Process of Institutionalization

Getting back to the actual process of institutionalization, Parsons argues that it has two levels: the structuring of patterned behaviors over time (this is the level we've been looking at) and individual internalization or socialization. Parsons understands internalization in Freudian terms. Freud's theory works like this: people are motivated by internal energies surrounding different need dispositions. As these different psychic motives encounter the social world, they have to conform in order to be satisfied. Conformity may be successful (well-adjusted) or unsuccessful (repressed), but the point to notice here is that the structure of the individual's personality changes as a result of this encounter between psychic energy and the social world. The superego is formed through these encounters. For Parsons, the important point is that cultural traditions become meaningful to and part of the need disposition of individuals. The way we sense and fulfill our needs is structured internally by culture. For Parsons, then, the motivation to conform comes principally from within the individual through Freudian internalization patterns of value orientation and meaning. As the same set of value patterns and role expectations is internalized by others, that cultural standard is said to be, from the point of view of the individual, institutionalized.

It is worth pointing out that Parsons argues that the content of the institutional solutions to societal needs doesn't matter. So, for example, it doesn't matter if a collective perpetuates itself biologically through the institution of family (however it is defined) or through an institutionalized hatchery like a chicken farm. What *is* important is that the perceived solutions are a set of highly ritualized behaviors that are seen as typical, belonging to particular settings (like church rather than school), and are believed to meet collective problems. I've diagrammed Parsons' bottom-up theory of institutionalization in Figure 10.3.

System Functions and Control

Systems needs: For quite some time, sociologists have been analyzing social institutions. One of the ways they have tried to understand these structures is through abstract analytical schemes. For example, we might say that societies all need religion, but why? What is it that religion does that society needs? Well, you might say, we all need to believe in something. Okay, but why? Obviously, I'm trying to push you to see things more abstractly, which is exactly what Parsons does.

Parsons argues that every system must meet certain needs in order to survive. This is where Parsons' grand theory comes in: he argues that these functions can be applied to any system (whether social, biological, personality, cultural, or any other). He proposes that every system has four needs: adaptation, goal attainment, integration, and latent pattern maintenance. Generally speaking, Parsons includes the three requisite functions of Spencer as well as the single requirement that Durkheim proposed. As we talk about the requisite functions, see if you can pick them out.

To get us thinking, let me ask you a question: Are you hungry? Maybe you aren't now, but sooner or later you will be, because every body needs food to live. Yet it isn't really food *per se* that you need. You need the nutrients that are in the food to survive. When you eat something, your body has a system that extracts the necessary resources from the food and converts it into usable things (like protein). So, your body doesn't really need a steak, it needs what is in it. Parsons calls this function **adaptation** because it adapts resources and converts them into usable elements.

Let's use a larger illustration. Every organism, society, or system exists within and because of an environment. For example, ducks are not found at the South Pole but penguins are. Each of these organic systems has adapted to a given environment and extracts from the surroundings what it needs to exist. It is the same with society. In order to exist, each and every society must adapt to its environment by inventing ways of taking what is needed for survival (like soil, water, seeds, trees, animals) and converting them into usable products (food, shelter, and clothing). Society must also move those products around so that they are available to every member (or at least most members). In society we call this subsystem the economy. The economy extracts raw resources from the environment, converts them into usable commodities, and moves the commodities from place to place.

I want you to be careful about something here. The economy and adaptation are not the same thing. The economy is *the subsystem in society that fulfills the adaptation need.* In the body, it's the digestive subsystem. Yet the digestive system and the economy are obviously not the same things. They fulfill the same function but in different systems. The reason I'm taking such pains here is that it is important to see that Parsons' scheme is very abstract and can be used to analyze any system, so I want you to be clear on how to apply it. (In addition, if your professor asks you to explain the adaptation function and you say that it is the economy, you'd be wrong.)

Every system also needs a way of making certain that every part is energized and moving in the same direction or toward the same goal. Parsons refers to this subsystem as **goal attainment**. In the human body, the part of us that activates and guides all the parts toward a specific goal is the mind. The mind puts before us certain goals, things that we need or want to do. We feel motivated to action because our mind invests emotion into these goals. Let's say that you have the goal of becoming the next Jimmy Hendrix. So you set about listening to all of the legendary rock guitarist's CDs, reading all the books about Hendrix's style, and practicing six hours a day. You also work a job and save your money so that you can buy the same kind of guitar and equipment that Hendrix used so you can sound just like him. You are motivated. Your mind has caught the image of yourself playing guitar and has controlled and coordinated your fingers, arms, and legs—in short, all your activities—to move you toward that goal. On the other hand, perhaps you aren't as motivated about school. After all, you're going to be a big rock star, so who needs school? So the different parts of your body are not energized and coordinated to meet the goal of doing well in school. In the body, it's the mind that coordinates all the different actions and subsystems to achieve a goal. In the social system, the institution that meets this need for goal attainment is government or polity (same meaning, different word).

> The concept of "integration" is a fundamental one in the theory of action. It is a mode of relation of the units of a system by virtue of which, on the one hand, they act so as collectively to avoid disrupting the system and making it impossible to maintain its stability, and, on the other hand, to "co-operate" to promote its functioning as a unity. (Parsons, 1954, p. 71n)

Systems also need to be integrated. By definition, systems do not contain a single part, but many different parts, and these parts have to be brought together to form a whole. Have you ever watched a flock of geese in flight? Rather than flying singularly in a haphazard manner, their actions are coordinated and integrated. The dictionary defines *integration* as meaning to form, coordinate, or blend into a functioning or unified whole; to unite with something else; and to incorporate into a larger unit. The geese are able to form into a larger unit mostly because of instincts. For human beings, it is a bit more complex. Humans generally use norms, folkways, and mores to integrate their behavior. Norms can be informal (like the norms surrounding our behavior in an elevator) or they can be formal and written down. Formal and written norms are called laws. Laws help to integrate our behaviors so that, rather than millions of individual units, we can function as larger units. Parsons refers to this function as **integration,** and in society that function is performed by the legal system. The legal system links the various components together and unites them as a whole. When, for example, Apple Computer crosses the boundaries of IBM, it is the legal system that makes them work together, even though they probably don't want to.

As should be apparent, polity and the legal system are intimately connected, because these two functions are closely related. In our bodies, for instance, the mind functions as the goal-attainment system and it uses the central nervous system to actually move the different parts of the body. But the mind and the central nervous system are two different things. A person can be completely paralyzed and still have full access to her or his mind; or the body can be in perfect working order with the mind completely gone. In the same way, polity and law are related but separate.

The final requisite function that Parsons proposes is **latent pattern maintenance.** Every system requires not only direct management, like that performed by a government, but also indirect management. Not everything that goes on in our bodies is directed through cognitive functions. Rather, some of these functions, like breathing, are managed and maintained through the autonomic nervous system, a subsystem that maintains patterns with little effort. Society is the same way. It is too costly to make people conform to social expectations through government and law; there has to be a method of making them *willing* to conform. For this task, society uses the processes of **socialization** (the internalization of society's norms, values, beliefs, cognitions, sentiments, etc.). The principal socializing agents in society are the structures that meet the requirement of latent pattern maintenance—structures such as religion, education, and family. (By the way, the word *latent,* from which Parsons gets his term, means not visible, dormant, or concealed.)

Parsons argues that these four requirements can be used as a kind of scheme to understand any system. When beginning a study of a system, one of the first things that must be done is to identify the various parts and how they function. Parsons' scheme allows us to categorize any part of a system in terms of its function for the whole. In Figure 10.4, I have diagrammed the way this analytical scheme looks. The four functions are noted by the initials AGIL. The larger box represents that system as a whole, which, of course, needs the four functions. Because they function as systems themselves, each of the four subsystems can be analyzed in terms of the same scheme. I've used adaptation in this case, but the same can be done with each of them.

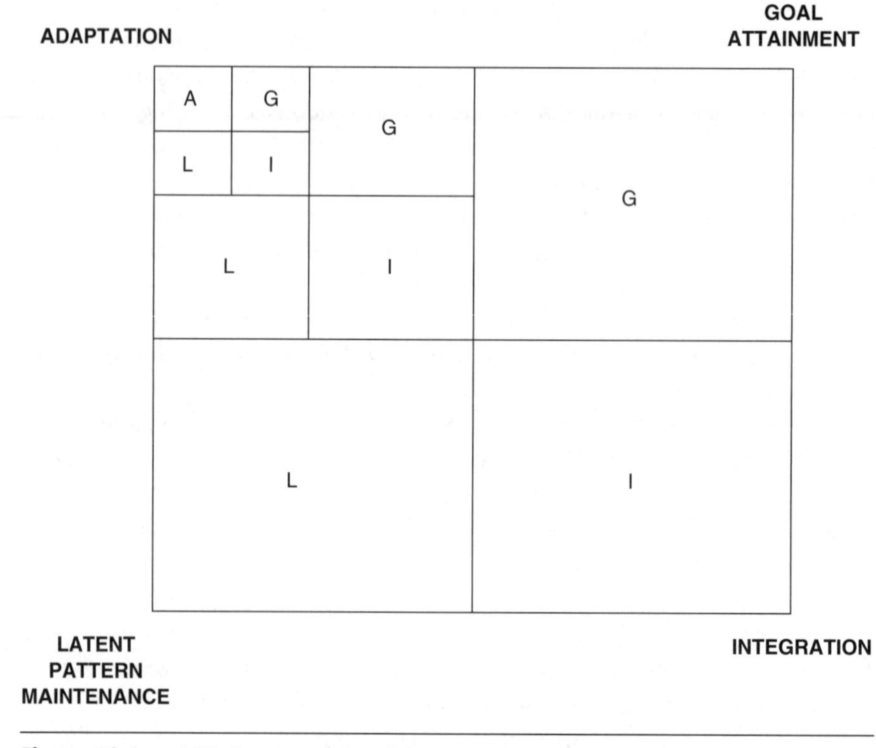

Figure 10.4 AGIL-Functional Requisites

Systemic relations: Further, Parsons gives us a way of understanding inter-institutional relations. This is important but little explored ground. What usually passes as institutional analysis is *within* an institutional sphere rather than between. For example, we might look at the institution of family in the United States and see that changes are occurring. Some of these changes are society-wide (like the increase in single-parent homes), some are the subject of much moral debate (such as whether or not to define gay couples as family), and some are present but not part of the public discourse (like an acquaintance of mine who introduced me to his wife, the mother of this children, and his girlfriend—three different women, all four people living happily in this arrangement). We can also look at marriage and divorce rates, birth and death rates, proportions of families under the poverty line, and so on.

These kinds of research agendas can be enlightening, but they are also quite limited. As we've noted, each subsystem is part of a whole and, as such, each is related to the other. If, for illustration, I put dirt in the fuel system of an automobile, it will affect the rest of the car, not just the fuel system itself. The same is true for society. If there are changes in one subsystem, those changes will ripple their way through all of society.

Family, in our society, is usually thought of as a married couple with 2.5 kids. Yet this model, called the nuclear family, has not always been the norm. In fact, it is a pretty recent model, historically speaking. Up through feudalism, marriage and family were far more important politically and economically. People got married to avoid wars or to seal economic commitments. As a result, the kinship structure was considerably more extensive and marriages were generally arranged because they were *socially* important. Today we think that marriage exists for the individual. Marriages in the United States are not arranged; we conceptualize marriage as existing principally for the individual and being motivated by love.

Thus, when we bemoan the loss of "family values," it is a historically specific set of values. These values came about because of changes in the rest of society. As institutions differentiated, the goal-attainment and adaptation functions were no longer dependent upon or related to family in the same ways. Bureaucratic nation-states emerged that were able to negotiate their interstate relations through treaty and war (using a standing army); the economy shifted to industrialized production and forced families to move from their traditional home to the city where most of the relationships that people have are not with or associated with family, as they were in traditional settings. Many other changes, such as the proliferation of capitalistic markets and the de-centering of religion, also influenced the definition, functions, and value of family.

The point that I'm trying to make is that for us to truly understand an institution, we must see it in its institutional context, in its relationships to other institutions. Parsons gives us a way to do just that using his AGIL scheme. He conceptualizes the institutions in the social system as relating to one another through "boundary interchanges." Each subsystem provides something for all the others. In Figure 10.5, I've pictured the boundary interchanges that Parsons gives us. What you will see is that each

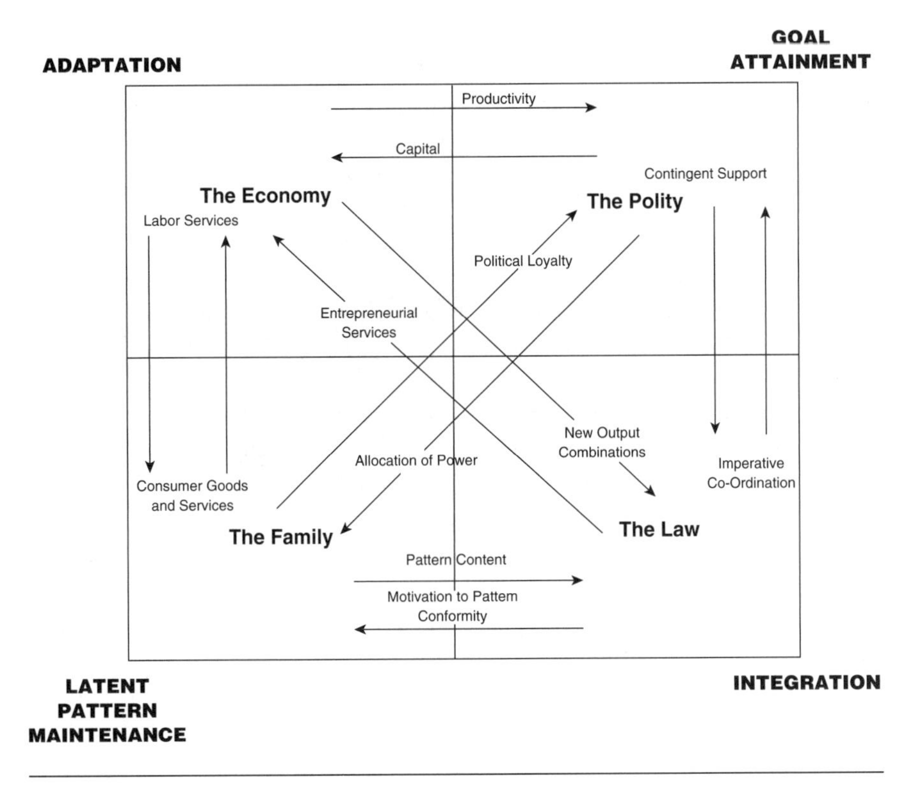

Figure 10.5 Interinstitutional Relations

relationship is defined in terms of what one subsystem gives to another. Both Spencer and Durkheim argued that as institutions differentiate, they become mutually dependent, but neither of them explicated the dependency. Here Parsons does that for us.

Let's look at family for a moment in the figure. Follow each of the arrows coming from family. These are the functions that it provides for the other subsystems. Through proper socialization, family provides political loyalty to the government; it provides a compliant pool of labor for the economy; and it influences the moral content of socialized patterns of norms that become law. I don't want to take us through each of these relations; you can do that on your own. The most important thing to glean is the idea of interinstitutional relations. In addition, while we may at some point become more sophisticated in our analysis of the associations, the place to begin is right where Parsons does: the functional dependencies.

Cybernetic hierarchy of control: Parsons develops an overall model of how the systems surrounding human life integrate. The model is called the general system of action or the **cybernetic hierarchy of control**. Cybernetics is the study of the automatic control system in the human body. The system is formed by the brain and nervous system and control is created through mechanical-electrical communication systems and devices. In using the term *cybernetic*, Parsons tells us that control and thus integration are achieved primarily through information. Also note that in cybernetics, control is achieved automatically, through what Parsons calls latent patterned maintenance.

I've outlined the control system in Figure 10.6. As you can see, the cybernetic hierarchy of control is understood through Parsons' AGIL system. (In the model, I have also expanded the social system to indicate what we have already seen: the social system is understood in terms of AGIL as well.) There are thus four systems that influence our lives: the culture, social, personality, and organic systems. The culture system is at the top, indicating that control of human behavior and life is achieved through cultural information. This emphasis on culture would obviously not be true for most animals. Regardless of the recent news and debates about apes being able to use and possibly share sign language, culture is not the primary information system for any animal other than humans. For most animals, information comes generally through sensory data, instinctual predispositions, and habitual patterns of action.

The position of culture at the top also indicates that it requires the most energy to sustain. As information flows from the top down, energy moves from the bottom up. Culture has no intrinsic energy. It is ultimately dependent upon the systems that are lower in the hierarchy for its existence. Without such energy, culture will cease to exist. For example, anthropologists and archaeologists know that a Babylonian culture existed at one time. That knowledge of past existence is itself part of our culture, but the Babylonian culture has long since died because its support mechanisms have passed away.

Culture is most immediately dependent on the social system for its existence. It is also dependent upon the personality system, because it is individual humans who internalize and enact culture. Since the personality system is dependent upon the organic system (the human mind needs the human body), culture is indirectly

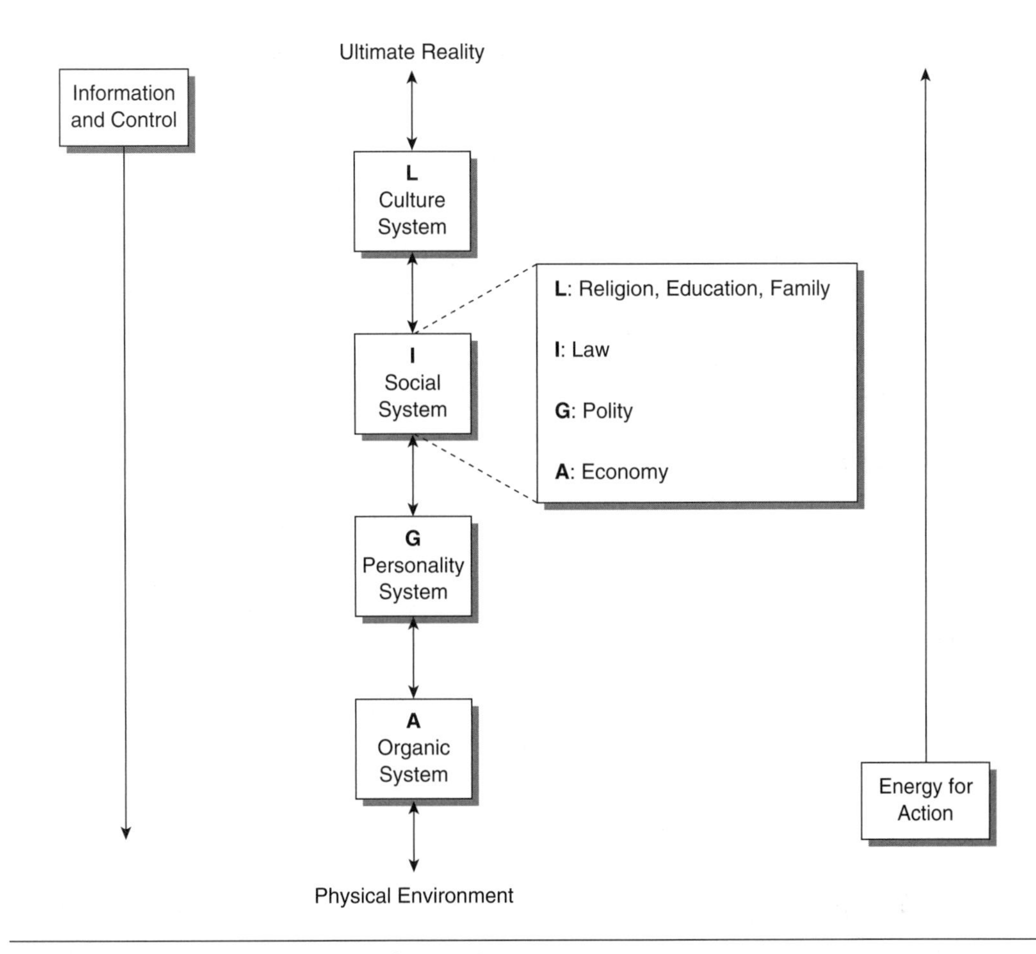

Figure 10.6 Cybernetic Hierarchy of Control

dependent upon it as well. I'd like to pause here and mention that recent theorizing argues that culture is also directly reliant on the organic body. Pierre Bourdieu (1984), for instance, argues that culture becomes embodied. There are not simply cognitive and emotional elements in culture. Culture also contains practices that form part of the way our bodies exist. Through our culture we develop tastes, dispositions, and automatic behaviors. For example, our taste in food is dependent upon our culture; and not only is our language cultural, but so is the way we speak it, as in regional accents.

Notice also that human life is contextualized by conditions of ultimate reality and the physical environment. Parsons never makes any comment about what ultimate reality is, but its understood existence is extremely important for the culture system. Remember that our most important values are framed in terms of ultimate truths, and these truths are religious in nature. Parsons, then, sees religion as an important influence on the culture system in general.

Overall, information moves down and energy moves up. Each system is embedded in and dependent upon the other—systems are reciprocally related to one another. One of the things that Parsons wants to point out with this kind of model is that differentiated, complex systems are dependent upon **generalized media of exchange** for facilitating communication and cooperation among and between the diversified parts. For example, in a complex society, each of the major structures has distinct goals, values, norms, and so forth. The capitalist economy has the goal of producing profit, while the education system has the goal of producing critical thinking. These value-oriented goals may at times clash, but a generalized media of exchange will tend to keep the system in equilibrium. Parsons offers language as the prototype of such generalized medium of communication and explicitly identifies money (from the adaptive subsystem), power (from goal attainment), and influence (integration) as other such media.

Social Change

Equilibrium: One of the critiques that is leveled against Parsons is that he only sees systems in equilibrium and his theorizing thus maintains the status quo. The criticism is not entirely correct. Parsons does assume that systems are in a state of **equilibrium**; that is, the forces of integration and disintegration are balanced. He feels that any social system worth studying would have a fair degree of permanence, and thus, "there must be a tendency to maintenance of order except under exceptional circumstances" (Parsons & Shils, 1951, p. 107). Parsons (1951) calls this tendency toward equilibrium the "first law of social process" (p. 205) and the "law of inertia" (p. 482). However, it is not the case that Parsons ignores social change. He actually has a notion of revolutionary change in addition to slow evolutionary change. Like Durkheim, Parsons argues that the principal dynamic of evolutionary change is differentiation. And like Spencer, Parsons sees that structural differentiation brings about problems of integration and coordination. Parsons argues that these problems would create pressures for the production of an integrative, generalized value-culture and a generalized media of exchange. Thus, like Durkheim, Parsons argues that culture is the most important facet of a complex social system. It is culture that provides the norms, values, and beliefs that allow us to interact, and it is culture that provides the information that the social system needs in general in order to operate.

Cultural strain: But the process of culture generalization, Parsons (1966) notes, may also bring about severe conflicts: "To the fundamentalist, the demand for greater generality in evaluative standards appears to be a demand to abandon the 'real' commitments" (p. 23). For example, in U.S. society, the call to return to family values is just such an issue. Societies that are able to resolve these conflicts move ahead to new levels of adaptive capacity through innovation. Others may "be so beset with internal conflicts or other handicaps that they can barely maintain themselves, or will even deteriorate" (Parsons, 1966, p. 23).

It is at this point that revolutionary change becomes more likely. Culture generally allows people and other social units (like organizations) to interact. It provides us with a language and value system. When people or organizations begin to value different kinds of things or to speak different languages, the situation is ripe for conflict. Parsons sees this kind of problem as a type of strain; *strain* is defined as a disturbance of the cultural expectation system. When we have different values, we do not know what to expect in an encounter. Strain always sets up re-equilibrating processes, but these processes may take a long time to reach balance and the system may be substantially different as a result. Change due to revolution occurs in two phases: 1) the ascendancy of the movement; and 2) the adoption of the movement as "setting the tone" for the society—the re-equilibrating process (this latter is the part that Marx and critical theorists leave out of their theories of revolutionary change).

Revolution: There are four conditions that must be met for a revolutionary movement to be successful. First, the potential for change must exist; Parsons refers to this potential as "alienative motivational elements." People become motivated to change the system as the result of value inconsistencies. These inconsistencies are inevitable and continually present in an empirical system of action, particularly one that has been generalized to incorporate a number of diverse groups, such as in the United States. For example, the term *equality* has been stretched to include groups not intended by the founding principals. The term has become more general; yet at the same time, the generality of the term sets up conflicts as more and more groups see themselves as disenfranchised, and others—the fundamentalists—see the generalization as movement away from truth. Second, dissatisfaction with the system is not enough to begin a revolutionary movement; the subgroup must also become organized. The organization of a group around a subculture enables members to evade sanctions of the main group, create solidarity, create an alternative set of normative expectations and sanctions, and it enables expressive leadership to arise.

Third, the organized group must develop an ideology that incorporates symbols of wide appeal and can successfully put forward a claim to legitimacy. The ability to develop an alternative claim to legitimacy is facilitated by two factors. One is that the central value system of large societies is often very general and is therefore susceptible to appropriation by deviant movements. The other factor is that serious strains and inconsistencies in the implementation of societal values create legitimacy gaps that can be exploited by the revolutionary group.

The fourth condition that must be met is that a revolutionary subgroup must eventually be connected to the social system. It is this connection that institutionalizes the movement and brings back a state of equilibrium. There are three issues involved: 1) The utopian ideology that was necessary to create group solidarity must bend in order to make concessions to the adaptive structures of society (e.g., kinship, education)—in other words, the revolutionary group must meet the reality of governing a social system; 2) the unstructured motivational component of the movement must be structured toward its central values and the movement must institutionalize its values both in terms of organizations and individuals; and 3) out groups must be disciplined vis-à-vis the revolutionary values that are now the new values of society.

Thinking About Modernity and Postmodernity

Parsons, science, and modernism: Parsons isn't specifically concerned with modernity. In fact, his position is decidedly in the direction of creating theories that transcend time and space and can be used to understand any society at any time. However, that position itself is characterized as modern by postmodernists. Modernity is not only defined in structural terms, as a period of time when industrialization, capitalism, and nation-states dominate, it is also defined by a particular kind of culture. We saw this specifically in our chapter on Durkheim. Durkheim's notion of a generalized collective consciousness is seen by postmodernists as a kind of grand narrative, a distinctly modern idea of cultural unification. Scientific theory that transcends time and place is also seen a grand narrative.

There are a couple of reasons for this position. First, science has a value base. This statement is a bit startling because science is supposedly value free. That is, science claims to be in pursuit of pure knowledge through the discovery of what empirically exists. Many postmodernists, however, argue that the objective stance of science is simply the technical arm of the ideology of European Western civilization. This brand of civilization defines progress in terms of technological advance, rather than humanistic ethics and morality, and it gives a select group of societies the ability to feel themselves superior to others. It's this combination that helped create the doctrine of manifest destiny in the United States and undergirded the colonization movements of Western European nations. If science is based on a value position (the universe is an object that is available for human use and control, rather than an equal part in the sacred universe that must be respected and honored), then it has no more basis to claim truth than any other knowledge system, and science is reduced to an ideological language in competition with other language games.

Another reason why science, and particularly scientific theory, is seen as part of the project of modernity is its belief in the cumulation of knowledge. *Cumulation* refers to a gradual building up and combining into one; and scientific theory is based on the cumulation of knowledge. Science progresses as theory is tested, modified, and restated. As Robert Merton (1967) explains, science is guided by the rule of "obliteration by incorporation." Each work builds upon and incorporates the work before it. The very meaning of scientific work is that each accomplishment will raise new questions and new problems to be solved. Each previous work should then be overshadowed and outdated by the next. Sir Isaac Newton embodied this scientific approach when he said, "If I have seen further it is by standing on the shoulders of Giants."

This kind of approach to knowledge and theory is exactly the one espoused by Parsons. *The Structure of Social Action,* Parsons' first major work, is explicitly a synthesis of four classic theorists: Marshall, Pareto, Durkheim, and Weber. A theoretical synthesis is a cumulative enterprise; it brings together different theories to form a whole. Parsons (1949) clearly states this kind of cumulation as his goal in *Structure:* "The keynote to be emphasized is perhaps given in the subtitle of the book; it is a study in social *theory,* not *theories*" (p. v). Parsons sees cumulation and obliteration by incorporation as the proper path for sociological theory.

The problem of logocentrism: So, what's wrong with theory cumulation? From a postmodern position, there are a couple of issues. Theory in this sense functions as a grand narrative. As theory is cumulated, the specifics of each theorist are discarded. The end result is a theory, or narrative, that presents itself as the only story worth telling. The problem is that there are many stories and the act of cumulation ignores them. We can take the theory book you hold in your hand as an illustration. In writing this book, I intentionally included the work of W.E.B. Du Bois, Harriet Martineau, and Charlotte Gilman. This attempt at inclusion is only a recent trend. It's been called "expanding the canon" of classical theory. The word *canon* has an ironic connotation—it refers to the third church council of Carthage, which in the year 397 decided which books could be considered scripture and which could not; the Bible was thus limited to the 39 books of the Old Testament and the 27 books of the New Testament. Until recently, sociological theory was canonized as well, being understood in terms of a restricted group of white men. The voices of Du Bois, Martineau, and many others were and are silenced. They were silenced for obvious value reasons: race and gender were not seen as important sociological concerns.

> Postmodernism criticizes the modernist notion that science itself . . . is a privileged form of reason or the medium of truth. It disputes the scientist claim that only scientific knowledge can be securely grounded. It takes issue with the unifying, consensus-building agenda of science. It contests the modernist idea that social theory has as its chief role the securing of conceptual grounds for social research. (Seidman & Wagner, 1992, p. 6)

From the postmodernist's position, the most insidious effect that canonizing the classics has is what Mark Gottdiener (1990) calls "logocentrism." (Notice the ironic connotation again: *logos* is the Greek term for *word*, which is a name used for Jesus in the New Testament.) *Logocentrism* refers to behind-the-scenes validation standards for sociological work. According to this point of view, in order for a new theory to be seen as valid, it must invoke one or more of the classical theorists. This idea implies that if I were to write a new theory, it would automatically be held suspect if it didn't refer to the work of Marx, Durkheim, Weber, or others in the accepted canon. If the legitimation of certain previous theorists over others is guided by values, as the history of theory indicates, then using a limited list of classics as a standard of legitimacy is in fact a political ploy by established theorists who want to control the discipline and maintain their privileged position. The postmodern stand on social theory is inclusive, not exclusionary.

Sociological and social theories: At this point in our discussion, I'd like to point out a distinction that some make between sociological theory and social theory. Before I begin to make comparisons, I want to first make a disclaimer: everything I'm going to say is generally but not necessarily *specifically* true. That is, the distinctions between sociological and social theory are broadly based and may not be entirely true about a specific theory that claims to be sociological or social. Generally speaking, then, sociological theory is understood to be oriented toward the discovery and explanation of the social factors that influence human feelings, thought, and behavior. The key elements are "discover" and "explain." For something to be discovered, it must first exist independently of the discoverer, as when Clyde Tombaugh discovered the planet Pluto on February 18, 1930. Thus, things that are discovered exist as objects. To explain this object implies that we are simply describing whatever it is as it exists; we aren't adding anything to the explanation, nor are we making any kind of value judgment about it. Sociological theory is oriented toward explaining how the social world works.

On the other hand, social theory is generally concerned with explaining and creating dialog around social issues. People who espouse social theory argue that the theorist ought to be politically and socially involved. To separate theory from the social and political situations of the theorist is to deny theory's source. Social theory emerges from the questions that the theorist generates from her or his experiences and the experience of others as social beings. Notice that social theory *emerges* through social experience; it is not involved in *discovery* because the theory and the theorist are intrinsically part of the issue. Social theory, then, doubts the ability of social actors to create an objective stand from which to gaze disinterestedly at society. Further, social theory is meant to be somewhat critical and it is intended for audiences larger than the scientific community, which usually constitutes the audience for sociological theory. Social theory is designed to incite social discourse.

Again, these are general distinctions, but they are informative and cause us to reflexively evaluate where we stand and the kind of theorizing in which we want to engage. In addition, sociological theory and social theory are not mutually exclusive, though they are rarely found in the same individual. A good example of someone who embodied some of both approaches is Karl Marx. He described the material dialectic in historical, mechanical terms. The material dialectic is the way society works—it is what drives history. At the same time, Marx advocated **praxis** (practically and politically applied theory), wrote political pamphlets, and headed political organizations.

Complex systems and neo-tribes: There is another point of the modernity/postmodernity issue that we can focus on with Parsons' theory. As we've seen, Parsons is a systems theorist. For Parsons, a system is a set of interrelated parts that as a whole are relatively self-sufficient. Such systems are comparatively stable, tend toward equilibrium, and have rather predicable results. Some people who work in the area of postmodernity argue that society no longer functions as such a system. Zygmunt Bauman (1992), for example, argues that as a result of de-institutionalization, people live in complex, chaotic systems. Complex systems differ from the mechanistic systems that Parsons talks about in that they are unpredictable and not controlled by statistically significant factors. In other words, the relationships among the parts are not predictable. For example, in a complex system, race, class, and gender no longer produce strong or constant effects on the individual's life or self-concept. Thus, being a woman, for instance, might be a disability in one social setting and not have any meaning at all in another; and race, class, and gender might come together in a specific setting in unique and random ways. Within these complex systems, groups are formed through unguided self-formation. In other words, we join or leave groups simply because we want to. And the groups exist not because they reflect a central value system, as Parsons would argue; instead, they exist due to the whim and fancy of their members and the tide of market-driven public sentiment.

The absence of any central value system and firm, objective, evaluative guides tends to create a demand for substitutes. These substitutes are symbolically, rather than actually or socially, created. The need for these symbolic group tokens results in what Bauman (1992, pp. 198–199) calls "tribal politics" and defines as self-constructing practices that are collectivized but not socially produced. These *neo-tribes* function solely as imagined communities and, unlike their premodern namesake, exist only in symbolic form through the commitment of individual "members" to the *idea* of an identity. But this neo-tribal world functions without an actual group's powers of inclusion and exclusion. Neo-tribes are created through the repetitive and generally individual or imaginative performance of symbolic rituals and exist only so long as the rituals are performed. Neo-tribes are thus formed through concepts rather than actual social groups. They exist as imagined communities through a multitude of agent acts of self-identification and exist solely because people use them as vehicles of self-definition: "Neo-tribes are, in other words, the vehicles (and imaginary sediments) of individual self-definition" (Bauman, 1992, p. 137).

Let me give you an example to illustrate what Bauman is getting at. During the 1960s, there was a youth-based social movement called hippies. Hippies had a very distinct culture. They were generally against war, practiced free love, and valued psychedelic drugs for their mind-expanding properties. Hippies often got together and interacted. They had sit-ins, love-ins, readings, protests, marches, concerts, and so on. And they developed cultural symbols that demarcated group boundaries—in particular the peace symbol and tie-dye tee-shirts. The only way to get a tie-dye tee-shirt was to either make one or have one made for you. And only hippies wore them back then. If you were a hippie, there was instant group recognition and solidarity when you saw someone wearing one. You knew what the person believed in and what she or he practiced. You knew how to talk with the individual and how to act toward her or him.

I saw one of my students the other day wearing a tie-dye tee-shirt. I asked where he had purchased it, and he told me at Wal-Mart. I saw the same student a week later wearing a Boston Marathon tee-shirt, which he said he'd bought over the Internet. He was neither a runner nor a hippie. I saw a professor I know at a bar last month. He was dressed as a cowboy, yet he's from New York City. All of these bits of clothing function as symbolic group tokens, and they create imagined communities—communities that exist in the moment, with no boundaries or ability to exclude or include. The professor became a member of a neo-tribe: groups that only exist because individuals are creatively defining themselves. He's not a cowboy, never was, and never will be. He will probably be at a college bar next week listening to modern rock and dressed post-grunge. And neither the rockers nor the cowboys can tell him to get out of the group, because there is no real group there.

Summary

- Parsons is the individual who is usually associated with clearly articulating a systems approach in sociology. This kind of theoretical method encourages us to see society in terms of system pressures and needs. Two issues in particular are important: the boundary between the system and its environment and the internal processes of integration. Parsons divides each of these into two distinct functions.

External boundaries are maintained through adaptation and goal attainment; internal-process functions are fulfilled by integration and latent pattern maintenance. Systems theory also encourages us to pay attention to the boundaries between subsystems, in terms of their exchanges and communication. Because relatively smart or open systems have goal states, take in information, and contain control mechanisms, they tend toward equilibrium. Parsons conceptualizes society as just such a system. In addition, because Parsons sees everything operating systemically, his theory is cast at a very abstract level and is intended to be applied to any and all systems.

- Parsons builds his theory of the social system from the ground up. He begins with voluntaristic action occurring within the unit act. Humans exercise a great deal of agency in their decisions; however, their decisions are also circumscribed by the situation and normative expectations. The normative expectations in particular are where human agency is most expressed and where culturally informed motives and values hold sway. These different motives and values orient the actor to the situation and combine to create three general types of action: strategic, expressive, and moral. People tend to interact socially with those who share their general types of action. As a result, interactions become patterned in specific ways, which in turn tend to create sets of status positions, roles, and norms. We may say that status positions, roles, and norms are institutionalized to the degree that people pattern their behaviors according to such sets and internalize the motives, values, and cultures associated with them.

- Different sets of institutionalized status positions, roles, and norms are clustered around different societal needs. Because society functions as a system, there are four general needs that must be met: adaptation, goal attainment, integration, and latent pattern maintenance. In complex, differentiated societies, these functions are met by separate institutional spheres. The different institutions are integrated through the system pressures of mutual dependency and generalized media of exchange. The social system itself is only one of four systems that surround human behavior. There are the cultural, social, personality, and physical systems, each corresponding to AGIL functional requisites. Because systems are dependent upon information, the culture system is at the top. Information flows from the top down, and the energy upon which culture is dependent flows from the bottom up. Parsons refers to this scheme as the cybernetic hierarchy of control.

- Systems tend toward equilibrium. They can, however, run amiss if the subsystems are not properly integrated. In the social system, this happens through cultural strain. As societies become more differentiated, the media of exchange must become more general. In this process, it is possible that some groups will seek to hold onto the dysfunctional culture. This case sets up a strain within the system, with some subsystems or groups refusing to change and other subsystems moving ahead. Motivation for social revolution is possible under these conditions. After people are motivated to change society, they must then create a subculture that can function to unite their group and create an alternative set of norms and values. This culture must eventually have wide enough appeal to successfully make a claim to legitimacy.

In a revolution, either side could win (the reformers or the fundamentalists), but in either case, certain steps are systemically required to reintegrate the system. After the revolution, the subgroup must produce a culture that can unite the system. Institutionalization occurs at this point as it does at any other time: through behaviors patterned and people socialized around a set of status positions, roles, and norms.

- Parsons' theory allows us to focus on some of the broader, more fundamental issues represented by postmodern theory. Postmodern theory proposes that society has changed at its core, in the way it works—society no longer functions as a predictable system. Society is better understood in postmodernity in terms of complex or chaotic systems. Simply put, this means that social elements can combine in nonlinear and undetermined ways. Bauman's notion of neo-tribes is a good example of this thinking. Postmodernism also challenges the way in which we construct knowledge. Thus, postmodernism isn't simply about the possibility that we may be living in a new kind of social world; it is also a critique of how we know what we know. Science and its methods are value laden; they came into existence in response to and support of the nation-state and Western colonialism. Science produces technologies of control, for both the physical and human environments, and is thus a part of modernist philosophy. The objectifying, linear way of knowing must be set aside for more subjective, language-based ways of knowing. Rather than giving truth-value over to one voice (science), we must give equal place to multiple voices. Rather than seeking to control the social world, we must reflexively place ourselves within the context and collectively work, through inclusion and dialog, to allow social worlds to emerge.

Building Your Theory Toolbox

Conversations With Talcott Parsons—Web Research

Unfortunately, as of the writing of this book, there are very few Web sources for Parsons, none of which provide "conversational" material.

Passionate Curiosity

Seeing the World (using the perspective)

- Parsons' primary point of view is that he sees things as a system. Recall what makes a system a system and analyze this society in terms of a system. Is this society a system? If so, in what ways? If not, in what ways does it not meet the criteria? Let's take it down a level: analyze the university you attend in terms of system qualities. Is it a system? What about your classroom? Is it a system in Parsonian terms? Can you analyze your friendship network in terms of systems? What about you? Do you exist as a system? If all these are systems, how are they linked together? Do you think systems analysis breaks down at any point?

- Can we understand globalization from a systems perspective? (I asked a similar question with Spencer, but Parsons' theory is more robust.)

Engaging the World (using the theory)

- According to Parsons, what is cultural strain? Why is cultural strain important? Under what conditions does cultural strain come about? Take a look at this society. Is it ripe for cultural strain? If so, why? What kinds of cultural strain can you identify? From a Parsonian approach, what are the effects we might expect?

- Earlier in this chapter, I asked you where you would begin if you were to put together a theory of society. I ask you again now: Where would you begin? Parsons begins with the unit act. Evaluate the unit act as a beginning point. What could be some benefits and drawbacks from such a starting point? Use Parsons' unit act analytical scheme (see Figure 10.1) to explain your behaviors at school today. Take the scheme and use it in at least five different settings (like school, home, shopping mall, crosswalk, beach, and so on). How does his scheme hold up? Were you able to analyze all of the behaviors equally well?

- Parsons talks about generalized media of exchange. In doing so, he is clarifying something that Spencer—and to a lesser degree, Durkheim—glossed over: the interdependency of institutions. Choose two institutions. What generalized media of exchange do you think exist between these two institutions? How would you go about determining if the media you propose are actually at work?

Weaving the Threads (synthesizing theory)

- One of sociology's abiding concerns revolves around the issue of social change. In the chapter on Durkheim, I asked you to consider Spencer and Durkheim's theories of social change and the problems of integration. In the chapter on Weber, I asked you to compare and contrast Marx and Weber on the issues of inequality and social change. How does Parsons' theory of revolution and change include the issues from the Spencer/Durkheim synthesis and the Marx/Weber synthesis? In other words, to what extent does Parsons' theory include both conflict and functional issues of social change? What does Parsons' theory leave out?

- Throughout these chapters, we have been weaving together a theory of religion. In the chapter on Simmel, I asked you to bring together the work of five theorists on religion. If you did, you would have a robust theory of religion in society. How does Parsons approach religion? What function does religion play in society, according to Parsons? In terms of the cybernetic hierarchy of control, where would something like religion fit? How does Parsons' approach compare to the other theorists?

- Spencer and Durkheim both propose requisite functions. Compare and contrast both Spencer's and Durkheim's lists with Parsons'. Where do Spencer's functions fit? Where does Durkheim's function fit? Does Parsons leave anything out?

- Another of sociology's abiding concerns is the relationship between the individual and society. Sometimes this problem is phrased in terms of agency (free will) versus structure (determination), and other times it is talked about as the micro–macro link. How are the individual and society related? How do the actions of people in face-to-face encounters get translated to macro-level structures? How do structures influence actors? Parsons doesn't answer all these questions, but he gives us one of our first detailed theoretical explanations of the micro–macro link. According to Parsons, how do the actions of people in face-to-face encounters get translated to macro-level structures? Does his theory seem reasonable to you? Can you think of how Durkheim or Weber or Simmel would explain this same phenomenon? How different or similar are their explanations compared to Parsons'?

Further Explorations—Books

Alexander, J. C. (Ed.). (1985). *Neofunctionalism.* Beverly Hills, CA: Sage. (Collection of articles and chapters that help define Neofunctionalism and Parsons' influence)

Hamilton, P. (1983). *Talcott Parsons.* New York: Tavistock. (Good, short, book-length introduction to the work and life of Parsons)

Lidz, V. (2000). Talcott Parsons. In G. Ritzer (Ed.), *The Blackwell companion to major social theorists.* Malden, MA: Blackwell. (Good, chapter-length introduction to Parsons' work and life)

Robertson, R., & Turner, B. S. (Eds.). (1991). *Talcott Parsons: Theorist of modernity.* Newbury Park, CA: Sage. (Leading social theorists use Parsons' theory to explain modernity, postmodernity, and globalization)

Treviño, A. J. (Ed.). (2001). *Talcott Parsons today: His theory and legacy in contemporary sociology.* Lanham, MD: Rowman & Littlefield. (Ten international scholars reappraise and extend Parsons' theory)

Further Explorations—Web Links

http://www2.pfeiffer.edu/~lridener/DSS/Parsons/parsbi01.html (Extensive biographical sketch of Parsons' life)

Further Explorations—Web Links: Intellectual Influences on Parsons

Sigmund Freud: http://www.ship.edu/~cgboeree/freud.html (Site maintained by Dr. C. George Boeree, professor of psychology at Shippensburg University; clear presentation of Freud's basic ideas and his life)

Max Weber: http://www.faculty.rsu.edu/~felwell/Theorists/Weber/Whome.htm (Site maintained by Frank W. Elwell at Rogers State University; site specifically oriented toward undergraduates and contains good explanations of some of Weber's concepts)

Émile Durkheim: http://www.relst.uiuc.edu/durkheim/ (The Durkheim Pages; contains reviews, summaries of Durkheim's work, timeline of Durkheim's life, glossary, and discussion threads that everyone can post to)

Bronislaw Malinowski: http://www.lse.ac.uk/lsehistory/malinowski.htm (Site maintained by the London School of Economics; short introduction to Malinowski)

Vilfredo Pareto: http://cepa.newschool.edu/het/profiles/pareto.htm (Site maintained by the Department of Economics, New School; extensive background and introduction to Pareto's theory)

Alfred Marshall: http://cepa.newschool.edu/het/profiles/marshall.htm (Site maintained by the Department of Economics, New School; extensive background and introduction to Marshall's theory)

CHAPTER 11

Theorizing Society

> *To theorize is to engage in the art of explanation. People who master the art know how to collect ideas, analyze content, and marshal conclusions in the most cogent reasoning possible.* (Mithaug, 2000, p. ix)

Final chapters are sometimes very difficult to write. We have come a long way and covered a lot of ground; how can I wrap up something like this? I would love for us to sit down and bring it all together. What did everybody say about religion and culture? What are the important dynamics of modernity? What kinds of things are leading us into the possibility of postmodernity? How do markets, money, communication technologies, transportation technologies, production, the division of labor, the centralizing of government, and so on—how do they all affect society and our lives?

As interesting and important as I think a discussion like that would be, I have a more important task in mind for this final chapter. In the opening chapter, we considered a couple of ideas surrounding theory and theorizing. We saw that theory is a perspective that is formed around two primary issues: assumptions about society, and values. We also saw that theory isn't simply what we know, it is something we do. Theorizing is an active, reflexive, and ongoing relationship between the mind of the theorist and the social environment. To theorize, we must actively think and ask questions. In this, our final chapter, I'd like to be a bit more proactive in passing on the torch of theory to you.

What to See—Provocative Possibilities

Ontology and values: This is a book about perspectives, and perspectives are about seeing: they are positions from which to view and attribute meaning, and they are made up of language, values, sentiments, norms, and beliefs. We've seen that there are two basic questions that contribute to our theoretical perspective about society: What is society (a question of ontology)? Phrasing this question differently, society exists, but how does it exist? What is the essence of society? Does society have an objective existence with its own set of laws? Or is society and everything therein an idea, a symbolic network through which we create and live by means of meaning? The second question is one of values; What should we do about society? What should our values be about this entity called society? Should we seek out those things that produce inequality and heartache and try to change them, or should we remove ourselves and simply describe society as it exists? The answers to these questions run on continuums that overlap at different places. Thus, in response to these two basic questions, we are able to create multiple and diverse perspectives that contain languages, values, feelings, beliefs, and scripts that tell us how to go about doing the business of sociology. Once produced, these perspectives tell us what to see and not see in the world around us.

The theorists we have considered in this book each have a perspective, a clear, definable way of viewing the social world. Some of them, like Durkheim, argue fervently for an objective, scientific approach to society. Others, like Weber, aren't so sure, and they approach social phenomena cautiously, being certain to pay attention to every nuance. Some, like Mead, see society emerging out of negotiated symbolic interaction; and others, like Schutz, see meaning created by the individual out of social stocks of knowledge and the backward glance of intentionality. Adding the value question, we get even more facets. Mead simply wants to describe how consciousness arises out of symbolic interactions; but Du Bois gives us a critical understanding of the same basic processes that Mead talks about—Du Bois sees that the culture of race creates different kinds of interactions and thus divergent consciousnesses. Marx is likewise critical and interested in consciousness, but he sees it being produced through structured economic relations, rather than culture or interaction.

Levels of analysis: Another defining issue of our theorists' perspectives is their level of analysis. As we've seen, social phenomena occur at three distinct levels: the macro level of historical, structural dynamics and relationships; the meso level of organizational structure; and the micro level of face-to-face encounters. Some of our theorisits are clearly linked to one level or another. Thus, for Spencer, the basic dynamics of social change and stability are at the level of entire societal populations; Gilman, like Spencer, sees the dynamics of society in terms of large-scale evolutionary development, but she, unlike Spencer, lends a critical perspective, concerned like Marx with a specific category of people—in this case, gender. On the other hand, Mead sees negotiated symbolic meanings behind social change and stability that occur only at the level of the interaction and self. Some other theorists are an interesting blend: while we generally think of Parsons as a macro theorist, he in fact begins at the level of the act and builds a theory of institutionalization—in other words, Parsons gives us a way of seeing the link between the macro and micro levels. People like Du Bois and Weber see the individual or micro level within historical milieus and then wonder about the effects of such historical processes upon consciousness and subjective experience.

In Table 11.1, I've begun a task I encourage you to finish. As you can see, I note the theorists' general perspective of society (objective/subjective), their position concerning values (critical/descriptive), and their level of analysis. After noting these issues, I've written a brief statement that I think encapsulates their perspective. The value of such a table is that it gives us a quick reference to specific perspectives. In turn, these perspectives give us eyes to see distinctive questions. The questions that each of our theorists asks are intrinsically linked to the kind of perspective each has. Accordingly, seeing the world through a Marxian perspective will enable you to ask questions that are distinctly different than if you see the world through Durkheimian eyes. *This kind of effort is fundamental to theorizing.* Theorizing begins with a base perspective—we will either construct this base reflexively and consciously, or our perspective will be a result of happenstance. Good theorizing can only be built on a consciously adopted base. Putting all of our theorists in such a table will allow you to not only be reflexive about your theorizing but flexible as well.

Flexible theorizing: Flexible theorizing comes out of the realization that perspectives are not right or wrong, they are either useful or not so useful. Why can't they be right or wrong? Because they are based on assumptions that are never tested, and thus, all assumptions (and perspectives) are taken on faith. Perspectives can't be right or wrong, but they can be more or less powerful in yielding and informing

Table 11.1 The Theorists' Perspectives

Theorist	*Society*	*Value*	*Levels*
Herbert Spencer	Objective	Descriptive	Macro
Spencer views society as progressively evolving due to the same objective elements (matter, force, motion) and dynamics (instability of homogeneous units, multiplication of effects, segregation) as the rest of the universe. For society, the basic elements revolve around population size and the dynamics around structural relations. Spencer's perspective is oriented toward the macro level; he gives us the basic tenets of functionalism and eyes to see the interconnections among structural units.			
Karl Marx	Objective	Critical	Macro → Micro
Marx views society as determined by economic relations and changing in response to dialectic elements within the economic structure. Marx's perspective is oriented toward the macro level in terms of dynamics, but for outcomes Marx is concerned with the micro level (consciousness); in doing so, he lays down the foundation for critical and conflict theorizing, and gives us eyes to see the effects of oppressive structures on human potential.			
Émile Durkheim	Objective	Descriptive	Micro → Macro
Durkheim sees society as a result of emotionally infused interactions; once born, the morally infused symbols that come from interaction take on an objective life of their own. The micro level is the basis for society, but macro-level cultural elements strongly inform what happens in interactions. Durkheim gives us eyes to see the issues surrounding cultural integration of society.			

certain kinds of theoretical questions. For instance, if you are interested in class structure and oppression, whose perspective would yield the most insightful questions and answers? Marx would be the most obvious answer. However, what if you want to ask questions about the intersection of race and class? To whom would you turn? In this case, Marx would not be as helpful as Weber and/or Du Bois. What if you are not so interested in how historical, structural relations between class and race come about, as in how race and class intersect and are produced in face-to-face interactions? In that case, you would turn to Mead's perspective. On the other hand, perhaps you don't think the macro structures of Marx or Weber are the real issues and you don't buy into Mead's reduction to meanings in interaction. Then where do you turn? Well, maybe the intersections between race and class actually operate as social forms that colonize the subjectivities of the people in interaction. Who can give you this perspective? Simmel, of course. Or maybe class and race crossroads are not constructed at the macro level of structure, or at the micro level of interaction, and not at the more meso level of social forms. What if they are part of the stocks of knowledge and relevance hierarchies that individuals use to construct meanings in We and Thou relations? The best person to turn to here is Schutz.

There is obviously an important, general point I want you to see in all this: *you can't begin to think about different possibilities unless you can take different perspectives.* The more perspectives with which you are familiar and comfortable, the better prepared you are to see multiple, provocative possibilities in the social world around you. Theory isn't stagnant; it is a living, breathing thing that moves us around a social phenomenon, granting us new insights into the world around us. How many different kinds of perspectives do you own? How many different kinds of lights can you shed on the world around us? What can you see, what can you hear? Now is the time for you to own your knowledge; that's why I'm asking you to complete the table, rather than my doing it for you. Don't be satisfied with only seeing part of the world, or having someone tell you what to see—make it your goal to own your own knowledge and to see the world in which you live more complexly by being able to use multiple perspectives.

In addition to perspectives, I've emphasized two other issues throughout these chapters: recurring themes and our historical era. The possibility that we are living in an age of transition is a stimulating idea. Most of our classical theorists lived in an age of transition: the birth of modernity. They became preoccupied with asking certain kinds of questions about their world. They were interested in religion, culture, the state and economy, equality and oppression, social integration, and change. One of the possibilities that I want us to consider is that society is changing, and with it the kinds of questions we ask. In particular, I've brought out certain features of postmodern theory, such as advertising, mass media, institutionalized doubt, and so on. At the same time, I've made it clear that many of these questions are or can be informed by the classical theorists. For example, Marx's theory of consciousness (the link between the mode of production and the way we see the world and ourselves) forms the basis of Jameson's version of postmodernism; and Weber's ideas of class and status groups informs Lash and Urry's postmodern theory. In addition, we've seen theories that may provide an answer to some of the typical postmodern problems, such as Mead's idea of fusing the I and the Me.

One of the things that I obviously want you to get ahold of—or better, have it get ahold of you—is our historical possibility: this may be the beginning of a new age, and I want that to fire your imagination. But I am also hoping for something a bit more mundane. Throughout this book, I've tried to model the activity of theorizing for you. As we have made our way through the theorists, I've given you opportunities at the end of the chapters to build your theory toolbox; one of the important ways to do so is by weaving together what the theorists say about the common themes. Asking you to "Compare and contrast Marx, Durkheim, and Simmel on religion" may sound like a test question, but it is so much more. Those kinds of questions are the very ways through which professional sociologists theorize. It's how I linked classical theory with the possibility of postmodernity. Theory is an activity that questions, compares, contrasts, and synthesizes. It is an activity that begins with reading.

How to See—Tools of the Trade

Reading well: The first step in theorizing is reading, which is what I've tried to introduce you to in this book. While I may be able to orient you to Durkheim's theory, there is nothing that substitutes for reading Durkheim. That's one of the reasons that these people's works are considered classic—repeated reading gleans new insights. However, chances are you don't know how to read. Oh, I imagine you can read well enough; otherwise you wouldn't have progressed this far in the book. But perhaps you don't know how to *read in a way that will remake your mind.* And that is the challenge and thrill of theory—if we really get what a Martineau or Mead is saying, the way we view the world will be changed forever. In reading theory, there are two tenets and several questions to keep in mind.

- TENET 1: Read the work repeatedly—at least three times. No, I'm not kidding; and no, you can't do a good job of understanding the theorist without immersing yourself in the work. Read with a notepad and pencil; each time you read through the work (or section of the work), write down observations and questions (and answers to your questions.).
- TENET 2: Use your imagination. Whatever else theorizing may be, remember that theoretical work always involves the imagination—the ability to see and understand and also to think outside the lines. Every theorist is one who has seen something that no one else has seen. Look at your world imaginatively; read the works using your imagination.
- QUESTIONS TO ASK OF THE WORK: The following questions are meant as starting issues only. Your understanding will not be complete if you simply answer these questions. Use these to help you ask more questions. Also keep in mind the sets of tensions we talked about earlier. Most classical theory will address at least some of these. In general, you want to discover the argument. Remember that these people are addressing an issue or question. What is it? Everything else in the work should revolve around it. After you've discovered the central point, find the stepping-stones that the author uses to lead us to the conclusion. In particular ask,

What are the general contours of the phenomenon? Every theory pertains to some specific phenomenon. We can always creatively extend a theory, but we first must know exactly what a theory is about. The largest defining issue here is the question the theorist is addressing. You cannot understand the theory unless you know the question. For example, Durkheim had two issues in mind when he wrote *The Elementary Forms of the Religious Life.* The first is obvious: he wanted to know what constitutes the basic elements of religion, regardless of when, where, or what kind of religion is under consideration. His second purpose was to uncover the source for the basic categories of knowledge that we use (like time and space)—he was arguing for a social epistemology (studying how we know what we know). Unless we keep in mind each of these purposes, we will be confused by some of what Durkheim is saying. In addition to the chief questions, here are some other ways to understand the general contours of the phenomenon:

- What is the level of analysis? Is the theory concerned with macro-, micro-, or meso-level phenomenon? Does the theory argue for bridges among or between these levels? Or, does the theory argue that those distinctions are false? What is presented in their stead?
- What are the theory's scope conditions? Identify the limits of the phenomenon to be described and provide its definition. Is the theory about the economic structure, the state, inequality, self/identity, authority structures, or something else? What are the other limits? Is it only concerned with traditional, modern, or postmodern societies?
- What are the temporal dimensions of the theory? How much time is involved for the processes to work? Do they diminish over time or increase?

What are the major concepts? As we've said, theory is an insightful, formal, and logically sound argument that is built from concepts and definitions. Concepts are the building blocks of all theory. Concepts tell us what to see and the definitions make our theory specific and replicable. So, every theory will have concepts and they are the single most important element. Under "concepts," I am also including analogy, types, and analytical frameworks.

- Look for repeated ideas. Are there concepts that appear over and over? Also be aware that concepts that are repeated often indicate more complex relationships than simple, linear ones.
- Look for linchpin concepts—those concepts without which the entire argument would fall apart. It is not the case that all the important concepts are repeated over and over. Sometimes they are only spoken of once or twice. But if you remove that concept, the work or the theory becomes incomprehensible (or at least impoverished).
- Look for concepts for which the author gives explicit definitions. Unfortunately it is not always the case that theorists intentionally give us exact definitions for their concepts. We oftentimes have to infer this element of theory. However, if the theorist does give us the definition, then we can be fairly certain that the concept is important.

- *Remember this:* theoretical concepts are technical terms, which means that they have special kinds of definitions. You can't define a theoretical term the way you would any other word. For example, to define Durkheim's concept of collective consciousness as culture or even as the norms, values, and beliefs of a society, falls well short of the mark. Generally speaking, you should look for four attributes of a theoretical concept. Here's a handy acronym to help you remember them—BEVO: **B**oundaries (What counts as an instance? What are the defining features?); **E**ffects (What is created? How does it influence the individual, group, or society?); **V**ariation (How does the concept vary? What produces higher or lower amounts?); **O**peration (How does it work? What are the internal processes?). Not every concept will have all of these attributes, but if it is a theoretical concept, it will have one or more of them.

Can the concepts/issues be made more abstract? If we are looking at or wanting to produce a scientific theory, or a theory that can be used in more than one setting, then the concepts have to be abstract. Actually, it is better to think of abstraction as a variable. Concepts are not simply empirical or abstract—they are more or less abstract. The more abstract a concept is, the more powerful, in terms of being usable in multiple settings (explaining, predicting, controlling).

What are the relationships among and between the concepts? Here again we must address the issue of the kind of theory we are reading or using. Remember that certain kinds of theories do not propose relationships (ideal types, analytical schemes, and so forth). So be aware that sometimes there will not be any relationships (like in symbolic interactionist theories). Determining whether or not there are relationships and their nature is extremely important. Often, the author will tell you straight out the kind of theory she or he is using, but many times the author won't. So you will have to search for the relationships. Occasionally, the relationships are explicitly laid out in models or propositions, but most of the time they are buried in the text.

- Look for such words and phrases as "results from," "increase in," "development of," "consequence," "doing away with," and the like.
- Keep in mind the kinds of relationships you might encounter: positive, negative, linear, curvilinear, additive, multiplicative, and so on.
- *Remember this:* theories of this sort explain how something works or comes about. They should read, then, like a mechanic's manual or recipe. A recipe will tell you exactly what you have to do to get a given result. My chocolate chip cookie recipe, for example, explicitly tells me what ingredients to use, what to do to those ingredients, when to put them in the oven, how hot to have the oven, how long to bake them, and so on. Marx's theory of class consciousness is very much like that. It has specific ingredients that come together and work in precise ways in order to produce class consciousness. In a paper or examination, when you write a scientific theory, it should have these qualities.

In the case of critical theory, what ideologies are being exposed? What are the relations of power? Which groups are systematically excluded from the social system and which are generally on top? What kind of society is being espoused? What are the values of the theory? What action is advised?

In the case of non-critical theory, what values are present but not reflexively acknowledged? As I mentioned, all theory is culturally based or informed (even science), and all culture is based on values. Even social scientific theories have values, even though they are generally not acknowledged.

Transforming theory: After reading (there really isn't an "after"—reading should be ongoing), there are a few approaches we can use in theorizing. We can *modify an existing theory* in some way. There are a number of methods we can use to make this modification. For example, we can expand or reduce a concept's definition. Take Durkheim's notion of ritual, for instance. By ritual, Durkheim generally meant patterned behaviors that create emotional effervescence. We could expand that idea by explicitly including the concept of dramatic enactment—the picturing of some important social event. By doing this, we will be able to see more of what is going on during a ritual like the Fourth of July. Fourth of July celebrations not only have the potential of increasing the level of experienced emotion, thus producing a sacred quality around the symbols used, they also emblematically picture certain events in American history.

Quite a bit of theoretical progress is made through expansion and contraction. Often, these changes are made through processes of disagreement. For example, we may disagree with Marx's theory of estrangement. As it stands, it is a critical theory of capitalism and alienation that occurs within the context of human beings cut off from species-being. Perhaps we don't buy into Marx's notion of species-being and maybe we don't want to condemn all of capitalism, but at the same time we do want to be able to talk about some of the alienating effects of working on an assembly line. In this case, we can argue against Marx's overall critique but accept and use his idea of alienation. But be aware that taking it out of Marx's context changes the meaning of the term—it constricts it in a way that makes it usable in our theory. Notice that in order to expand or contract a concept, *we must first understand completely what the theorist originally meant.*

We can also *synthesize theories.* To synthesize means to combine or put together. So we can put together elements from different theories to form something new. That's basically what George Herbert Mead did. He took ideas from William James (self and consciousness), Wilhelm Wundt (general will), and Charles H. Cooley (looking glass self), and formed what became known as symbolic interactionism. This approach to theorizing is fairly common. The contemporary theorist Anthony Giddens draws elements from structuralism, functionalism, phenomenology, dramaturgy, symbolic interactionism, and psychoanalytic theory to form his approach, called structuration theory.

We can synthesize the theories we have looked at as well. Let's say that we want to understand how the self is formed in modernity and what its unique characteristics might be. None of the theorists that we deal with in this book try to answer those questions directly. However, we can take Mead's theory of self-formation (role-taking and generalized other) and combine it with elements of Simmel's theory of the metropolis and mental life (urbanization, rational group membership, cognitive stimuli, blasé attitude) to form a theory that would answer those questions.

Theory modification and synthesis are the two main ways in which theory is produced and "progresses." One of the main reasons for this is the amount of work that has been done up to this point. You've heard the saying, "there is nothing new under the sun." That may not be exactly true, but it is certainly the case that we have explicated much of the social world and reinventing the wheel is a meaningless effort.

Grounded theorizing: Theory can also be produced through what is called "grounded theory." **Grounded theory** (and approaches like it) is the way in which interpretive theory is shaped. It begins at the ground level, *in situ,* in the context of human action and interaction. As much as possible, people who use a grounded theory approach attempt to come to a social phenomenon with as few preconceptions as possible. They want the situation and the participants to speak to them. They are interested in how the actors themselves view, interpret, and act within their own context.

One of the problems with grounded theory is that it is based on the exclusion of all preexisting concepts. Many researches feel that getting rid of all concepts is not possible. To help solve this problem, Derek Layder (1998) proposes an intermediary approach that he calls "adaptive theory." This approach recognizes the worth in being grounded in the situation, but also acknowledges that existing theoretical ideas and frames have value (and are really tough to get out of your head). Layder recommends that we move back and forth between the situation and theory. The adequacy of the theory, then, is measured in two ways. In subjective adequacy, the researcher is concerned with how well the concepts reflect the lived experiences of the people in the context. The researcher constantly asks the participants about the concepts and definitions. Do they make sense to the subjects? The second kind of adequacy is analytic. Here the researcher needs to tie the subjectively adequate concepts to the theoretical literature. In using this approach, then, concepts are developed in an ongoing manner and in dialog with the subjects, the context, and existing theory.

My primary motivation in writing this book was to inspire provocative thinking about the social world. I hope I have planted a seed or two. Ideas drive the human world. Every major change and development in history has been fostered and fueled by ideas; and behind the great ideas of human epochs are theories. Social and sociological theory is part of this great tapestry of thought. The people we have looked at in this book are among its most skilled weavers. Yet, while we may stand in awe of such thinkers as Weber and Martineau, theorizing is not just for the great. You, too, can enter this creative stream. And who knows? There might be a group of students reading your books a hundred years from now.

References

Aggar, B. (2004). *Syllabus for Sociology 5341: Contemporary Social Theory.* University of Texas at Arlington. Available: http://www.uta.edu/huma/agger/5341.html

Alexander, J. C. (1985). Introduction. In J. C. Alexander (Ed.), *Neofunctionalism.* Beverly Hills, CA: Sage.

Alexander, J. C. (1988a). Parsons' 'Structure' in American sociology. *Sociological Theory, 6,* pp. 96–102.

Alexander, J. C. (Ed.) (1988b). *Durkheimian sociology: Cultural studies.* Cambridge, UK: Cambridge University Press.

Allan, K. (1998). *The meaning of culture: Pushing the postmodern critique forward.* Westport, CT: Praeger.

Allan, K., & Turner, J. H. (2000). A formalization of postmodern theory. *Sociological Perspectives, 43*(3), 363–385.

Anderson, P. (1998). *The origins of postmodernity.* London: Verso.

Barthes, R. (1967). *Elements of semiology* (A. Lavers & C. Smith, Trans.). London: Jonathan Cape. (Original work published 1964)

Baudrillard, J. (1993). *Symbolic exchange and death.* Newbury Park, CA: Sage.

Baudrillard, J. (1994). *Simulacra and simulation* (S. F. Blaser, Trans.). Ann Arbor: The University of Michigan Press. (Original work published 1981)

Bauman, Z. (1992). *Imitations of postmodernity.* New York: Routledge.

Bellah, R. N. (1970). *Beyond belief: Essays on religion in a post-traditional world.* New York: Harper & Row.

Bellah, R. N. (1975). *The broken covenant: American civil religion in time of trial.* New York: Seabury Press.

Bellah, R. N. (1980). *Varieties of civil religion.* San Francisco: Harper & Row.

Bellah, R. N., Madsen, R., Sullivan, W. M., Swidler, A., & Tipton, S. M. (1985). *Habits of the heart: Individualism and commitment in American life.* New York: Harper & Row.

Bellah, R. N., Madsen, R., Sullivan, W. M., Swidler, A., & Tipton, S. M. (1991). *The good society.* New York: Knopf.

Berger, P. L. (1967). *The sacred canopy: Elements of a sociological theory of religion.* New York: Doubleday.

Berger, P. L., & Luckmann, T. (1966). *The social construction of reality: A treatise in the sociology of knowledge.* Garden City, NY: Doubleday.

Bourdieu, P. (1984). *Distinction: A social critique of the judgment of taste* (R. Nice, Trans.). Cambridge, MA: Harvard University Press.

Cassirer, E. (1944). *An essay on man.* New Haven, CT: Yale University Press.

Charon, J. (2001). *Symbolic interactionism: An introduction, an interpretation, an integration.* Upper Saddle River, NJ: Prentice Hall.

Chodorow, N. (1989). *Feminism and psychoanalytic theory.* New Haven, CT: Yale University Press.

Collins, R. (1986). *Max Weber: A skeleton key.* Newbury Park, CA: Sage.

Collins, R. (1988). *Theoretical sociology.* Orlando, FL: Harcourt Brace Jovanovich.

Collins, R. (1998). *The sociology of philosophies: A global theory of intellectual change.* Cambridge, MA: Belknap Press of Harvard University Press.

Comte, A. (1896). *The positive philosophy of Auguste Comte* (H. Martineau, Trans.). London: G. Bell & Sons.

Coser, L. (1956). *The functions of social conflict.* Glencoe, IL: The Free Press.

Coser, L. A. (2003). *Masters of sociological thought: Ideas in historical and social context* (2nd ed.). Prospect Heights: IL: Waveland Press.

Csikszentmihalyi, M. (1990). *Flow: The psychology of optimal experience.* New York: Harper & Row.

Du Bois, W. E. B. (1996a). The souls of black folk. In E. J. Sundquist (Ed.), *The Oxford W. E. B. Du Bois reader.* New York: Oxford University Press. (Original work published 1903)

Du Bois, W. E. B. (1996b). Darkwater. In E. J. Sundquist (Ed.), *The Oxford W. E. B. Du Bois reader.* New York: Oxford University Press. (Original work published 1920)

Du Bois, W. E. B. (1996c). The propaganda of history. In E. J. Sundquist (Ed.), *The Oxford W. E. B. Du Bois reader.* New York: Oxford University Press. (Original work published 1935)

Du Bois, W. E. B. (1996d). In black. In E. J. Sundquist (Ed.), *The Oxford W. E. B. Du Bois reader.* New York: Oxford University Press. (Original work published 1920)

Durkheim, É. (1938). *The rules of sociological method* (S. A. Solovay & J. H. Mueller, Trans.; G. E. G. Catlin, Ed.). Glencoe, IL: The Free Press. (Original work published 1895)

Durkheim, É. (1951). *Suicide: A study in sociology* (J. A. Spaulding, & G. Simpson, Trans.). Glencoe, IL: The Free Press. (Original work published 1897)

Durkheim, É. (1957). *Professional ethics and civic morals* (C. Brookfield, Trans.). London: Routledge.

Durkheim, É. (1961). *Moral education: A study in the theory and application of the sociology of education* (E. K. Wilson, Trans.). New York: The Free Press. (Original work published 1903)

Durkheim, É. (1984). *The division of labor in society* (W. D. Halls, Trans.). New York: The Free Press. (Original work published 1893)

Durkheim, É. (1993). *Ethics and the sociology of morals* (R. T. Hall, Trans.). Buffalo, NY: Prometheus. (Original work published 1887)

Durkheim, É. (1995). *The elementary forms of the religious life.* New York: The Free Press. (Original work published 1912)

Engels, F. (1978a). The origin of the family, private property, and the state. In R. C. Tucker (Ed.), *The Marx-Engels reader.* New York: Norton. (Original work published 1884)

Engels, F. (1978b). Speech at the graveside of Karl Marx. In R. C. Tucker (Ed.), *The Marx-Engels reader.* New York: Norton.

Flyvbjerg, B. (2001). *Making social science matter: Why social inquiry fails and how it can succeed again* (S. Sampson, Trans.) Cambridge, UK: Cambridge University Press.

Fromm, E. (1961). *Marx's concept of man.* New York: Continuum.

Geertz, C. (1973). *The interpretation of cultures.* New York: Basic Books.

Gergen, K. J. (1991). *The saturated self: Dilemmas of identity in contemporary life.* New York: Basic Books.

Giddens, A. (1986). *The constitution of society.* Berkeley: University of California Press.

Giddens, A. (1991). *Modernity and self-identity: Self and society in the late modern age.* Stanford, CA: Stanford University Press.

Gilman, C. P. (1975). *Women and economics: A study of the economic relation between men and women as a factor in social evolution.* New York: Gordon Press. (Original work published 1899)

Gilman, C. P. (2001). *The man-made world.* New York: Humanity Books. (Original work published 1911)

Goffman, E. (1967). *Interaction ritual: Essays on face-to-face behavior.* New York: Pantheon.

Gottdiener, M. (1985). Hegemony and mass culture: A semiotic approach. *American Journal of Sociology, 90,* 979–1001.

Gottdiener, M. (1990.) The logocentrism of the classics. *American Sociological Review, 55*(3), 460–463

Gramsci, A. (1971). *Selections from the prison notebooks.* London: Lawrence & Heinemann. (Original work published 1928)

Grathoff, R. (Ed). (1989). *Philosophers in exile: The correspondence of Alfred Schutz and Aron Gurwitsch, 1939–1959* (J. C. Evans, Trans.). Bloomington: Indiana University Press.

Hall, R. J., & Neitz, M. J. (1993). *Culture: Sociological perspectives.* Englewood Cliffs, NJ: Prentice Hall.

Harrison, F. (1913). Introduction. In H. Martineau (Trans.), *The positive philosophy of Auguste Comte.* London: G. Bell and Sons.

Hart, J. A. (1993). *Rival capitalists: International competitiveness in the United States, Japan, and Western Europe.* Ithaca, NY: Cornell University Press.

Havens, J. J., & Schervish, P. G. (2003). *Why the $41 trillion wealth transfer estimate is still valid: A review of challenges and questions.* Retrieved August 31, 2004, from http://www.bc.edu/research/swri/meta-elements/pdf/41trillionreview.pdf

Hewitt, J. P. (1998). *The myth of self-esteem: Finding happiness and solving problems in America.* New York: St. Martin's Press.

Hochschild, A. R. (1983). *The managed heart: Commercialization of human feeling.* Berkeley: University of California Press.

Hofstadter, D. R. (1985). *Metamagical themas: Questing for the essence of mind and pattern.* New York: Basic Books.

James, W. (1890). *The principles of psychology.* New York: Holt.

Jameson, F. (1984). *The postmodern condition.* Minneapolis: University of Minnesota Press.

Jencks, C. A. (1977). *The language of post-modern architecture.* New York: Rizzoli.

Kennickell, A. B. (2003). *A rolling tide: Changes in the distribution of wealth in the U.S., 1989–2001.* Retrieved August 27, 2004, from http://www.federalreserve.gov/pubs/feds/2003/200324/200324pap.pdf

Langer, E. (1989). *Mindfulness.* New York: Perseus.

Lash, S., & Urry, J. (1987). *The end of organized capitalism.* Madison: University of Wisconsin Press.

Layder, D. (1998). *Sociological practice: Linking theory and social research.* London: Sage.

Lemert, C. (2000). W. E. B. Du Bois. In G. Ritzer (Ed.), *The Blackwell companion to major social theorists.* Malden, MA: Blackwell.

Lengermann, P. M., & Niebrugge-Brantley, J. (1998). *The women founders: Sociology and social theory, 1830–1930.* Boston: McGraw-Hill.

Lengermann, P. M., & Neibrugge-Brantley, J. (2000). Early women sociologists and classical sociological theory: 1830–1930. In G. Ritzer (Ed.), *Classical sociological theory* (3rd ed.). Boston: McGraw-Hill.

Lidz, V. (2000). Talcott Parsons. In G. Ritzer (Ed.), *The Blackwell companion to major social theorists.* Malden, MA: Blackwell.

Luckmann, T. (1991). The new and the old in religion. In P. Bourdieu, & J. S. Coleman (Eds.), *Social theory for a changing society.* Boulder, CO: Westview Press.

Lukács, G. (1971). *History and class consciousness: Studies in Marxist dialectics* (R. Livingstone, Trans.). London: Merlin. (Original work published 1922)

Marshall, G. (1998). *A dictionary of sociology* (2nd ed.). Oxford: Oxford University Press.

Martineau, H. (1966). *Society in American.* New York: AMS. (Original work published 1837)

Martineau, H. (2003). *How to observe morals and manners.* New Brunswick, NJ: Transaction Press. (Original work published 1838)

Marx, K. (1977). *Capital: A critique of political economy* (Vol. 1; E. Mandel, Trans.). New York: Random House. (Original work published 1867)

Marx, K. (1978a). Economic and philosophic manuscripts of 1844. In R. C. Tucker (Ed.), *The Marx-Engels reader.* New York: Norton. (Original work published 1932)

Marx, K. (1978b). The German ideology. In R. C. Tucker (Ed.), *The Marx-Engels reader.* New York: Norton. (Original work published 1932)

Marx, K. (1979). Contribution to the critique: Introduction. In R. C. Tucker (Ed.), *The Marx-Engels reader.* New York: Norton. (Original work published 1844)

Marx, K. (1995). Economic and philosophic manuscripts of 1844. In E. Fromm (Trans.), *Marx's concept of man.* New York: Continuum. (Original work published 1932)

Marx, K., & Engels, F. (1978). Manifesto of the Communist Party. In R. C. Tucker (Ed.), *The Marx-Engels reader.* New York: Norton. (Original work published 1848)

Maryanski, A., & Turner, J. H. (1992). *The social cage: Human nature and the evolution of society.* Stanford, CA: Stanford University Press.

Mead, G. H. (1925). The genesis of the self and social control. *International Journal of Ethics, 35,* 251–277.

Mead, G. H. (1934). *Mind, self, and society* (C. W. Morris, Ed.). Chicago: University of Chicago Press.

Mead, G. H. (1938). *The philosophy of the act.* Chicago: University of Chicago Press.

Merton, R. K. (1967). *On theoretical sociology: Five essays, old and new.* New York: The Free Press.

Mills, C. W. (1959). *The sociological imagination.* New York: Oxford University Press.

Mithaug, D. E. (2000). *Learning to theorize: A four-step strategy.* Thousand Oaks, CA: Sage.

Oakes, G. (1984). Introduction. In *Georg Simmel on women, sexuality, and love.* New Haven, CT: Yale University Press.

O'Brien, J., & Kollock, P. (2001). *The production of reality: Essays and readings on social interaction* (3rd ed.). Thousand Oaks, CA: Pine Forge Press.

Orwell, G. (1946). *Shooting an elephant and other essays.* New York: Harcourt.

Parsons, T. (1949). *The structure of social action* (2nd ed.). New York: The Free Press.

Parsons, T. (1951). *The social system.* London: The Free Press.

Parsons, T. (1954). *Essays in sociological theory* (Rev. ed.). New York: The Free Press.

Parsons, T. (1961). Culture and the social system. In T. Parsons (Ed.), *Theories of society: Foundations of modern sociological theory* (pp. 963–993). New York: The Free Press.

Parsons, T. (1966). *Societies: Evolutionary and comparative perspectives.* Englewood Cliffs, NJ: Prentice Hall.

Parsons, T. (1990). Prolegomena to a theory of social institutions. *American Sociological Review, 55*(3), 319–339.

Parsons, T., & Shils, E. (Eds.). (1951). *Toward a general theory of action.* Cambridge, MA: Harvard University Press.

Parsons, T., & Smelser, N. J. (1956). *Economy and society: A major contribution to the synthesis of economic and sociological theory.* New York: The Free Press.

Pirsig, R. M. (1981). *Zen and the art of motorcycle maintenance: An inquiry into values.* New York: Bantam.

Ritzer, G. (2004a). *The McDonaldization of society* (Rev. ed.). Thousand Oaks, CA: Pine Forge Press.

Ritzer, G. (2004b). *The globalization of nothing.* Thousand Oaks, CA: Pine Forge Press.

Ritzer, G., & Goodman, D. (2000). Introduction: Toward a more open canon. In G. Ritzer (Ed.), *The Blackwell companion to major social theorists.* Malden, MA: Blackwell.

Roy, W. (2001). *Making societies.* Thousand Oaks, CA: Pine Forge Press.

Schutz, A. (1944). The stranger: An essay in social psychology. *The American Journal of Sociology, 49,* 499–507.

Schutz, A. (1962). *Collected papers I: The problem of social reality* (M. Natanson, Ed.). The Hague, Netherlands: Martinus Nijhoff.

Schutz, A. (1964). *Collected papers II: Studies in social theory* (A. Brodersen, Ed.). The Hague, Netherlands: Martinus Nijhoff.

Schutz, A. (1966). *Collected papers III: Studies in phenomenological philosophy* (I. Schutz, Ed.). The Hague, Netherlands: Martinus Nijhoff.

Schutz, A. (1967). *The phenomenology of the social world* (G. Walsh & F. Lehnert, Trans.). Evanston, IL: Northwestern University Press.

Schutz, A., & Luckmann, T. (1973). *The structures of the lifeworld* (Vol. 1). Evanston, IL: Northwestern University Press.

Seidman, S., & Wagner, D. S. (1992). Introduction. In Steven Seidman and David G. Wagner (Eds.), *Postmodernism and social theory: The debate over general theory.* Cambridge, MA: Blackwell.

Simmel, G. (1950). *The sociology of Georg Simmel* (K. H. Wolff, Trans. & Ed.). Glencoe, IL: The Free Press.

Simmel, G. (1959). *Essays on sociology, philosophy, and aesthetics [by] Georg Simmel [and others]: Georg Simmel, 1858–1918* (K. H. Wolfe, Ed.). New York: Harper & Row.

Simmel, G. (1971). *Georg Simmel: On individuality and social forms* (D. N. Levine, Ed.). Chicago: University of Chicago Press.

Simmel, G. (1978). *The philosophy of money* (T. Bottomore, & D. Frisby, Trans.). London: Routledge and Kegan Paul. (Original work published 1907)

Simmel, G. (1984). *Georg Simmel on women, sexuality, and love* (G. Oakes, Trans.). New Haven, CT: Yale University Press. (Original work published 1858–1918)

Simmel, G. (1997). *Essays on religion* (H. J. Helle, Trans. & Ed.). New Haven, CT: Yale University Press.

Smith, A. (1937). *Inquiry into the nature and causes of the wealth of nations.* New York: Modern Library. (Original work published 1776)

Smith, D. E. (1987). *The everyday world as problematic: A feminist sociology.* Boston: Northwestern University Press.

Snow, D., & Anderson, L. (1992). *Down on their luck: A study of homeless people.* Berkeley: University of California Press.

Spencer, H. (1896). *First principles* (4th ed.). New York: D. Appleton and Co. (Original work published 1862)

Spencer, H. (1954). *Social statics: The conditions essential to human happiness specified, and the first of them developed.* New York: Robert Schalkenbach Foundation. (Original work published 1882)

Spencer, H. (1961). *The study of sociology.* Ann Arbor: University of Michigan Press. (Original work published 1873)

Spencer, H. (1967). *The evolution of society* (R. L. Carneiro, Ed.). Chicago: University of Chicago Press. (Original work published 1876–1896)

Spencer, H. (1975a). *The principles of sociology* (Vol. 1). Westport, CT: Greenwood Press. (Original work published 1876–1896)

Spencer, H. (1975b). *The principles of sociology* (Vol. 2). Westport, CT: Greenwood Press. (Original work published 1876–1896)

Spencer, H. (1975c). *The principles of sociology* (Vol. 3). Westport, CT: Greenwood Press. (Original work published 1876–1896)

Timasheff, N. S. (1967). *Sociological theory: Its nature and growth* (3rd ed.). New York: Random House.

Tucker, R. C. (1978). *The Marx-Engels reader.* New York: Norton.

Turner, J. H. (1985). *Herbert Spencer: A renewed appreciation.* Beverly Hills, CA: Sage.

Turner, J. H. (1988). *A theory of social interaction.* Stanford, CA: Stanford University Press.

Turner, J. H. (1993). *Classical sociological theory: A positivist's perspective.* Chicago: Nelson.

U.S. Census. *Household shares of aggregate income by fifths of the income distribution: 1967–2001.* Retrieved August 27, 2004, from http://www.census.gov/hhes/income/histinc/ie3.html

Wallerstein, I. (1980). *The modern world system* (Vols. 1 & 2). New York: Academic Press. (Original work published 1974)

Warner, W. L. (1959). *The living and the dead.* New Haven, CT: Yale University Press.

Weber, M. (1948). *From Max Weber: Essays in sociology* (H. H. Gerth, & C. Wright Mills, Trans. & Eds.). London: Routledge and Kegan Paul.

Weber, M. (1949). *The methodology of the social sciences* (E. A. Shils, & H. A. Finch, Trans. & Eds.). New York: The Free Press. (Originally published 1904, 1906, 1917–1919)

Weber, M. (1961). *General economic history* (F. H. Knight, Trans.). New York: Bedminster. (Originally published 1923)

Weber, M. (1968). *Economy and society* (G. Roth, & C. Wittich, Eds.). Berkeley: University of California Press. (Originally published 1922)

Weber, M. (1993). *The sociology of religion* (E. Fischoff, Trans.). Boston: Beacon Press. (Originally published 1922)

Weber, M. (2002). *The Protestant ethic and the spirit of capitalism* (S. Kalberg, Trans.). Los Angeles: Roxbury. (Originally published 1904–1905)

Wuthnow, R. (1987). *Meaning and moral order: Explorations in cultural analysis.* Berkeley: University of California Press.

Wuthnow, R. (1995). The editors talk with Robert Wuthnow. *Perspectives: The Theory Section Newsletter 18*(1), 1–2.

Zerubavel, E. (1991). *The fine line: Making distinctions in everyday life.* New York: The Free Press.

Finding Your Way Through the Maze

An Annotated Glossary of Terms

Absolute surplus labor: From Marx's labor theory of value; a method of increasing profit by increasing the number of hours a worker works.

Act: According to Mead, because human action is meaningful and social, it is complex. A single action has four stages: impulse, perception, manipulation, and consumption. The distinctly human elements are found in the second and third stages, which are both based on symbols and meaning.

Action theory: In general, any theory that begins and is concerned with individual, social action rather than structure. For *Weber*, action is social insofar as the individual takes into account other people. Weber is specifically concerned with rational action (in his typology—traditional, affective, value-rational, instrumental-rational—only value and instrumental action are truly rational). *Parsons* is interested in the conditions under which action takes place (the choice of means and ends constrained by the physical and social environments). Parsons' theory of the social system is built upon his notion of voluntaristic action.

Adaptation: One of four subsystems in Parsons' AGIL conception of requisite needs; the subsystem that converts raw materials from the environment into usable stuffs (in the body, the digestive system; in society, the economy).

Affectual action: From Weber's typology of social action; action that is motivated by emotion.

Agency: Agency refers to the freedom to choose one's own behaviors or to independently influence a course of events. Agency is one half of a central debate in sociology focusing on the relative influence of structure over individual choice.

Alienation: The word itself means separation from; it also implies that there is something that faces us as an unknown or alien object; for Marx, there are four different kinds of alienation: alienation from one's own species-being, alienation from the work process, alienation from the product, and alienation from other social beings.

Altruistic suicide: Part of Durkheim's typology of suicide; the word *altruistic* means to have uncalculated devotion to the interests of others; altruistic suicide, then, is motivated by excessively high group attachment.

Analogy: Analogy is used in **hermeneutic theory** to explain through reference and comparison. For example, Goffman's dramaturgy explains social interaction through the analogy of the stage.

Analytical frameworks: Analytical frameworks or models are theoretical schemes that help us to understand a given social situation. They contain abstract concepts and explicit definitions, but do not propose relationships among the concepts. Weber's ideal types and Parsons' unit act are good examples.

Anomic suicide: Part of Durkheim's typology of suicide; *anomie* is to be without rules or norms; thus, anomic suicide is motivated by extremely low normative regulation.

Anomie: Used by both Durkheim and Simmel; literally, to be without norms or laws. Because human beings are not instinctually driven, they require behavioral regulation. Without norms guiding behavior, life becomes meaningless. High levels of anomie can lead to anomic suicide.

Anonymization: The process of becoming anonymous. Schutz argues that other people become increasingly anonymous to us the more we use typifications to understand them. Anonymization is characteristic of "they" relations.

Ascetic religions: Part of Martineau's typology of religion; ascetic religions are those that use abstinence for proof of holiness; a society wherein this type of religion dominates will be characterized by an ever-expanding inventory of sin and an emphasis on outward restraint rather than inward ethics.

Assumptions: Assumptions are elements of cultural knowledge systems that are believed to be true but never tested. All knowledge systems are based on assumptions. They allow the system to work. For example, science assumes that the universe is empirical and then conducts empirical, scientific research.

Bi-polarization of conflict: From Marx; it is the process through which conflict is limited to two parties. As an overall principle, the more bi-polarized a conflict, the greater will be its intensity and overtness.

Blasé attitude: The word *blasé* means to be uninterested in pleasure or life. According to Simmel, the blasé attitude is the typical emotional state of the modern city dweller and is the direct result of the increased emotional work and anomie associated with diverse group memberships, and increased cognitive stimulation due to the rapidness of change and flow of information.

Bourgeoisie: In Marxian theory, the owners of capital.

Bracketing: For most phenomenologists, consciousness is the only phenomenon of which human beings can be certain. In order to observe consciousness purely, phenomenologists bracket or set aside all that they know or presume about the world.

For Schutz, it is exactly that which phenomenologists want to bracket that constitutes the lifeworld and the subject of sociology.

Bureaucracy: From Weber, a system of organizing people and their behavior that is characterized by the presence of written rules and communication, job placement by accreditation, expert knowledge, clearly outlined responsibilities and authorities, explicit career ladders, and an office hierarchy. The purposes of the bureaucratic form are to rationalize and routinize behavior. Its superiority is purely utilitarian.

Bureaucratic personality: An extension of Weber's theory, the result of extensive use of bureaucratic methods for organization; the bureaucratic personality is characterized by rational living, identification with organizational goals, reliance on expert systems of knowledge, and sequestered experiences (experiences that are removed from social or family life and placed in institutional settings).

Capital: The means of production (technology, buildings, labor) that are not owned by the people who are using the means (workers); money used for investment, the purpose of which is to create more capital (from Marx and Weber).

Civil religion: An extension of Durkheim's idea of the importance religion has for the creation and maintenance of the collective consciousness. In civil religion, the role that religion plays in the collective consciousness is carried out by and through the state.

Class: A social structure built around issues of economic production. For *Marx*, class is the only structure that matters and it is specifically defined by the ownership of the means of production. In capitalist societies, class bifurcates into owners (bourgeoisie) and workers (proletariat). For *Weber*, class is one of three systems of stratification, status and power being the other two. Weber also defines class more complexly than does Marx. According to Weber, one's class is defined by the probability of acquiring the goods and positions that are seen to bring inner satisfaction. Class position is determined by the control of property or market position, both of which may be positively, negatively, or medially possessed.

Class consciousness: Being aware that class determines life chances. Class consciousness is one of the prerequisites to social change in Marxian theory.

Collective consciousness: The collective representations and sentiments that guide and bind together any social group. According to Durkheim, the collective consciousness takes on a life and reality of its own, particularly in response to high levels of ritual.

Commercial class: Part of Weber's typology of class; property and commercial classes are differentiated by the way each obtains money: the property class through property ownership and the commercial class through controlling a specific market niche, like professors and knowledge.

Commodification: The process through which material and nonmaterial goods are turned into products for sale. From this point of view, nothing is by its nature a product for sale; goods must be placed in markets in order to be commodities. In Marxian theory, commodification is seen as an ever-increasing force in capitalism.

Commodification of labor: According to Marx, the historical process through which human labor became a good that is bought and sold within a market; equating human creative labor with money.

Commodity fetish: From Marx; in commodity fetish, workers fail to recognize the human factor in products. Creative production is the distinctive trait of humanity. Therefore, all products have an intrinsic relationship to the people who make them. However, in capitalism, the product is owned and controlled by another, and the product thus faces the worker as something alien that must be bought and appropriated. In misrecognizing their own nature in the product, workers also fail to see that there are sets of oppressive social relations in back of both the perceived need and the simple exchange of money for a commodity.

Common emotional mood: One of the three variables that make up Durkheimian ritual performance. *Common emotional mood* expresses the degree to which the participants in any social interaction are emotionally oriented toward the interaction in the same way.

Common focus of attention: One of the three variables that make up Durkheimian ritual performance. *Common focus of attention* measures the degree to which the participants in any social interaction are cognitively oriented toward the same idea.

Compounding: In Spencerian theory, compounding refers to increases in the level of the population in a society brought about by adding other distinct populations through conquest or migration. Compounding can be compared to the normal population growth that occurs through a higher birth than death rate.

Concepts, abstract: One of the three elements out of which scientific theory is constructed. If something is abstract, it exists apart from any particular object or specific instance. Because two of the goals of scientific theory are to predict and control, the ideas, terms, or concepts must be general or abstract so that they can be used in multiple settings.

Consciousness: Humans are not simply aware, they are aware that they are aware. For example, we not only feel pain, we are aware of that pain and give it meaning (think about the relationship between pain and gender). Consciousness, then, is reflexive awareness. *Mead* argues that consciousness is the result of language. *Marx,* on the other hand, argues that consciousness has a material base—the effect of creative labor is a product that mirrors our nature back to us.

Coordination and control: A Spencerian systems problem that arises due to increases in structural differentiation. As social structures become different from one another, coordinating and controlling their activities becomes a problem that the system needs to solve. Problems in coordination and control are typically solved through increasing the power of government.

Co-presence: One of the three variables that make up Durkheimian ritual performance. *Co-presence* is measured by the physical proximity of the members of any social interaction.

Credential inflation: An extension of Weber's theory of bureaucracy, it is the result of extensive use of bureaucratic methods for organization; because bureaucracies rely heavily upon educational credentials for job placement, the level of required credentials tends to inflate over time as more and more of the populace completes higher levels of education.

Credentialing: From Weber, the use of credentials, rather than experience or social connections, for job and career placement. Credentialing results in a new class of workers—experts and professionals who traffic in knowledge.

Critical theory: In sociology, theory that is directed toward changing the existing society in some way. It is contrasted with descriptive theory that seeks only to describe things as they are. Critical theory generally deconstructs social relationships or arrangements in order to reveal the underlying ideology. Marxian theory is a good example of critical theory.

Culture: In general, culture is the system of meaning that societies hand down from generation to generation. For *Weber*, culture is defined as that part of the universe upon which human beings place value, meaning, and significance; in Weber's scheme, culture may at times lead social change. For *Durkheim*, culture is the fundamental source of society and integration; it is made up principally of social categories and sentiments. When held by society as a whole, it is the collective consciousness; culture that is specific to special groups is particularized and may represent a threat to the collective consciousness and social integration. *Parsons* considers the culture system to be at the top of the societal cybernetic hierarchy of control and to consist of value hierarchies, belief systems, and expressive symbols. *Simmel* sees culture divided into two specific types: objective and subjective—subjective culture can be completely known and experienced by the individual but objective culture cannot. *Du Bois* sees that culture can be used to oppress a disenfranchised group. Culture can be hegemonic or ruling and can exclude the experiences of others. In particular, culture may be used to misrepresent a group, create stereotypes and default assumptions, and to leave out the history and identity of the disenfranchised from the national narrative. This use of culture not only provides legitimation for prejudice, it also creates double consciousness for the oppressed.

Culture generalization: According to Durkheim and Parsons, as a society becomes more diversified, a more general or abstract culture (collective consciousness) is needed to hold it together and facilitate exchanges and interactions.

Cybernetic hierarchy of control: In Parsonian theory, the idea that all systems are controlled by information; the subsystem that controls and provides needed information for the rest of the system (in the social system, culture).

Deconstruction: A method of inquiry most intimately associated with post-structuralism. In that application, language is assumed to represent or refer only to itself rather than to an external reality or truth. It assumes multiple interpretations of an expression of language, like a text, and interprets language based on the philosophical, political, or social implications rather than the author's intention. In general, it refers to any critical analysis that assumes that statements can contain ideology, and seeks to uncover those elements of cultural oppression.

Definition of the situation: Part of symbolic interactionist theory, it refers to the primary meaning given to a social interaction. The definition of the situation is important because it implicitly contains identities and scripts for behavior. Because the definition of the situation is a meaning attribution, it is flexible and negotiable—in other words, people can change it at a moment's notice and with it the available roles and selves.

Disenchantment: A process that Weber sees as generally accompanying modernity and specifically the use of bureaucracies; in general, *enchantment* refers to a holistic view of a world that is mysterious and uncertain; *disenchantment* is the companion of rationalization—as humans relate more rationally to the physical and social environments, they are seen as less mysterious and more controllable.

Distributive system: One of Spencer's three requisite functions (the other two are regulatory and operative). The distributive structure moves needed supplies among the various parts of a system. In society, such a function is played by markets, roads, mail, the Internet, and so on.

Division of labor: The way in which work is divided in any economy. It can vary from everyone doing similar tasks to each person having a specialized job. This concept is an important one for most theories of modernity. In previous epochs, labor was more holistic in the sense that the worker was invested in a product from beginning to end. Thus, a shoemaker made the entire shoe. One of the distinctive traits of modernity is the use of scientific management, or Fordism, to divide work up into the smallest manageable parts. The division of labor is an important variable for Durkheim, Simmel, Marx, and others. For *Durkheim,* the division of labor creates specialized cultures for each group that may in turn threaten the cohesiveness of the general culture; for *Simmel,* the division of labor increases the level of objective culture in any society and it trivializes the meaning surrounding products; for *Marx,* the division of labor is understood as potentially separating people from species-being (natural or forced; material versus mental).

Double consciousness: Du Bois' argument that members of a disenfranchised group have two ways to understand and be aware of themselves: as disenfranchised and as full members of society. These two awarenesses war with and negate one another other so that the disenfranchised are left with no true consciousness.

Durée: From phenomenology, durée is the pure duration of unmarked time; time and experience as it naturally flows from one event and moment to the next without break. Meaning, by definition, breaks the natural flow of durée into discrete bits.

Dynamic model: Dynamic models are theoretical statements that spatially picture the relationships among and between the concepts. Dynamic models are different than analytic models in that they contain relationships (hence, they are dynamic or moving). The use of space in the diagram is important in these models as time is usually depicted as moving left to right (thus, two concepts appearing at the same place in the model occur at the same time). Concepts are related through the use of arrows and + (positive) and – (negative) relationships. Models are used to simplify theories and make them explicit.

Dynamic, moral, or physical density: An important variable in Durkheim's understanding of modern society. Dynamic/moral/physical density refers to the intensity and frequency of interactions among people. It is facilitated by urbanization and results in higher levels of competition and division of labor.

Dysfunctional: Describes any effect that upsets the equilibrium of a system or one of its parts; the effect of any social structure or process that tends to reduce the level of integration; the term was specifically introduced to functional analysis by Robert K. Merton.

Effervescence: Durkheim's term for the production of collective emotion. Effervescence is the product of repeated, intense ritual performance; it creates high levels of sacredness or morality when associated with a symbol or norm.

Egoistic suicide: Part of Durkheim's typology of suicide; ego refers to a single individual, thus egoistic suicide is one that is motivated by extremely low group attachment; intrinsic within this view is the assumption that the self is a social rather than personal entity.

Emergence: To *emerge* means to rise from or come out into view, like steam from boiling water. Emergence is a central idea in Meadian and symbolic interactionist theories, and it is a way to understand how meaning is created. Emergence is based on the idea that meaning is not intrinsic to any object, but is a function of social interaction: the meaning of a sign, symbol, nonverbal cue, social object, and so on comes out of (emerges from) social interaction. In this perspective, meaning is thus a very supple thing and is controlled by people in face-to-face interactions, rather than being determined beforehand by a structure of meaning.

Empirical: If something is empirical, it is discerned through one or more of the five senses. Empirical knowledge relies on direct experience and testing, rather than logic or belief. For example, a brick falling from a building is empirical, but the existence of God is not.

Empiricism: A philosophical base for science, empiricism is the assumption that all that may be known may be experienced through one or more of the five senses.

Enlightenment: A period of European history covering the seventeenth and eighteenth centuries, and characterized by the assumption that the knowledge of most worth is based on reason and observation (science) rather than faith (religion); it resulted in the hope that humans could control not only the physical universe but society as well; it also resulted in individualism and utilitarianism; also known as the Age of Reason.

Equilibrium: A concern of functionalist theory, equilibrium refers to a state of balance between integrative and disintegrative forces; it implies a social system that maintains certain constancies of patterns relative to its environment; the idea is that social systems must maintain a balance, whether internally or between itself and its environment.

Exchange: For Simmel, exchange is the social form that determines value, and value is based on sacrifice and scarcity, both of which are most directly determined by the immediate exchange relationship. Even though value is felt subjectively, it is determined by the totality of elements in the exchange (how scarce the exchange object is and how much someone is willing to sacrifice for it). Further, Simmel argues that exchange is the basic social form because it is based on reciprocity.

Exchange-value: From Marx; the value for which a product is purchased (compare to use-value).

Exploitation: From Marx, the measurable difference between what a worker gets paid and the amount of product she or he produces.

False consciousness: According to Marx, false consciousness is consciousness that is false in its very foundations. True consciousness is founded on species-being: recognizing the intrinsic human nature of creative production. Any idea of humanity apart from species-being is laid on a false foundation.

Fatalistic suicide: Part of Durkheim's typology of suicide; fatalism is the belief that all things are determined by fate or are already determined; in fatalism, the person's agency is severally reduced; thus, fatalistic suicide is motivated by excessively high normative regulation (control of behaviors through norms).

Flow: A term coined by Mihaly Csikszentmihalyi to describe non-reflexive experience. For humans, most situations are experienced in two ways: the immediate experience itself and our ideas about the experience. Human consciousness is thus divorced from the actual experience. We mentally sit back and observe the situation in order to evaluate and direct it. In flow, the second part of experience is not there. The situation is purely experienced as it unfolds.

Game stage: One of the three stages in Mead's theory of self-formation. The game stage is the second stage and it corresponds to the time when children are able to take the role of multiple others separately and can understand rule-based behavior.

Generalized media of exchange: A concern for functionalist theory, specifically Durkheim and Parsons, generalized media of exchange refers to symbolic goods that are used to facilitate interactions across institutional domains; thus, values and symbols that allow different structures to relate to one another; the goods and services exchanged by different institutions.

Generalized other: One of the three stages in Mead's theory of self-formation. The generalized other also refers to sets of perspectives and attitudes indicative of a specific group or social type with which the individual can role-take. In the formation of the self, the generalized other represents the last stage in which the child can place herself in a collective role from which to view her own behaviors. It is the time when the self is fully formed as the person takes all of society inside.

Goal Attainment: One of four subsystems in Parsons' AGIL conception of requisite needs; the subsystem that motivates and guides the system as a whole (in the body, the mind; in society, government).

Grand narratives: A postmodern concept that refers to national stories that encompass and encase diverse groups and cultures. According to many postmodern thinkers, these stories are losing their power in technically advanced capitalistic societies as people are expressing more localized cultures and identities.

Grand theory: A theory that claims to explain all phenomena, whether social, physical, or biological, through one set of concepts; Spencer and Parsons are considered grand theorists.

Grounded theory: Theory that is constructed apart from any preconceived ideas or expectations. In grounded theory, the researcher puts aside any theory that she or he thinks might apply to the data and allows the "data to speak for themselves," rather than basing her or his work on an existing abstract or formal theory; symbolic interaction is a type of grounded theory.

Gynaecocentric theory: Originated with Lester F. Ward and used by Gilman; gynaecocentric theory argues that women—not men, as was and is commonly assumed—are the general species type for humans: it is through women that the race is born and the first social connections are created; in gynaecocentric theory, men and women have essential sex-specific energy and the dominance of the masculine energy in society makes its evolutionary path unbalanced.

Hermeneutic theory: The word *hermeneutic* is from a Greek word meaning "to interpret or translate." Thus, the goal of hermeneutic theory is to understand and convey the cultural meanings and practices of one group to another. It focuses on culture rather than structure.

History as ideology: From Du Bois' critical perspective, history is told from the point of view of the ruling class in order to sustain and legitimate society.

Hyperreality: A term in postmodern theory, hyperreality refers to an image that presents itself as real but is not based on any reality. This idea is founded on the notion that signs, symbols, and images at one time corresponded to and represented social reality. A clear example is art: at one point in time, almost all of the artistic paintings visually represented real social scenes; in postmodern or abstract art, the representation is not of real scenes, but is at best an image of an idea that is about real social life. This divorce from true representation is even more pronounced in mass media and entertainment (such as Disneyland and Las Vegas) that is driven by advertising and commodification. These images are simulacrum (images that represent other images) and create a hyperreal or virtual world rather than a real one.

Ideal types: One of Weber's ways of creating objective knowledge in the face of the subjective world of culture; ideal types are purely conceptual constructions that are used as measuring rods against which empirical instances may be compared.

Idealism: A school of philosophy that argues that the objects perceived by humans are actually ideas contained in the perceiving mind. For idealists, ideas are ultimately real. Plato, Kant, and Hegel are noteworthy examples.

Ideology: In Marxian theory, ideology is any system of ideas that blinds us to the truth of economic relations. In a more general sense, ideology is any set of norms, values, and beliefs that undergird and legitimate a culture or specific knowledge system.

Industrial societies: Part of Spencer's typology of society; generally characterized by less governmental control and increased economic production and innovation.

Industrialization: In general, industrialization refers to the process through which machines replace direct human manipulation of objects in work. Industrialization is also an intrinsic part of what we mean by modernity. As such, it carries with it other factors such as high division of labor, rationalization, scientific management, urbanization, markets, and so on. Industrialization is specifically a concern for Marx.

Instability of homogeneous units: One of Spencer's three forces of evolution (the other two are segregation and multiplication of effects). Since evolution is motivated by survival of the fittest, evolution moves forward because groups of similar and unconnected entities are intrinsically unstable.

Institutionalization: The process through which institutions are created; in Parsonian theory, institutions are the result of two different paths: 1) building up of actions through identifiable modes of orientation, types of action, interactions among like-minded people, and the resulting norms, values, and status positions that constitute social structure; 2) the internalization of society's cultures and structures.

Institutions: Enduring sets of norms, values, beliefs, and status positions that are recognized as collectively meeting some societal need.

Instrumental-rational action: Part of Weber's ideal typology of social action; instrumental-rational action is behavior that is guided by means–ends considerations.

Integration: In general, the degree to which subsystems or units within a larger system work collectively for the common good; in social systems, integration is often related to cultural consensus, particularly in the work of *Durkheim.* In *Parsons,* integration refers to one of four subsystems in his AGIL conception of requisite needs, the subsystem that regulates the activities of the system's diverse members (in the body, the central nervous system; in society, the law).

Intentionality: According to Schutz, intentionality is a mental perception that involves intentional objects, intended behaviors, and thus, relations between the ego (self) and sets of intentional objects and behaviors. Each object and behavior is perceived in a group or matrix of other objects and behaviors. Intentionality creates meaning; meaning cannot exist apart from intentionality. It is subjective (takes place within the individual) and always retrospective (meaning can only be given to past experiences).

Interaction: Interaction is the intertwining of individual human actions. According to Mead, interaction is the three-part process through which meaning emerges: the presentation of a cue, the initial response to the cue, and the response to the response. At that final point, meaning may be said to have emerged. However, that meaning can itself become an initial cue for further interaction, staring the process over again.

Intersubjectivity: One of the principal problems that sociology must explain, according to Schutz. Intersubjectivity refers to the sense that people share similar internal worlds. Intersubjectivity is, by its very nature, something that we must either assume or create a sense of. The feeling that we share inner worlds and experiences is chiefly produced through language. Language, because it is social, creates a sense of shared inner worlds.

Iron cage of bureaucracy: A result of extensive use of bureaucratic methods for organization; Weber argues that once in place, bureaucracies are difficult if not impossible to replace because rulers cannot easily get rid of the need for expert knowledge or a professionally trained staff, and professionals tend to perpetuate their job status and make knowledge secret in order to control the organization and avoid public scrutiny, thereby creating an *iron cage of bureaucracy.*

Language: A system of signs through which people communicate meaning and instrumental goals. Language is abstract (it transcends space and time), the repository of social experience (the chief way in which we organize and pass down meaning and activities), and is reflexive (it makes reference chiefly to itself and creates basically the same response in the sender and recipient). Language is specifically a concern for Mead and Schutz.

Latent function: An element of functional theorizing specifically introduced by Robert K. Merton; the hidden or unacknowledged function of any social structure that leads to integration and equilibrium.

Latent pattern maintenance: One of four sub-systems in Parsons' AGIL conception of requisite needs; the subsystem that indirectly preserves patterns of behavior that are needed for survival (in the body, the autonomic nervous system; in society, education, religion, and family).

Legitimation: Legitimations are stories through which power and social relations are given institutionalized and moral basis.

Licentious religions: Part of Martineau's typology of religion; licentious religions are so called because in their beginning phases they lack moral constraints: gods and humans alike participate in lewd behavior. In the latter phases, this kind of religion still focuses on immoral behavior: more and more actions are defined as sin and are easily atoned for through ritual. In a society, licentious religions produce a condition with little concern for humanistic morals, and place a strong emphasis on ritualized behaviors.

Lifeworld: According to Schutz, the lifeworld is made up of the sets of assumptions, beliefs, and meanings against which the individual judges and interprets everyday experiences.

Looking glass self: Charles Horton Cooley's concept of self-formation, which is based on emotion, as compared to Mead's theory, which is cognitively based. There are three phases to Cooley's theory: the individual imagines how she or he appears to another person, then imagines the evaluation that other makes, and finally the individual has an emotional response to the evaluation: either pride or shame.

Macro: Large-scale social facts and processes; generally seen as lying outside the ability of an individual to influence. Population size and institutions are good examples. All of the factories, markets, suppliers, and distribution systems make up the macro level of the economic institution.

Magic: Defined by Weber as the manipulation of forces through direct means; compare to religion, which is symbolic.

Markets: In general, markets refer to any spheres wherein goods and services are exchanged for money. As such, they are social mechanisms for the distribution of commodities. Markets are not the only way through which this social need could be met; the government-controlled economies of socialism are another example. Thus, markets are more than sterile mechanisms; they are themselves social factors. In other words, markets have their own processes (such as susceptibility to expansion and trade cycles) and effects (such as rationalization and objectification). Markets play an important role in the theorizing of Marx, Weber, and Simmel.

Material dialectic: Marx's theory of historical change: every economic system contains within it contradictions that will eventually lead to its demise. History thus moves cyclically and because of structural tensions. The only economic system that doesn't contain such contradictions is communism, because it of all systems expresses humanity's species-being.

Materialism: In general, materialism presupposes that only concrete qualities exist. Thus, all the facts in the universe, including those connected to humankind like history and the mind, are causally dependent upon physical processes and may in fact be reducible to them. However, when the term is used with reference to Marx, it takes on a specialized meaning. Rather than the physical properties of the universe, Marx's materialism refers to the material of production. This is a powerful philosophical move. Marx is obviously moving away from idealism, but he isn't advocating a brute materialism either. For Marx, the material fabric of human reality is economic production.

Matter, force, and motion: Spencer's three basic elements of the universe. In theorizing society, matter refers to population and force, and motion refer to the movement of populations through space.

Means of production: The way in which commodities are produced. However, in Marx's hands the concept comes to denote quite a bit more. Because the basic social fabric is the economy, relations of production are inherent in the means of production. Thus, the organization of the means of production—such as capitalism, feudalism, and socialism—determine how we will relate to the self and others.

Mechanical solidarity: Part of Durkheim's typology of society: mechanical solidarity refers to high levels of group cohesiveness and normative regulation.

Meso: Mid-level social phenomenon, such as an economic organization (like IBM).

Micro: Face-to-face social interactions; the social encounter where all the actors are physically present.

Militaristic societies: Part of Spencer's typology of society. Militaristic societies are characterized by a high level of governmental control.

Mind: According to Mead, the mind is a social entity constituted through behavior: the internalized conversation of linguistic signs and symbols. The mind is socially necessary because of its place in human behavior.

Misrecognize: A central Marxian idea: people fail to see or recognize the relations of production within a commodity or means of production. Misrecognition is a function of ideology.

Moderate religions: Part of Martineau's typology of religion; moderate religions emphasize the overall well-being of humanity through self-actualization and education; a society that has a high level of this type of religion will have a culture that stresses freedom and true equality and broadmindedness.

Modernity: The idea of modernity is constructed against the notion of traditional society, and more recently against postmodernity. Modernity is defined by the movement from small local communities to large urban settings, a high division of labor, high commodification and use of rational markets, and large-scale integration through nation-states. Rather than traditional and religious authority, modernity is characterized by individualism, rationality, bureaucracy, secularization, and alienation. In general, the defining moments of modernity are the rise of nation-states, capitalism, mass democracy, science, urbanization, and mass media; the social movements that set the stage for modernity are the Renaissance, Enlightenment, Reformation, American and French Revolutions, and the Industrial Revolution. To understand the transition from traditional societies to modern societies, some sociologists constructed social typologies, such as Tönnies's "gemeinschaft" (community relations distinguished by moral, homogeneous culture and informal cooperation) and "gesellschaft" (individual, self-interested competition, and heterogeneous culture), and Durkheim's mechanical and organic solidarity.

Modes of orientation: The beginning element in Parsons' theory of social action and institutionalization; modes of orientation are the different values (cognitive, appreciative, and moral) and motivations (cognitive, cathectic, and evaluative) that people bring into a situation; they result in identifiable types of action (strategic, expressive, or moral) that in the long run pattern interactions across time and space and result in the taken-for-granted norms, roles, and status positions that constitute social structure and society.

Money: Money is the generalized media of economic exchange. In barter, things are equated with other things, but the use of money creates a general system through which all products are valued. *Simmel* sees the use of money as having both positive and negative effects: increases in personal freedom, rationalization and calculability, the number and extent of exchange relations, continuity between groups and individuals, trust in the national system; decreases in emotional attachment to things and people, and moral constraint. For *Marx*, money has no socially redeeming qualities. Money is yet another abstract system that stands between the worker and her or his creation. Rather than seeing vibrant social relations and connections in products, people simply see money. Money comes to stand for the product and the worker's labor; and eventually money comes to stand in the place of self and other.

Moral boundaries: An analogous way of understanding morality: we can think of morals as fences surrounding behaviors, with some being forbidden or fenced off. In this case, the picture that we want to bring to mind is the fences or boundaries surrounding a group. Some groups have a strong moral basis and high or dense fences that keep members in and strangers out. Others have permeable boundaries and people can move somewhat freely in and out of a group. Both Durkheim and Simmel are specifically concerned with moral boundaries.

Morals and manners: Martineau's categorization of two primary issues in society: morals are the ideals that a society claims it wants to live up to, and manners are the actual behaviors. Martineau argues that the best way to measure the success of any society is to hold it up to its own standard (morals) and see how close its behaviors come (manners).

Morbid excess in sex distinction: Sex distinctions are those physical and visual cues that make the sexes different from one another; Gilman argues that the sex distinctions in humans have been carried to a harmful extreme due to women no longer living in the natural environment of economic pursuit, but rather, living in the artificial environment of the home that has been established by men.

Multiplication of effects: One of Spencer's three forces of evolution (the other two are instability of homogeneous units and segregation). As structures become dissimilar, they become subject to different kinds of forces that in turn multiply the differences among the structures. For example, education used to be a function of family, but because of population forces, education broke off and became a separate structure. Thus, family and education are now subject to different kinds of pressures that increase the differences between them.

Nation-states: A defining feature of modernity; a nation-state is a socially diverse collective that occupies a specific territory, creates a common history and identity, and its members sees themselves as sharing a common fate. Nation-states are bureaucratically organized and emphasize mass democracy.

Natural attitude: Part of Schutz's theory of meaning, the natural attitude is characterized by being wide awake, suspending doubt, engaged in work, and assuming intersubjectivity. In the natural attitude, we unthinkingly use categories, typifications, roles, social recipes, and skills until we experience a problem in our routine.

Natural division of labor: Marx's idea that all people have natural abilities and sensitivities. In socially connected economies, these natural abilities correspond to the division of labor: everyone labors in those areas that they like and for which they have the natural gifts. Compare to forced division of labor.

Natural signs: According to Mead, a natural sign is one that has a natural relationship to that which it signifies. For example, smoke is a natural sign of fire. Compare to symbols which are abstract and arbitrary.

Necessary labor: From Marx's labor theory of value, the amount of money needed to reproduce the worker; the standard of living wage. Compare to surplus labor.

Neo-positivism: Social research and theorizing that is characterized by the primacy of quantification and statistics; compare to historic positivism that accepted a wide range of empirical measures (like historical comparison).

Non-evaluative: Different kinds of knowledge have different goals and functions. Some cultural systems are evaluative; that is, they evaluate whether things are right or wrong (like most forms of religion). Science, on the other hand, claims to be non-evaluative; that is, it asserts that it is only concerned with pure knowledge and doesn't make moral or value claims.

Normative specificity: Norms are behaviors that have some sanction attached to them. The sanctions can be positive (reward) or negative (punishment), and they can be formal (like laws) or informal (generally understood). Normative specificity, then, refers to the degree to which behaviors are specifically regulated (explicit norms for specific behaviors). Normative specificity is high when a group is regulating a large proportion of a person's behaviors. We find this issue in Durkheim, Simmel, and Parsons' theories.

Norms: Positive or negative sanctions attached to social behaviors.

Objective culture: A prime concern of Simmel's; objective culture refers to the amount of culture with which an individual has to contend that she or he cannot grasp in its entirety, cannot entirely use, and in which she or he has no emotional investment.

Ontology: A branch of philosophy that is concerned with how things exist. An obvious example is that rocks and people exist differently—one is biological and the other is geological. But ontology is concerned with less obvious, more fundamental questions, in particular, how categories exist. In our case, we are concerned with how society exists: What is its ontological source? Does society exist as an object made up of social facts and structures? Or, does society exist symbolically through the way we create meaning around the idea of society?

Operative function: One of Spencer's three requisite functions (the other two are regulatory and distributive). The operative function supplies everything that the larger system needs in order to work (in the body, the digestive system; in society, the economy).

Organic motivation: From Simmel's theory of the web of group affiliations, motivation for group membership that is based on either family ties or previously established social relations.

Organic solidarity: Part of Durkheim's typology of social integration; organic solidarity is created through mutual dependency, generalized culture, and intermediate (between the level of the state and the level of face-to-face interaction) groups and organizations. Organic solidarity is the way in which modern societies are integrated.

Organismic analogy: A way of understanding how society functions by comparing it to organisms. The organismic analogy is specific to functionalist theorizing.

Particularized culture: Culture that is peculiar to a small group within a larger collective. According to Durkheim, particularized culture may in the long run reduce the influence of the collective consciousness.

Perspectives: Perspectives are positions from which to view and attribute meaning. They are made up of language, values, sentiments, norms, and beliefs. The perspective we take determines what we see (or don't see) and the meaning we give it.

Phenomenology: A school of philosophy developed by Edmund Husserl that argues that consciousness is the only experience or phenomenon of which humans can be certain. It seeks to discover the natural and primary processes of consciousness apart from the influence of culture or society. Husserl hoped to create "a descriptive account of the essential structures of the directly given," that is, the immediate experience of the world apart from preexisting values or beliefs. Social phenomenology, on the other hand, argues that nothing is directly given to humans; we experience everything through stocks of language, typifications, and so on. Social phenomenology, then, seeks to explain the phenomena of the social world as it presents itself to us. This approach is distinct from most sociology in that phenomenologists reduce phenomena to their simplest terms. For example, a structural sociologist will study racial inequality and its effects. Race is simply a given in this kind of approach. A phenomenological approach will take race itself as the most basic problem to explain: How is it that race can be experienced as a given, as something that can be taken for granted, as an intersubjective reality? How does "race" present itself to us in just such a way as to appear real, taken for granted, and intersubjective?

Play stage: One of the three stages in Mead's theory of self-formation. The play stage is the initial stage wherein the child first sees herself as a social object separate from her behaviors. In the play stage, the child literally takes on the role of one significant other and acts and feels toward herself as the other does.

Positivism: The philosophy that assumes that the social world is subject to invariant laws and that human beings can thus control their world and life course.

Postindustrial: A society wherein the economy has moved from having its base in industry and manufacturing to service and knowledge. In a postindustrial society, knowledge is more important than property, and professionals and technicians are the most important social type. The idea of postindustrial society was popularized by Daniel Bell in his book, *The Coming of Post-Industrial Society.*

Postmodern: Describes the period of time contrasted with and coming after modernity. The beginning of this period is variously dated from the late 1940s to the mid 1960s. However, there is quite a bit of debate over this term. Some sociologists reject the notion that we are in a different social era and claim that the social processes that define modernity are still at work. Other sociologists argue that the idea of postmodern isn't quite right, but that we aren't still modern—we are late-modern. Theory that is based on the idea of late-modernity argues that modern processes have become extended to their logical conclusion. On the other hand, postmodernists argue that a clear break has occurred; institutional relations, culture, and identities have fragmented past the point of cohesion due to such things as increasing markets, advertising, transportation and communication technologies, de-centralized global capitalism, and institutionalized doubt. Postmodern social theory is characterized by contradiction, ambiguity, and paradox, particularly with reference to culture and identity—personal identities are increasingly important in a postmodern era, yet they are also less stable; and culture is more salient, yet meaning is less certain.

Pragmatism: A school of philosophy that argues that the only values, meanings, and truths that humans hold onto are the ones that have practical benefits. These shift and change in response to different concrete experiences. Pragmatism forms the base for many American social theories, most specifically symbolic interactionism.

Praxis: From Marx, practically applied critical theory; praxis is the practice through which the social world can be changed, beginning first with self and then extending to others.

Primary groups: From Cooley and symbolic interactionism, primary and secondary groups vary by length of time of association, purpose, degree of involvement, and intimacy. Primary groups stay together for long periods of time and tend to be open, honest, and emotionally based. Primary groups will have stronger influence on individual thoughts, feelings, and behaviors than secondary groups will.

Primitive communism: Marx argues that this kind of communism was practiced in primitive societies; corresponds to the natural division of labor.

Private property: According to Marx, private property, rather than being a natural consequence of economic production, is an effect of alienation. The worker must first be alienated from species-being and the creative work of her or his hands before the product can belong to another as private property.

Problem of routinization: From Weber, the problem of routinization occurs after the charismatic leader dies. Because charismatic authority is rooted in an individual, when that individual is gone, the authority must be made routine, either through traditional or rational-legal authority.

Proletariat: In the Marxian scheme of capitalism, the proletariat is the working class—the ones who do not own the means of production and must work to survive.

Property class: From Weber, class position that is determined through the ownership of property; this idea contains Marx's notion of class as ownership of the means of production as well as other types of property ownership, such as rentiers.

Proposition: A theoretical statement that proposes a relationship between two or more concepts or ideas. For example, the level of emotional energy that any symbol can elicit is a positive function of the level of previous ritual performance focused on that symbol.

Punitive law: From Durkheim, law that is oriented toward punishing offenders. Punitive law is associated with mechanical solidarity and functions to confirm a group's moral basis and reinvest it with emotional effervescence.

Race-preservation: Part of Gilman's evolutionary scheme; race-preservation stimulates the natural selection for skills that promote the general welfare of the collective; however, because women have been removed from the natural environment by male-dominated society, race-preservation skills are deemphasized and the evolutionary system is out of balance.

Rational capitalism: Part of Weber's ideal typology of capitalism; in rational capitalism, profit is pursued for its own sake (money invested to make more money to invest) and social relationships are based on rational, utilitarian equality (people are treated equally so that all may be used for profit).

Rational motivations: From Simmel's theory of the web of group affiliations, motivation for group membership that is based on free choice rather than family ties or previously established social relations.

Rationalization: According to Weber, rationalization is the main defining dynamic of modernity. It is the process through which spirituality, tradition, moral values, and affective social ties are replaced by rational calculation, efficiency, and control.

Reciprocity of perspectives: From Schutz, the assumption that we hold mutual points of view. This is one of the basic assumptions that allow culture to be taken for granted as reality. It's based on two other assumptions: the interchangeability of standpoints and irrelevant differences.

Reflexive construction: The idea that all cultural reality is created and has no direct reference to anything other than itself.

Regulatory function: One of Spencer's three requisite functions (the other two are operative and distributive). The regulatory function regulates or governs the overall operation of the system (in the body, the central nervous system; in society, the system of government).

Reification: In Marx, ideas that do not naturally spring out of species-being can only appear real through reification (making something appear real that isn't); all ideology is reified; the furthest reach of reification is the idea of God.

Relational statements: That part of a theory that explains how concepts vary. For example, in the following proposition, the italicized words are the relational statement: The level of emotional energy that any symbol can elicit *is a positive function* of the level of previous ritual performance focused on that symbol. It's the presence of relational statements that constitute the difference between dynamic (or causal) and analytical theoretical schemes.

Relations of production: In Marxian sociology, every economic system implies a specific set of social relationships. The means and the relations of production go together. The existence of the bourgeois and the proletariat social classes is created through capitalism. The important point is that economic systems are not simply rational ways of producing goods and services, they create social relationships.

Relative surplus labor: From Marx, surplus labor that is increased through industrialization.

Relevance structures: A hierarchically arranged structure of contexts and situations that is important for the individual. According to Schutz, relevance structures determine the importance we will place on any object or person in an encounter.

Religion: According to *Spencer,* religion functions to justify existing social structures (particularly those dealing with inequality), to legitimate the regulatory function (the state), and to infuse values and morals with supernatural power. Spencer, like Weber, traced the evolution of religion from its earliest forms of ancestor worship, to polytheism, pantheism, and eventually monotheism. The driving forces in the evolution of religion are compounding of diverse populations, principally through war, and the need to legitimate an increasingly segmented and bureaucratized state. *Marx:* Based on his notion of species-being, Marx sees religion as the highest form of ideology and reification. In religion, humans have separated themselves from their basic nature of species-being and the material basis for knowledge. Ideologically, religion serves the ruling class by shifting the workers' attention away from this life and its problems to a better hope in the next life. *Durkheim:* The centerpiece of Durkheim's definition of religion is the distinction between the sacred and profane. Durkheim argues that sacredness is created through intense human interactions; thus, anything can be made sacred. Once high levels of emotional energy have been produced and reproduced through rituals, a collective consciousness forms that constitutes the foundation of society and takes on independent effects of its own. Since it is the moral power of the collective consciousness that principally brings about social solidarity, Durkheim argues that some form of religion is always functional for a society. For *Simmel,* religion itself is an objective categorical system that overextends its rightful boundaries and is often at odds with the natural religious impulse that is characteristic of humans. Humans are born with a religious impulse and transcendence is also part of our social makeup. Religion is seen as originating in individuals and legitimately extending no further than experiences that can be shared holistically with others. The institution of religion, on the other hand, is a system of categories that comes to be part of objective rather than subjective culture. For *Weber,* religion is defined in contrast to magic. Magic is based in a naturalistic view of the universe and is concerned with direct manipulation of natural forces; religion, on the other hand, is symbolic and is concerned with rituals that dramatically enact abstract truths. Because of its symbolic nature, religion is functional for society because it provides abstract, symbolic ties that can relate diverse groups to one another. Weber thus sees the evolution of religion specifically joined to the ways in which people organize themselves as society, particularly the form of government, and connected to the economic concerns of its practitioners and leaders (professionalization).

Requisite needs: The different things that a system or society needs in order to function. Most functionalists propose system needs: *Spencer* proposes three (operative, distributive, and regulatory); *Durkheim* gives us basically one (collective consciousness); and *Parsons* gives us four (adaptive, goal attainment, integration, and latent patterned maintenance).

Restitutive law: From Durkheim, law that is oriented toward restoring offenders back to the group. Restitutive law is associated with organic solidarity and functions to enforce a group's non–morally-based laws and confirm the group's rationalized approach to collective life.

Rituals: Patterned sequences of behavior that recreate high levels of co-presence, common emotional mood, and common focus of attention. In Durkheim's scheme, rituals function to reinvigorate a group's moral and sacred base.

Role conflict: According to Simmel, role conflict occurs when behavior expectations of two or more roles that an individual holds clash with one another.

Roles: Scripts for behavior that are associated with a status position.

Role-taking: For Mead, the process through which an individual places her or himself in the role of the other person in order to see her- or himself from that position. Role-taking is the chief mechanism for the creation of the self.

Routinize: Typically associated with bureaucracies, to routinize behavior is to make it patterned and predicable; generally accomplished through written rules and job descriptions.

Science: A knowledge system based on the assumptions that the universe is empirical, operates according to law-like principles, and that those principles can be discovered and used by human beings to explain, predict, and control. Scientific knowledge is characterized as empirical and non-evaluative.

Scientific theory: Scientific theory is a formal and logically sound argument explaining some empirical phenomenon in general or abstract terms. In principle, all scientific theory is testable and is judged by its explanatory power.

Secondary groups: From Cooley and symbolic interactionism, primary and secondary groups vary by length of time of association, purpose, degree of involvement, and intimacy. Secondary groups are groups that do not last long, that tend to be goal- rather than emotion-based, and people in the groups tend not to reveal personal matters.

Segregation: One of Spencer's three forces of evolution (the other two are instability of homogeneous units and multiplication of effects). The verb *to segment* means to cut off or divide. Thus, segregation is the process through which social units (in the case of society) are isolated from one another. As they are detached, they become subject to different kinds of environmental forces.

Self: According to symbolic interactionist theory, the self is a perspective, a conversation, and a story. The self is a perspective in the sense that it is the place from which we view our own behaviors, thoughts, and feelings. It is an internal conversation through which we arrive at the meaning and evaluation we will give to our own behaviors, thoughts, and feelings. This conversation is ongoing and produces a story we tell ourselves and others about who we are (the meaning of this particular social object). The self is initially created through successive stages of role-taking and the internalization of language. The self is continually produced, given meaning, and furnished with stability or flexibility through patterns of interactions with distinct groups and generalized others.

Self-preservation: Part of Gilman's evolutionary scheme; self-preservation motivates the natural selection of skills that protect the individual; self-preservation and race-preservation skills ought to balance one another, however, because women have been removed from the natural environment by male-dominated society, the self-preservation skills of women are overemphasized and the evolutionary system is disequilibrated.

Sexuo-economic: Gilman's characterization of the human system when women are forbidden to labor in the economy; in such a society, the structures of economy and family are confounded, the sex distinctions between men and women are accentuated and unbalanced, sex itself becomes pathologically important to people, women become consumers par excellence, and men are alienated from their work.

Significant gestures: Mead's way of distinguishing between the gestures that humans and animals use. Significant gestures call out the same response in the sender and receiver; they are reflexive.

Simple, compound, and doubly compound societies: Spencer uses these terms to refer to the effects of population growth on the regulatory function (government). Simple societies are those that cooperate with or without a central governing figure. If a governing chief is present, there is only one level of power; that is, the people relate directly to the ruler. In compound societies, there is an additional level of government between the people and the chief; in doubly compound societies there are several layers of administration. For example, in the United States, there are many layers of government between the people and the president; the same is true between line workers and the CEO of General Motors.

Sociability: According to Simmel, a social form that is purely social, without any other purpose other than establishing and experiencing social connections with others.

Social differentiation: From Durkheim, the level of cultural diversity in any society: people become different from one another in response to increases in the division of labor and structural differentiation.

Social evolution or social Darwinism: From Spencer, the belief that certain social forms are selected for their ability to better survive in the environment. In its most extreme versions—eugenics and manifest destiny—it proposes that certain races or groups of people are inferior and others superior.

Social fact: Durkheim's term to describe the effects that the collective consciousness or moral culture has. Durkheim uses this term to underscore the argument that culture has the same attributes as any empirical object: it exists external to and coercive of the individual. The existence of social facts is used to argue in favor of a scientific approach to understanding society.

Social form: From Simmel, a patterned mode of interaction through which people obtain goals (examples: dyad/triad, conflict, exchange, sociability); forms exist prior to the interaction and provide rules and values that guide the interaction and contribute to the subjective experience of the individual; forms also imply social types—types of people that occupy positions within a social form (examples: stranger, adventurer, competitor, miser).

Social institutions: For most sociologists, social institutions are collective moral sets of predetermined meanings, legitimations, and scripts for behavior that resist individual agency and are perceived to meet the survival needs of a society. However, for *Mead* and symbolic interactionists, institutions are not macro-level, top-down forms of social control; rather, institutions are organized activities and attitudes that the individual carries around within. Meadian theory allows for a greater sense of individual agency, as these attitudes and activities may or may not be used by the individual in interaction. *Parsons*' theory of institutionalization bridges both levels. For Parsons, institutions are macro-level entities that become internalized through the formation of the Freudian superego.

Social morphology: A Durkheimian term that refers to the physical properties that correspond to the way in which groups are dispersed in space. It is literally the body of society. Social morphology is generally defined through the level of population density and structural differentiation.

Social objects: According to Mead, social objects are anything that we call attention to, name, and attach legitimate lines of behavior to in interaction.

Social solidarity: Durkheim's term for the level of integration in a society, measured by the subjective sense of "we-ness" individuals have, the constraint of individual behaviors for the group good, and the organization of social units.

Social structure: Sets of connections among people that are mediated by status positions, roles, norms, values, perspectives, and beliefs. These structures enable and constrain human behavior.

Social system: Society as a set of interrelated parts that function together to create integration and equilibrium; systems can be smart or dumb (ability to take in and adjust to information) and open or closed (exposure to external forces); systems may also contain feedback and feedforward effects as well as mechanisms that produce equilibrium; according to Parsons, social systems may be large (entire societies) or small (face-to-face interactions), are distinguished by boundaries between the system and its environment, and are controlled cybernetically.

Socialization: The process through which the norms, values, perspectives, languages, and status positions of a society are internalized by the individual; socialization is ultimately successful when the individual feels that the cultures and structures of society are natural and originate from within. Socialization is a specific concern of Parsons'.

Society: A basic, though abstract, concept in sociology. It is abstract because it cannot in itself be empirically discerned. What society is and where and how it exists are thus important issues for social theory. For *Durkheim,* society is made up of social facts that are known through moral power. Society is thus the collective consciousness, which responds to its own rules and effects and exerts power over individual conduct. For *Simmel,* society is found in the social forms we use to pattern goal-oriented behavior. Individuals must get their needs met through collective organization; it is thus in our best interest as individual actors to reduce uncertainly by creating and using a small number of social forms through which all interaction can be molded. For *Mead,* society is an emergent phenomenon that consists of sets of attitudes and legitimated behaviors that people may or may not use in any specific interaction. For *Spencer and Gilman,* society is a system of differentiated structures that meet collective needs. For *Marx,* society is formed through the dialectical relations found in the economy. For *Martineau,* society is the steward of human happiness: it exists chiefly to create and safeguard the natural law of happiness.

Solipsism: The philosophical idea that everything is reduced to the individual—the self can know nothing but its own state.

Species-being: Literally, species existence and awareness; in particular, how humans are conscious of being human. According to Marx, every species is unique because of the way it, as a biological organism, physically exists. Humans exist and survive through creative production. Thus, we become conscious of our own existence and the world around us through material production and the economy.

Spirit of capitalism: Weber's term to capture the cultural values and beliefs that undergird rational capitalism. There are at least three values and beliefs in the spirit of capitalism: life should be rationally organized to maximize profit; economic work is the most important thing we can do with our time and we must be diligent in it; and things of true value are quantifiably measured.

Status positions: Social positions that are hierarchically arranged according to different levels of honor or esteem; for Weber, status is one of the three systems of stratification; objective status positions are socially expressed by groups maintaining a distinctive way of life, or level of education, or specific types of careers or family traditions and history.

Stocks of knowledge: According to Schutz, stocks of knowledge are the cultural information that is available to any group. They include language, dialects, roles, norms, beliefs, and so on. They are "what everybody knows" and thus aid in the taken-for-grantedness of cultural reality.

Structural differentiation: The process through which the behaviors associated with social networks of roles, norms, and status positions are acted out in different places and at different times. In societies with high levels of structural differentiation, the requisite functions are carried out in distinct and separate institutions. Both Spencer and Durkheim are concerned with structural differentiation.

Structural interdependency: The level to which differentiated structures in a society must depend upon one another for goods and services. Structural interdependency is an important concept in functional theories of modernity; it provides a structural source of integration. Both Spencer and Durkheim are concerned with structural interdependency.

Subjective culture: From Simmel, a body of culture that is small enough for one individual to grasp, experience, use, and emotionally embrace.

Substructure: In Marx's scheme, the substructure is the economy and it provides the base for everything else in society.

***Sui generis*:** A Latin term meaning "of its own kind." Durkheim uses it to express his idea of the independent existence of society. In saying that society is a reality *sui generis,* Durkheim is claiming that society exists separately from its interactive or morphological base; as such, it obeys laws peculiar to it and it can create its own independent effects.

Superstructure: From Marx, the superstructure includes all of society other than the economy (the substructure). In other words, everything in society is built upon the foundation of the means and relations of production.

Surplus labor: According to Marx, surplus labor is the amount of labor a worker performs for which she or he does not get paid. It's the difference between the worker's pay and the value of the products the individual produces. This difference is where profit is created. Surplus labor is equivalent to the level of exploitation.

Survival of the fittest: The mechanism in Spencerian and Darwinian evolution that naturally selects certain forms for survival and others for extinction; the criteria for selection is competition for survival during specific environmental changes.

Symbols: Symbols are words, ideas, or objects that contain complex, abstract, and arbitrary meanings. For *Durkheim,* symbols can also be infused with high levels of effervescence or emotional energy. For *Mead,* symbols are organized sets of responses and attitudes.

Taken-for-grantedness: From Schutz, taken-for-grantedness becomes an issue only because human reality is assumed to be constructed. If human reality is constructed, then it must taken for granted in order for it to appear real. To seriously question a reality is to begin to deconstruct it. Thus, one of the concerns of phenomenology or social constructivism is finding and explaining the ways in which people are able to hide from themselves the truth that they are in fact constructing those things that appear objectively real.

They-orientation: In Schutz's theory, "they" are types of people who never appear but nonetheless form a backdrop to our interactions. "They" are utterly typified and anonymous.

Thick descriptions: Minute descriptions of a social phenomenon. Hermeneutic theory and research uses thick descriptions because of the assumption that all human behavior must be understood within its context.

Thou-orientation: In Schutz's scheme, thou-relations are people that we can observe but with whom we do not interact; in thou-relations, others are more subjectively available to us than those in they-orientation—we can thus attribute some individuality to the typification.

Traditional action: Part of Weber's ideal typology of social action; traditional action is behavior that is motivated by custom and long-practiced routines.

Traditional capitalism: Part of Weber's ideal typology of capitalism; in traditional capitalism, profit is pursued in order to live a comfortable life, and social relationships are based on emotional, traditional ties (some people therefore could not be exploited because they were in-group members).

Types: A categorical scheme that allows us to understand a social phenomenon by comparing and contrasting it with others that are like and unlike it.

Typifications: From Schutz, social types or categories through which we understand ourselves and others. People appear to us as variations of the average expectations associated with a social type. Those people who are closest to us (we-orientation) have the greatest variation and most individuality, and those that are furthest removed (they-orientation) are the most typified and most anonymous.

Unit act: Parsons' term to denote the entire set of conditions under which actions take place (the choice of means and ends constrained by the physical and social environments).

Urbanization: The process through which more and more of a given population moves from rural settings to the city. A distinct process of modernity that is associated with industrialization and capitalism, it results in such things as higher levels in the division of labor, increases in the use of money and the size and velocity of exchanges and markets, rational versus organic group memberships, overstimulation, increasing social diversity, and so on. Urbanization is specifically a concern of Simmel's.

Use-value: According to Marx, use-value is determined by the use or utility any commodity has. It is expended or consumed through use. Understood in relation to exchange-value.

Value-rational action: Part of Weber's ideal typology of social action; value-rational action is behavior that is motivated and makes sense based on some system of values (like religion).

Verstehen: One of Weber's methodological tools; German word that literally means "to understand"; in every circumstance, people are motivated either by emotional or intellectual reasons. Intellectual reasons can be observed and understood through ideal types; emotional reasons must be understood through empathy.

Web of group affiliations: Simmel's term to describe social networks: the number, frequency, and intensiveness of relationships among people. Simmel is particularly concerned with why people join groups (organic verses rational motivations).

We-orientation: In Schutz's theory, "we" are people that are immediately present, with whom we share reciprocal stocks of knowledge. Because "we" are present with one another, we can check our typifications against the subjective experience of the other—"we-relations" are the only time that typifications can get modified.

Index

About the Author

Kenneth Allan (Ph.D., University of California, Riverside) is Associate Professor of Sociology at the University of North Carolina at Greensboro. His research areas include theory, culture, and self—but his intellectual passion is the production of human reality; he is the author of *The Meaning of Culture: Moving the Postmodern Critique Forward.* When he isn't preoccupied with human reality, he enjoys playing guitar and spending time with his best friend and life partner, Jen.